Materials Science
and Metallurgy

Materials Science and Metallurgy

SECOND EDITION

Herman W. Pollack
*Orange County Community College
Middletown, New York*

Reston Publishing Company, Inc.
Reston, Virginia
A Prentice-Hall Company

Library of Congress Cataloging in Publication Data

Pollack, Herman W.
  Materials science and metallurgy.

  Includes bibliographical references and index.
  1. Materials. 2. Metals. I. Title.
TA403.P5■ 1977   620.1'12   76-50037
ISBN 0-87909-480-X

© **1977 by**
Reston Publishing Company, Inc.
A Prentice-Hall Company
Reston, Virginia 22090

10   9   8   7   6   5   4   3   2   1

Printed in the United States of America.

To my sisters,
Ann Moskowitz and Helen Dandes

Contents

Preface

This second edition of *Materials Science and Metallurgy* contains suggestions for revision made by faculty that have used and are using this text. The objects set forth in the preface to the first edition still prevail. This text provides a ready source of information of materials and their characteristics so necessary in engineering analysis and design.

Approximately half of the text deals with ferrous and nonferrous materials. The remainder of the text deals with non-metals. Thus, Chapters 1 through 8 deal with the manufacture, structure, physical properties of metals, and the analysis and interpretations of diagrams and curves. Chapters 9 through 12 deal with tool steels, nonferrous metals, and corrosion. Inorganic materials, cements, plastics, rubber, and wood are dealt with in Chapters 13 through 17.

As stated above, many suggestions by faculty who have used this text are included in this revision. If the suggestions help the student understand the content of this text, they were included. If necessary, entire sections have been rewritten. Chapter 1 was expanded to include the production of aluminum, magnesium, copper, nickel, lead, and tin. The periodic table has been added to Chapter 2 so that the instructor has a ready reference to the orbital populations in atoms. It is hoped that with the rewriting of sections 5-4 and 5-5, the student will more readily understand the physical properties of steel. Much of the chapter on corrosion (12) has been rewritten and expanded. Chapter 15 has been reworked and the production of rubber has been added to Chapter 16.

In addition, many more questions and numerical problems, drawings, and photomicrographs have been added. Tables have been included wherever possible. The student may use the "permissions" at the end of the tables for further investigation.

Once again there are people who need to be thanked for their contributions, efforts, and patience. Special recognition and thanks are extended to all those busy faculty who, after using the first edition, took the trouble to offer suggestions to improve it. As always my secretary, Mrs. John Westeris, did a magnificent job with the materials which I gave her. Finally, I wish to thank Mr. Matthew Fox at Reston Publishing Company for all these years of faith.

Herman W. Pollack

Acknowledgements

The author would like to acknowledge the following organizations who generously provided illustrations, photographs, and tables for this book.

United States Steel—Carnegie Illinois Steel Corp., Pittsburg, Pa.
Carilloy Steels, 1948
Appendixes 8.1, 8.2, 8.10
Tables 3.3, 8.1, 8.2
Making, Shaping and Treating of Steels
Figures 1.11(a), 1.12(a),(b), 1.13, 3.9(a),(b),(c),(d), 5.2(a),(b),(c), (d),(e),(f), 7.1, 7.2(b), 7.3(a),(b),(c), 8.a(c), 8.6(a)

American Iron and Steel Institute, 1963
Tool Steels Steel Products Manual
Table 9.1
The Making of Steel, 2nd edition, 1964
Figures 1.4, 1.14, 1.15, 1.16, 1.17(a),(b), 1.18, 1.19

American Society for Metals, Ohio
Tool Steels, 3rd Edition
Table 9.2 and Figure 7.2(a)
Heat Treating, Cleaning, Finishing, Vol. 1 and 2, 8th Edition
Tables 9.3, 9.14, 9.12, 10.9, 10.10, 11.1, 11.2, 11.3, 11.5, 11.6, 11.7, 11.8, 11.9

Aluminum Association, New York City
Aluminum Standards and Data, 2nd Edition, 1969
Tables 10.2, 10.3

McGraw-Hill Publishing Co.
Tool Engineering Handbook, 2nd Edition, 1959 (ASTME)
Tables 10.2, 10.5
Materials Handbook, Earl Parker
Tables 14.4, 15.1, 15.2, 17.1, 17.2

John Wiley Publishing Co. — Colin Carmichael
Kent's Mechanical Engineering Handbook, 12th Edition
Tables 13.3, 13.6, 13.8, 16.1

Portland Cement Association
The Design and Control of Concrete Mixes, 11th Edition, 1968
Tables 14.6, 14.10

American Concrete Institute
Table 14.2

Technical Tutor
Table 16.3

Addison Wesley Publishing Co., Mass.
Elements of Material Science, Van Vleek
Figure 15.20

Bausch and Lomb Instrument Division, Rochester, N.Y.
Figure 3.1(a),(b)

Ametex Technical Equipment, East Moline, Mich.
Figure 3.14

Wilson Instrument Co., Bridgeport, Conn.
Figures 3.15(a), 3.16, 3.17

International Textbook Co., N.Y. — Murphy
Figures 1.6, 14.5

American Institute of Mining and Metals
Figure 8.11(e)

Magniflux Corp., Chicago, Ill.
Figure 3.19(a)

Shore Instrument Co., Jamaica, N.Y.
   Figure 3.18

Prentice-Hall Publishing Co., Englewood Cliffs, N.J.
   Figures 3.21(a), (b)

American Smelting and Refining Co.
   Figure 11.4(b)

International Nickel Co., N.Y.
   Figures 6.2(a),(b),(c),(d)

Van Nostrand Publishing Co.,—Varney and Clark
   Figures 10.3(b),(c),(d),(e), 14.3(a)

Bethlehem Steel Co., Inc. Bethlehem, Pa.
   Figure 1.9

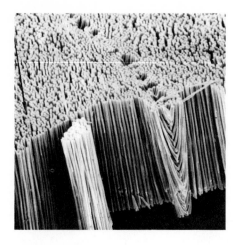

Photograph 1. Scanning electron micrograph of NiAL-34-Cr eutectic with NiAL matrix removed to show Cr rods. Arrow indicates direction of solidification.

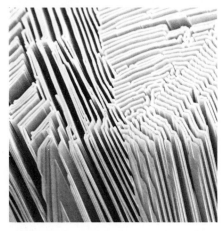

Photograph 2. Cr(MO) plates in NiAL-Cr (1.0 MO) eutectic. Longitudinal and transverse sections are revealed.

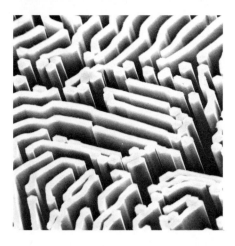

Photograph 3. NiAL-Cr (0.6 MO) eutectic with NiAL matrix removed to show faceted Cr(MO) rods and plates.

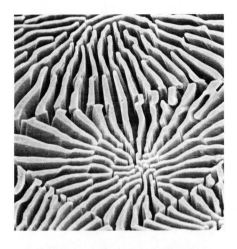

Photograph 4. Cells of Cr(MO) plates in NiAL-Cr (6.0 MO) eutectic.

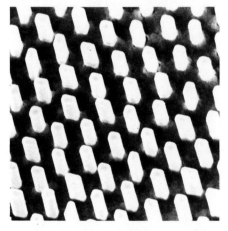

Photograph 5. NiAl-9MO eutectic with NiAL matrix removed to reveal MO rods.

Photograph 6. NiAL-34Cr eutectic showing faulted Cr rods. Arrow indicates direction of solidification.

1 | The Production of Ferrous and Nonferrous Materials

1-1 IRON ORE TO PIG IRON

The ancient name for iron is "star metal," probably because the first iron was extracted from meteors. Copper, tin, zinc, gold, and silver were used long before iron; these metals appeared near the surface of the earth and were easily accessible to ancient man. The impurities in the ore had to be separated from the iron. The process separating one from the other probably was discovered by accident. The earliest known appearance of iron is a single piece contained in the Great Pyramid (2900 B.C.). Tools made of iron appeared first in Palestine (1350 B.C.), where an iron furnace was found that dates back to about 1200 B.C. Metal appears also to have been used in ancient China and India about 2000 B.C. The art of smelting seems to have developed somewhere between the year 1350 and 1100 B.C. Prior to that, iron probably was "meteor" iron. Cast iron seems to have made its appearance in China in about 200 B.C.

The blast furnace seems to have been developed from the hearth (Fig. 1-1) by increasing its height. The process consisted of putting the fuel, flux, and ore in at the top and blowing air into the furnace at the bottom. The end product from the ancient hearth was wrought iron, but these "new" blast furnaces produced high-carbon iron.

The modern blast furnace (Fig. 1-2) is constructed from steel and lined with a highly heat resistant brick. Openings (tuyeres) in the side of the shell, and through the brick, permit air to be forced into the furnace at about 15 pounds per square inch (psi). This air enters the furnace at about 1100°F and aids combustion.

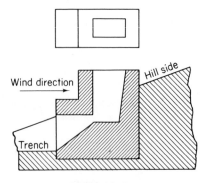

Figure 1-1

Ores that contain iron are magnetite, Fe_3O_4; hematite, Fe_3O_3; taconite; siderite, $FeCO_3$; and limonite, FE_2O_3 plus water. Magnetite is about 65 per cent iron and very strongly magnetic. Hematite is about 50 per cent iron and is capable of being magnetized. Taconite is very strongly magnetic and contains about 30 per cent iron. It also contains a great deal of silica, which must be removed before the ore can be processed. Siderite is about 50 per cent iron, whereas limonite is about 60 per cent iron.

A *charge* consists of iron-bearing materials, fuel, and a flux. The iron-bearing materials may be iron ore, mill scale, iron or steel scrap, and Bessemer or open-hearth slag. The fuel used is coke. The flux is limestone (CaO), dolomite, or both. Dolomite is a stone-bearing limestone composed mainly of calcium–magnesium carbonate.

An air blast is blown into the bottom of the converter to which the charge has been added, which burns part of the fuel, melts the iron, and burns off oxygen. Sometimes a fuel is added to the blast of air in the form of gas, oil, or powdered coal.

The ratio of each component of the charge is about two-thirds iron ore and other iron-bearing materials, one-fourth coke, and about one-twelfth flux in the form of limestone and/or dolomite. For each ton of pig iron produced this represents about one and three-quarters tons of iron-bearing materials, one-half ton of fuel, and one-fourth ton of flux. In addition, about two tons of air is used.

As the charges work their way down to the bottom of the furnace, carbon monoxide is formed by the combustion of air and coke. The temperature of the charge increases to about 3000°F, which is enough to melt the ore.

Slag is a fluid impurity made up of coke ash and other undesirable components. The chemical composition of slag is such that it also controls the sulfur content of the pig iron produced. In former years, slag was dis-

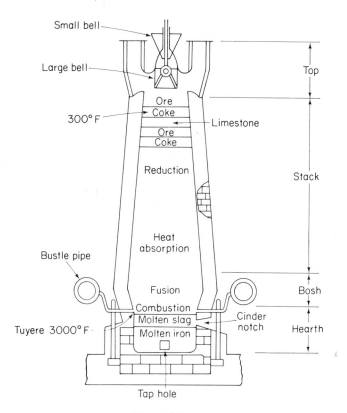

Figure 1-2

carded as an impurity. Recently, it has been used in various grades of paint, concrete, insulating and roofing materials, and fertilizer.

The end product produced by the charge discussed above is 1 ton of *pig iron,* ⅓ ton of slag, 3 tons of furnace gases, and a small amount of moisture and flue dust. Although the constituents in the pig iron may vary considerably, they are approximately 4 per cent carbon, 2.5 per cent silicon, 1.0 per cent sulfur, 1.5 per cent manganese, and 1.0 per cent phosphorus.

Pig iron may be either basic or acid, depending upon the phosphorous level. Since the acid furnace* for making steel cannot reduce the phosphorous content, raw materials low in phosphorous must be used for the few acid Bessemer furnaces in use. Since basic slag will reduce phosphorous, the basic furnace can accept the pig iron as it comes from the blast furnace.

*Acid and basic furnaces are discussed in subsequent sections of this chapter.

Nearly all blast furnace iron is used in its liquid state by the steel-makers. Less than 10 per cent of the iron goes to foundries in the solid pig state to be remelted and used to make castings, or refined in subsequent processes to make steel.

1-2   CAST IRON

To obtain a more uniform composition, pig iron is further refined to achieve more precise percentages of the necessary component elements. The *cupola* [Fig. 1-3(a)] is somewhat like a small blast furnace. It consists of a long steel shell lined with refractory firebrick. The bottom of the cupola is made from two semicircular hinged doors. A sand bottom, pitched toward the pouring spout, called the *breast hole,* forms the crucible in which the molten iron is collected.

The *charge* of coke, pig iron, limestone, and scrap steel is put into

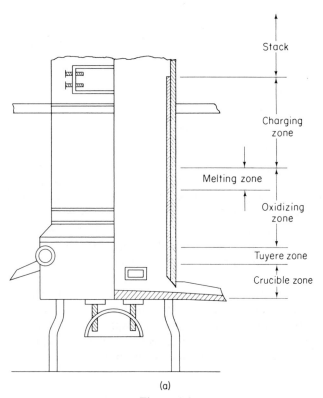

Stack

Charging
zone

Melting zone

Oxidizing
zone

Tuyere zone

Crucible zone

(a)

Figure 1-3

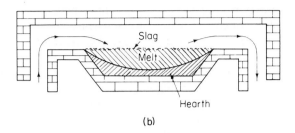

(b)

Figure 1-3 (continued)

the furnace through a charging door at the top of cupola. As in the blast furnace, the coke acts as a fuel, the pig iron supplies the component that makes up the end product (in this case, cast iron), the scrap steel is used as a constituent control, and, as before, the limestone acts as a slag former.

It is not the function of the cupola to change the components, but rather to purify the iron and produce a more uniform end product without the addition of components to the melt. The components of the end product cast iron are approximately 4 per cent carbon, 1.25 per cent silicon, 1 to 2.5 per cent manganese, 0.04 per cent sulfur, 0.06 to 3 per cent phosphorus, and about 90 per cent iron. There have been instances when nickel has been added to the melt to produce an alloy cast iron.

In a *reverberatory furnace* the charge is melted in a hearth that is shallow in comparison to its length and in which the roof is relatively low. Both the hearth and roof are built from refractory brick. The flame, entering from the sides, preheats the roof and the charge. Heat generated by the flame and radiation from the heated roof keep the metal fluid. Figure 1-3(b) shows a reverberatory furnace used to make cast iron. A charge of pig iron, scrap, and limestone is placed in the hearth. The heat from above forms a protective layer of slag and melts the iron.

A *duplexing process* is one that uses cast iron from the cupola, which is then transferred to an electric furnace. The refinement of the cast iron in an accurately controlled electric furnace produces high structural qualities in the resulting cast iron.

Recently, attempts have been made to go from the ore directly to the making of iron or steel without using the blast furnace as an intermediate step. These *direct reduction processes* aim at producing iron steel or a spongy iron that can be reprocessed into steel. For sponge iron, the iron ore need not be molten before it is processed into iron or steel. The use of the blast furnace to first melt iron ore so that it may be processed into steel is referred to as the *indirect reduction process*. The intermediate processing stage between the ore and the iron is sponge iron. The process of hammering sponge iron into implements was used before the blast furnace was developed.

Three reduction processes for producing sponge iron are currently in use or being developed and are worth discussing. They are the *kiln reduction process,* the *retort reduction process,* and the *fluidized bed process.*

The *kiln reduction process* (Fig. 1-4) is a method for making sponge iron that seems to show promise for direct large-quantity production. The aggregate of high-grade ore pellets and dolomite is brought to the reduction kilns, sorted to a uniform size, and stored in bins. The raw materials are fed directly and continuously onto a conveyor and then into the kiln.

These raw materials move through a *gas-*fired rotating furnace. The kiln is about 115 ft. long, 7½ ft. in internal diameter, and has a firebrick wall about 8 in. thick. The kiln rotates as the conveyor moves very slowly through it. It takes about 5 hours at about 2000°F for the aggregate to make the trip through the kiln. The kiln slopes about 2 degrees from the horizontal.

Once through, the aggregate drops through an opening into a cooler-kiln mounted below the main kiln. This cooler-kiln is 65 ft. long and 5 ft. in diameter. The input half of this kiln is lined for about 28 ft. The output portion is not lined and has a water spray directed against its outer shell. This kiln also slopes about 2 degrees from the horizontal to facilitate movement. The end product is sponge iron.

The *retort reduction process* uses reformed natural gas (73 per cent

Figure 1-4

hydrogen, 16 per cent carbon monoxide, 7 per cent carbon dioxide, and 4 per cent methane), which is preheated to 1600°F, injected into the retorts, and forced down through the charge. The partially spent gas is pumped out the bottom of the first retort into the second retort to preheat its charge.

Thus, during a full cycle, the four stages of the process include charging, primary reduction, secondary reduction, and cooling. The end product is sponge iron.

The *fluidized bed processes* use the heat of partial combustion of preheated air and natural gas for the reducing reactions. The reducing gas is composed of about 21 per cent carbon monoxide, 38 per cent nitrogen, and 41 per cent hydrogen. The reduced iron is pressed into briquettes and stored or shipped. This is a batch process and operates at about 1500°F and atmospheric pressure.

In some instances a semibatch process is used at 900°F and 500 psi pressure in conjunction with a *triple fluidized bed*. After about 5 hours in the reducing cycle, the lower bed is about 90 per cent reduced. The product is then removed and briquetted. The upper two beds are dropped and a new charge is put into the top. Purified gas is passed through the three beds, and the cycle is repeated.

A third method is a *two-stage* continuous operation that uses hydrogen as a fuel. Iron ore is preheated to 1700°F. In the first phase, Fe_2O_3 is reduced to FeO at about 1300°F. The gas in this initial phase is the spent gas from the preceding second phase. The original gas enters the second (final) phase. This original reducing gas, 85 per cent hydrogen, enters this phase at 30 psi, having been preheated to 1550°F. It is here that the end product of the initial phase, FeO, is reduced to Fe at about 1200°F.

Many additional variations of the direct process are being investigated.

1-3 TYPES AND PROPERTIES OF CAST IRON

Various types of cast iron structures may be developed by either controlling the amount and kinds of alloys added, or by controlling the temperature and cooling rate. Upon cooling, part of the carbon combines with the iron, forming iron carbide. The remaining carbon precipitates out in long slivers as free powdered carbon (graphite), shown in the photomicrograph of Fig. 1-5.

Gray Cast Iron (Fig. 1-5). If the rate of cooling is decreased (slower cooling), the crystals grow larger than normal and more graphite precipi-

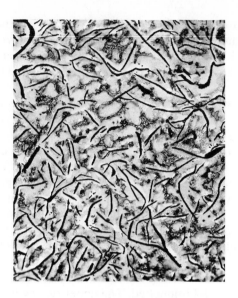

Figure 1-5

tates out, causing longer slivers to form. Since these graphite slivers are very weak, the cast iron itself is very brittle. The free graphite gives the cast iron its dark appearance.

The greater the amount of carbon in the structure, the greater the production of graphite. Since gray cast iron is essentially graphite flakes embedded in a matrix of steel—iron and iron carbide—the control of the relative amounts of iron and iron carbide to the amount of graphite formed is very important. Alloying elements that increase the formation of graphite flakes by aiding flake growth or by the dissociation of iron carbide are called *graphite formers*. Nickel, molybdenum, copper, silicon, chromium, and vanadium are such graphite formers.

The gray cast iron in general use is composed approximately as follows: carbon, 4 per cent; silicon, 2 per cent; manganese, 1.5 per cent; sulfur, 0.5 per cent; phosphorus, 1 per cent.

One to $3\frac{1}{2}$ per cent silicon is included to control the physical property characteristics of cast iron. For high-strength gray cast iron, the carbon content is kept on the low side and the silicon on the high side. Silicon in high percentages also imparts acid-resistant qualities to cast iron.

Nickel is used in amounts of 4 to 15 per cent to control graphitization, graphite flake size, strength, wear, and heat and corrosion resistance. Chromium up to about 0.5 per cent acts as a strong carbide former, thus producing a fine-grain, high-strength cast iron. Molybdenum, in amounts up to 1 per cent, is a mild carbide former. Vanadium is a strong carbide former, whereas copper is a mild stabilizer.

Other elements usually found in cast iron to achieve certain properties are manganese, phosphorus, and sulfur. Manganese in excess of 1.3 per cent increases the resistance to wear, hardness, and strength, but decreases machinability. Phosphorus in amounts of less than 0.5 per cent yields cast iron that does not have the brittle characteristics exhibited when greater percentages are used. When added to the melt in small quantities, it increases fluidity. Sulfur in amounts of 0.15 per cent and over acts as a carbide stabilizer. In the presence of manganese, this stabilizing effect of the carbides is reduced considerably. Since carbides are very hard substances, one method for controlling the physical properties of cast iron is accomplished by the addition of appropriate amounts of sulfur and manganese.

The gray cast iron in general use has low impact resistance and lacks ductility. On the plus side, it is relatively cheap to produce, has excellent fluidity and a relatively low melting point, and therefore is a good casting material. It has good machinability characteristics, compressive strength qualities, and abrasive hardness. The internal graphite flakes provide good damping qualities to gray cast iron.

Gray cast iron is classified according to the American Foundrymen's Association so that the minimum tensile strength of a particular structure is designated by a number. Thus a class 25 cast iron indicates a 25,000 psi minimum tensile strength.

White Cast Iron (Fig. 1-6). When molten cast iron is cooled very rapidly, almost all the carbon goes into solid solution as carbide and very little into graphite. Since carbides are very hard, this type of iron is very hard and brittle. When molds are lined with cold iron to produce selective rapid chilling, the resulting white cast iron is referred to as *chilled iron.*

Malleable Cast Iron (Fig. 1-7). Castings that have compositions of 2 to 3 per cent carbon, 1 per cent silicon, 0.5 per cent manganese, and 0.1 to 0.2 per cent phosphorus and sulfur when poured will develop a structure of white cast iron. If this casting is reheated to approximately 1600°F for

Figure 1-6

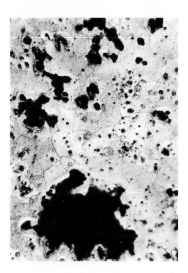

Figure 1-7

about 24 to 48 hours, the carbide slowly dissociates into small nodular forms of carbon, instead of the flakes discussed earlier. This is referred to as *temper carbon.* If the process is stopped before the combined carbon dissociates, the structure is called *mottled cast iron.* If the process is allowed to proceed, the end result is a large number of small nodules of graphite in a matrix of pure iron. This type of malleable iron is called *ferritic malleable cast iron.* It should be noted that the dissociation of carbon takes place entirely in the solid state, whereas, in gray cast iron, carbon dissociates from the melt. The structure of the latter type of malleable cast iron yields the following mechanical properties: 50,000 psi tensile strength, 20 per cent elongation, and 35,000 psi yield strength.

By controlled heat treatment the graphite nodules may be caused to form in a matrix of $Fe-Fe_3C$. The mechanical properties of this structure — *pearlitic malleable cast iron* — are considerably higher than ferritic malleable cast iron: 75,000 psi tensile strength, 10 per cent elongation, and 65,000 yield strength.

Ductile or Nodular Cast Iron (Fig. 1-8).   Nodular cast iron forms while the process of solidification takes place. This is accomplished by inoculating the melt with magnesium or cerium. Other elements may act as nodularizing agents, but magnesium and cerium are the most effective.

This spherical carbon microstructure has high shock-resistant properties and a high modulus of elasticity. Its tensile strength is in the neighborhood of 70,000 psi, with about 10 per cent elongation, and a Brinell hardness number of approximately 200.

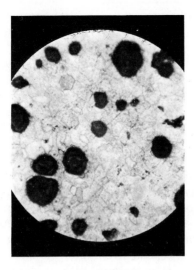

Figure 1-8

1-4 WROUGHT IRON AND ITS PROPERTIES

The Aston process is probably the only method used today in the United States for making wrought iron, which is a mixture of pure iron and silicate slag. Pig iron is melted in a cupola and refined in a Bessemer converter to a high degree of purity. This purified iron at a temperature of 2800°F is poured into a large ladle and moved to a slowly oscillating processing "cup." A composition-controlled slag at a temperature of 2400°F is moved into such a position that the molten iron can be poured through it. Since this slag is cooler, it causes the iron to solidify under rather violent conditions. This violence distributes the slag throughout the iron.

The fluxing action of the slag causes a spongy mass to form. This spongy material is processed by rolling and pressing. The latter squeezes out the excess slag and fuses the spongy material into a bloom and then into billets.

Since the iron should approach commercial purity, the resulting percentages of impurities remaining in the structure are all very low. Thus carbon is rarely higher than 0.03 per cent; silicon, 0.07 to 0.15 per cent; sulfur, under 0.02 per cent; phosphorus, 0.25 per cent; manganese, under 0.10 per cent. The slag in the iron is about 3 per cent by weight.

The mechanical properties of wrought iron are essentially the same as pure iron. The inclusion of the slag in the structure has a decided effect upon the ductility properties of wrought iron. A better distribution of the slag increases ductility. Up to a point, ductility is increased when the material is reworked because the slag is thus better distributed, refined, and rolled into finer threads throughout the structure.

The longitudinal ductility is somewhat lower than in steel because of the slag in the structure. The transverse ductility is very much lower than its longitudinal strength. A history of the direction of rolling is therefore important when the metal is to be fabricated. The following are some mechanical properties of wrought iron: tensile strength, 50,000 psi (longitudinal), 38,000 psi (transverse); elongation, in 8 in., 20 per cent (longitudinal), 5 per cent (transverse).

Wrought iron can be easily cold worked, forged, or welded. It is used to make pipe, corrugated sheet, iron bar products, grills, etc.

The microstructure of wrought iron produced by the Aston process is shown in the photomicrograph of Fig. 1-9. The iron silicate slag fibers — elongated and gray — dispersed throughout the refined iron matrix (white) and the fiber orientation in the direction of rolling can be seen. These long fibers result from repeated rolling of the material, which increases the desirable mechanical properties, such as ductility, in the structure. Large grain sizes, nonuniformity, distribution of the grain, or slag distribution are not desirable.

1-5   PRODUCTION OF STEEL

The production of steel is essentially a process of the purification of pig iron and the introduction of impurities in the form of alloys to impart the desired characteristics to the end product, steel. The type of furnace lining used is one important means of controlling this purity. Elements such

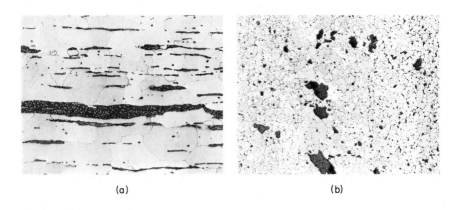

(a)   (b)

Figure 1-9

as carbon, manganese, sulfur, phosphorus, and silicon must all be eliminated, or very drastically reduced. The desired elements are then added.

The union of iron and carbon produces plain carbon steel. The addition of molybdenum, manganese, chromium, nickel, tungsten, etc., imparts special characteristics (not necessarily in the order listed) of deep hardening, strength, heat resistance, abrasion resistance, corrosion resistance, etc.

As has been indicated, the lining is an important aspect in the purification of iron. The two linings used are the *acid lining* and the *basic lining*. Carbon, silicon, and manganese may be removed by using either lining.

The removal of phosphorus and sulfur is accomplished with lime, which forms a basic slag. This basic slag will *not* react with a basic lining, and therefore the phosphorus and sulfur are removed from the iron.

Furnaces lined with silica are classified as having acid linings. Acid linings cannot remove phosphorus; consequently, pig iron low in phosphorus must be used.

Oxygen is also used to oxidize impurities, which are removed in the form of a slag. Only sulfur is not oxidized but must be removed by some type of basic slag process.

The oxygen is blown under pressure either through the molten iron, as shown in Fig. 1-10(a), or over the surface of the melt, as shown in Fig. 1-10(b), or onto the surface of the melt, as shown in Fig. 1-10(c). Two additional processes revolve the furnace while it is at an angle of about 30 degrees from the horizontal [Fig. 1-10(d)], or completely on its side, as shown in Fig. 1-10(e).

The bottom blown method [Fig. 1-10(a)] was first used in the *acid Bessemer* process for making steel. This process requires high-purity iron ore. At present, the steel made by this process is used for making free-machining bars, wire, steel castings, flat-rolled steel, seamless pipe, and metal to be used in the duplex process to be described later.

The *acid Bessemer converter* [Fig. 1-11(a)] is a pear-shaped steel vessel with a silicon lining. The tuyeres are made from carefully processed fire-clay tubes.

Either solid pigs (rarely used anymore) or molten metal from the blast furnace is poured directly into the converter when it is in the position shown in Fig. 1-11(b). In this position the air is turned on, and the converter is rotated into an upright position, as shown in Fig. 1-11(c).

While the converter is in this position, the *silicon blow* takes place. The silicon and the manganese burn during this blow. The temperature increases rapidly. Scrap steel is added to control this temperature increase. After about 4 minutes, the flame lengthens into the *carbon blow*. Whereas the initial flame is short and transparent, the carbon-blow flame is long and a brilliant yellow. After a time, the flame suddenly changes to

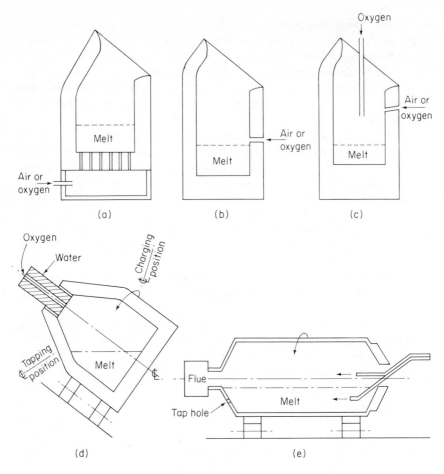

Figure 1-10

a deep red. This is an indication that the carbon content has reached a certain level. This stage of the purification is referred to as the *end point.* If the air is shut off at this point, the end product is said to be *young blown* steel. If the air is permitted to continue to blow through the melt, more carbon burns out and the end product is referred to as *full blown steel.* The air is finally shut off.

After the air has been shut off, a decision may be made to add materials such as silicon, manganese, or aluminum to the melt. These materials act as deoxidizers and produce *killed steel.* They combine with the oxygen in solution and thereby prevent blowholes from forming in the solid metal. Gas-free structures are referred to as "killed."

Once the process is completed, the converter is tilted to an almost

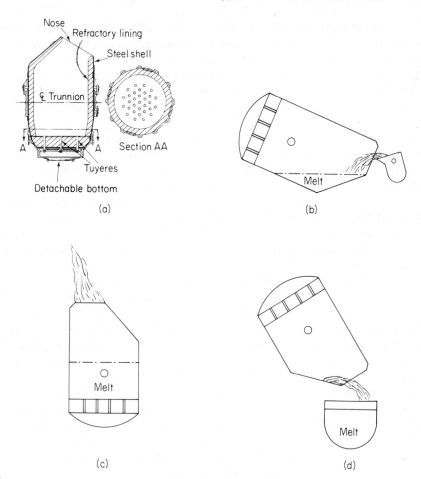

Figure 1-11

vertical position [Fig. 1-11(d)]. The melt is poured into a crucible very carefully so that the slag which flows out with the melt will float to the surface and form a protective coat for the liquid metal. The deoxidizers and carburizers are added, and the crucible is moved to the ingot molds, where they are filled. The crucible is moved to slag pits, where it is tipped vertically on its end so that the remaining slag may drop out.

The *basic Bessemer converter,* not generally used in the United States, is a process that requires a basic lining and limestone as a slag former. Thus it is capable of removing phosphorus and sulfur as already indicated. The end product has a low silicon content but an unacceptable level of nitrogen for certain applications. The level of nitrogen produces steel of lower ductility and higher strength — steel capable of strain aging.

The process partially desulfurizes the melt before pouring it into the converter. This is done by adding soda ash as a slag former. After the slag has formed, it is skimmed off. The converter is then rotated into a horizontal position, and lime is charged into the opening. The vessel is tilted further and the melt is added. The air blow is started, the converter is righted, and scrap is added as needed to control the temperature of the blow.

The blow proceeds in a similar manner to the acid converter. After the blowing is completed, the slag is removed by tilting the converter and desired amounts of ferromanganese are added. The melt is then poured into crucible cars, which transport it to the ingot molds.

The *basic oxygen process* [Fig. 1-10(c)] employs a vertical pipe, inserted through the mouth of a basic-lined Bessemer-shaped converter. Blowing oxygen into the top of the converter results in an iron of very high purity. The charge consists of iron ore that has a low phosphorus and a high manganese content.

The furnace is charged with molten iron and scrap while it is in a horizontal position. It is righted, and a lance is inserted into the mouth of the converter to a predetermined height above the melt. The oxygen is forced out from the end of the lance at about 150 psi. The oxygen, striking the surface of the hot metal, causes the formation of iron oxide and carbon monoxide. This causes violent reactions, which make the melt boil and thus accelerate the refinement.

Slagging materials are added shortly after the oxygen is turned on. A hood covers the mouth of the converter. All reactions take place inside the converter, and the products of combustion are carried away through the hood and up the stack after passing through a filter system.

A variation of the oxygen process just described is the *double-slag process*. This is used to refine molten iron that has a high phosphorus content. The process utilizes a powdered lime, which is blown into the melt with the oxygen. The first slag that results is rich in phosphorus and is removed from the converter. A second slag forms. The refined metal is removed through the taphole. The second slag is retained for the next heat.

The *Kaldo process* [Fig. 1-10(d)] employs a Bessemer-shaped converter with a solid bottom. The converter is rotated to a position about 20 degrees back from the vertical position, where a chute charges the lime and ore into its mouth. The converter is then rotated about 20 degrees to the opposite side of vertical and is charged with the melt. It is then rotated to its operating position and is water cooled; the lance is inserted, and the fume hood is placed into position. The converter starts to revolve and the oxygen, 95 per cent pure, is turned on. The phosphorus and carbon are burned out simultaneously. Iron ore and scrap are used as temperature regulators.

The *rotor process* [Fig. 1-10(e)] is used to refine high-phosphorus hot metal. It uses a rotary furnace that is designed to rotate slowly about its central axis. The furnace may also be tilted toward either end. That is, when tilted toward the entry end, it receives the charge of molten metal, solid iron ore, and slag formers. The vessel is then tilted back to horizontal and caused to revolve, and two lances are inserted. The lance that enters the melt injects high-purity oxygen directly into the molten iron. The other lance injects low-purity oxygen into the furnace over the molten metal.

Upon completion of the refinement, the furnace is tilted toward the discharge end, the tap is opened, and the metal is extracted. A waste gas flue is attached to the discharge end to carry off the gaseous products of combustion.

The *basic open-hearth furnace* [Fig. 1-12(a)] is the most commonly used furnace for making steel. The acid open-hearth furance is rarely used today. The furnaces are set up in rows, with the charging doors raised about 20 ft. above the ground. The tap sides of the furnaces are at ground level.

On the charging side of the furnaces and in front of the doors are boxes filled with the charge. The charging machines, mounted on wide gage tracks, lift these charging boxes, thrust them into the furnaces, unload, and withdraw to be refilled.

Hot metal may also be fed into the furnace to compensate for any charge deficiencies by using an overhead crane or an electric car.

Pouring platforms are installed on the output side of the furnace at a level convenient for pouring the molten metal into ingot molds. These molds are mounted on mold cars. Once filled, they are hauled away.

The furnace consists of a large shallow hearth housed in a low-roof refractory enclosure, as shown in Fig. 1-12(b). The charge is placed in the hearth and heated by a flame, which is blown over the top of the melt, first from one side and then from the other side. This flame also heats the low roof so that the heat level is also maintained by radiation. Thus the open-hearth furnace is reverberatory in that it employs radiation to maintain the heat.

The furnace also has *"checkers"* built in both ends. These checkers are stacked to permit the hot gases of combustion to pass through them, which give up their heat to the brick. Once heated, the burner above these heated checkers turns on, and the flow of air in the furnace reverses so that the incoming air of combustion now passes through these heated checkers. This preheated air helps maintain the high heat needed in the combustion chamber.

The tapping hole and spout are shown in the middle and to the rear of the furnace. Their position above the ground permits tapping directly into ingot molds mounted on flatcars.

The charge consists of scrap steel, molten pig iron, and iron ore to

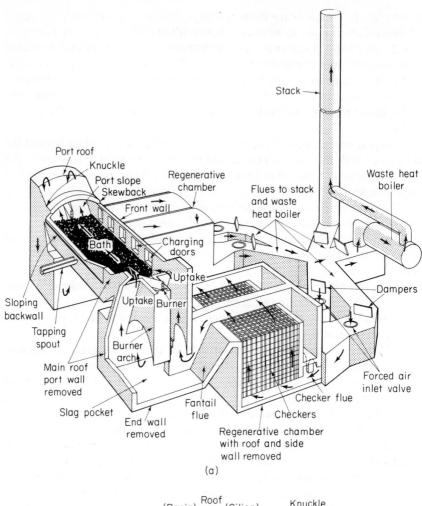

(a)

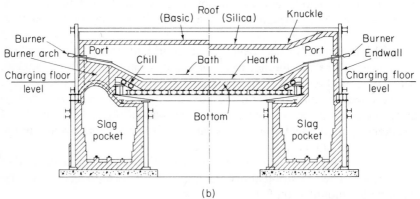

(b)

Figure 1-12

control the carbon content. When solid scrap and solid pig iron are used, limestone is used as a fluxing agent.

Oxygen lances are now used to hasten the removal of carbon and to shorten the meltdown period, which starts at about the same time the charge is inserted into the furnace. This may be done by lances inserted either through the roof of the furnace through the slag and into the melt, or directly underneath the burner tubes.

The operation of the open hearth consists essentially of five steps: the meltdown, hot metal addition, ore boil, lime boil, and refining period.

As soon as the first scrap has been added, oxygen is injected to quickly melt *(meltdown)* the solid scrap and maintain a high temperature and oxidization process. As part of the charge melts, molten pig iron is added to the hearth. This is the *hot metal addition* step.

Very soon thereafter silicon and manganese are removed by the formation of oxides. They become part of the slag. While this reaction is progressing, the carbon has begun to oxidize and is becoming more agitated, with the release of carbon monoxide. This is called the *ore boil*. The more violent turbulence results when carbon dioxide is released. Once the carbon content is reduced, the boil decreases.

A much more violent boil then results from the calcination of the lime, the *lime boil*.

After these processes have been completed, the *refining phase* begins. During this period the phosphorus and sulfur content is lowered, the carbon is eliminated, and the metal is processed to a condition suitable for tapping.

The tap hole is opened, and the metal is allowed to flow into large ladles. At this point, the addition of alloys takes place to recarburize and deoxidize the melt. Once the appropriate amount of slag covers the melt, the ladle is lifted by a crane, and the metal is poured into ingot molds.

The *duplex process* consists of a combination of the Bessemer and basic open-hearth processes. The melt is first processed in the Bessemer converter until the carbon, manganese, and silicon have been oxidized. The steel is then charged into the basic open-hearth furnace, where the phosphorus and whatever carbon is left are reduced by iron oxide and lime. The desired alloys are then added to the melt to produce the end product.

In the *triplex process* the electric furnace is added as a third step to the Bessemer and the basic open-hearth furnaces. This process is very expensive and is rarely used except to produce very specific characteristics.

1-6 ELECTRIC FURNACES

Another method used to produce steels is the *basic (or acid) electric arc furnace.* These furnaces are used to produce high-grade alloy and tool steels, such as tool and die steels, bearing steels, stainless steels, and high-heat-resistant steels.

The electric current's sole purpose is to produce heat. Heat may be generated directly when the arc is permitted to pass through the melt. Heat that is generated by a resistance may also be applied indirectly. The latter method is not used to melt steel, but is used for heat treating materials.

Indirect arc heating is used in electric furnaces where the arc is generated between the electrodes above the melt. Heating is accomplished as a result of radiation. In the direct arc method the arc jumps from electrode to the melt and then to another electrode. A third type of arc furnace uses the arc generated between one electrode, the melt, and another electrode at the bottom of the furnace. The resistance of the refractory material at the bottom completes the process.

The basic electric arc furnace (Fig. 1-13) is a circular steel shell lined with refractory brick. It is mounted on trunions to permit tilting for pouring off the finished melt. The dome may be removed so that charging takes

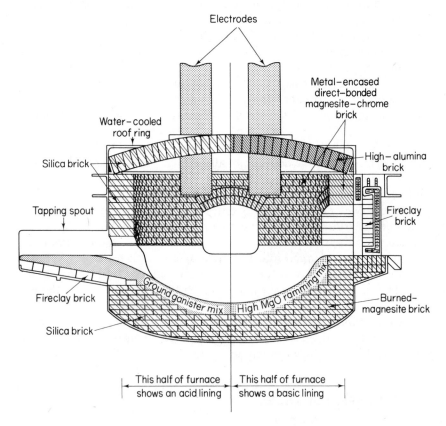

Figure 1-13

place through the top of the furnace. Once charging is completed, the dome is replaced. Carbon or graphite electrodes are inserted through the dome to the proper height above the charge. The low-voltage, high-amperage current is controlled by an operator. As the amperage increases, the arc length increases, and the electrodes must be raised.

Careful separation of the scrap used is important because of the necessity of controlling the type and kinds of alloys that get into the finished product. *Double slagging* permits a better control of the end-product alloys.

The *meltdown* period is the beginning of the refining process. A pool of molten metal forms on the hearth as a result of the heat generated, of radiation from the crucible floor, generated by the arc, and as a result of the resistance to the current flow in the scrap.

During the meltdown period carbon, manganese, silicon, and phosphorus are oxidized. The oxygen needed during this oxidizing period comes from the charge, the air, or from an oxygen lance.

When the melt has been completed, some slag is removed; lime and additional ore may be added, and the steel may be poured. In the double-slag method, all the slag must be removed before the second slag is permitted to form.

The second slag which forms is strongly reducing and contains calcium carbide. Since this carbide cannot exist in the same slag with oxides, it returns reducible oxides to the melt. As soon as the calcium carbide is formed, chromium, manganese, vanadium, iron, and tungsten oxides are added to the melt for direct reduction.

As soon as all the desired materials are added, the slag is shaped up. Then ferrosilicon, as a deoxidizer, and aluminum, for grain size control, are added. Once ready, the furnace is tilted so that the metal runs off first, followed by the slag, which forms a protective coat.

The *induction furnace* (Fig. 1-14) is an electric furnace in which the primary current is carried by a coil of copper tubing, circling the shell of the furnace. Almost immediately after the current is sent into the coil, a strong secondary current is generated in the charge. The high flux density generated causes these very strong secondary currents, which, because of the resistance within the charge to the passage of this secondary current, heat the charge. The heat starts at the outside of the charge and passes very quickly to the center.

Melting is very fast. The melt is thoroughly mixed by the turbulence present, which produces a homogeneous metal. This turbulence is not conducive to the formation of either basic or acid slags used for refining metals. As soon as the charge becomes clear, the slag present is skimmed off, and the desired alloying elements are added. The current is increased so that the pouring temperature may be achieved.

If slag is to be used, the current is reduced to cut down on this mix-

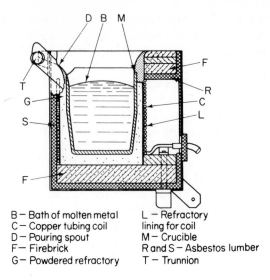

B — Bath of molten metal   L — Refractory
C — Copper tubing coil   lining for coil
D — Pouring spout   M — Crucible
F — Firebrick   R and S — Asbestos lumber
G — Powdered refractory   T — Trunnion

Figure 1-14

ing effect. A constant recycling of molten material from the bottom to the top of the crucible brings more and more of the melt into contact with the slag, which results in purification of the resulting metal.

Once the melt has reached the pouring temperature, the metal is poured into ladles or directly into molds.

These furnaces have an economic advantage over other types of furnaces in that they are less costly and therefore several furnaces may be tied to one frequency oscillator. Another factor is that they melt metal much faster than other types of furnaces; consequently, the losses of elements due to oxidation are less. As already noted, another advantage of the induction furnace is the stirring action created by the electrical input, which yields a very homogeneous material.

A much refined process is the *vacuum induction* method (Fig. 1-15). One method is to enclose the induction furnace wholly inside a large water-cooled sphere. An opening is provided for charging under vacuum conditions. Other openings may be provided for viewing the melt or measuring its temperature. The vacuum is maintained at very low pressures.

This method provides the following possibilities: percentages of reactive elements may be included; hydrogen, nitrogen, and oxygen may be held at very low levels; freedom from contamination by air may be obtained; the composition of the melt may be easily controlled.

The *consumable electrode vacuum arc melting process* (Fig. 1-16) consists of melting an electrode inside a water-cooled evacuated copper

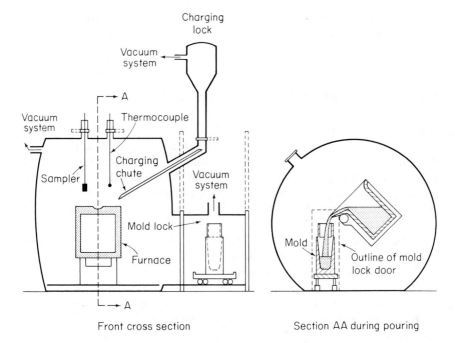

Front cross section

Section AA during pouring

Figure 1-15

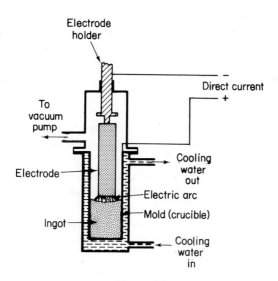

Figure 1-16

chamber. The composition of the electrodes is controlled. Therefore, when the electrode is consumed, it produces a very sound, homogeneous, clean ingot.

1-7 INGOTS AND CONTINUOUS CASTING

Ingots [Fig. 1-17(a)] are roughly square in cross section, and have rounded corners and tapered and corrugated sides to facilitate removing them from the molds. Shrinkage of the metal as it solidifies may create *pipes* in the upper part of the ingot. Another condition that may exist as a result of solidification is *segregation*. This happens when alloys such as carbon, sulfur, or phosphorus accumulate at the top section, where solidification takes place last.

The combination of carbon with oxygen to form gases causes turbulence. If this oxygen is not removed by the addition of ferrosilicates or aluminum, the process continues during solidification.

When there is a negligible amount of gas formed, the ingot steel that results is said to be *killed*. The structure is very uniform because it has been strongly deoxidized. A pipe forms in the extreme upper section. This pipe is removed easily. However, some segregation may take place in killed steel. This steel is most suitable for forging, piercing, carburizing, and heat treating.

Rimmed steel is at the other extreme and results when the melt is slightly deoxidized. At the boundary between the solid and melt, gas is produced. As a result, the melt solidifies as carbon-free steel. This carbon-free rim may grow until the center of the ingot solidifies, or it may be stopped at any time. The low-carbon rim creates a very ductile steel that may be rolled into a very sound structure. Segregation may also take place in this steel.

Semikilled steel is intermediate in deoxidization between killed and rimmed steels. Shrinkage is held to a minimum, and composition is more uniform than may be found in rimmed steel, but segregation is greater than in killed steel.

Capped steels are those that have had a deoxidizer added to stop the rimming at some desired point after solidification has started.

Figure 1-17(b) shows the various degrees of oxidization.

The *continuous casting process* (Fig. 1-18) was developed to roll steel directly from the melt, thus bypassing the slabbing or billeting processes.

A preheated bottom pouring ladle takes the melt from an electric furnace to the casting machine. The ladle is lifted to the casting floor, the nozzle is opened, and the melt is allowed to flow into a refractory-lined

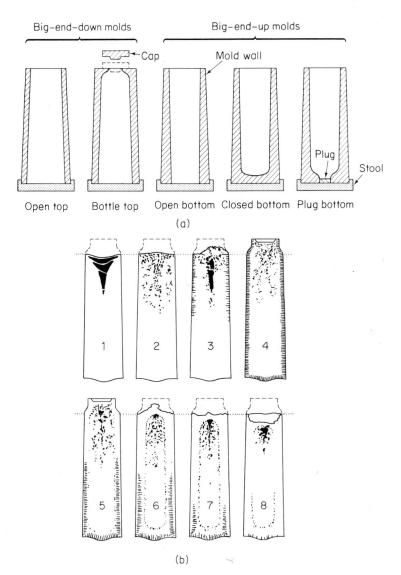

Big-end-down molds   Big-end-up molds

Cap

Mold wall

Plug

Stool

Open top   Bottle top   Open bottom   Closed bottom   Plug bottom

(a)

1   2   3   4

5   6   7   8

(b)

Figure 1-17

turndish. This reservoir allows the melt to enter a water-cooled mold. A "starting" bar cools the end of the liquid stream, so that, when the bar is pulled down, the liquid above solidifies and is also capable of being pulled down. The hot billet is supported by rollers on all sides and cooled as it proceeds down the machine.

At the point shown in Fig. 1-18 a clamp fastens the oxyacetylene

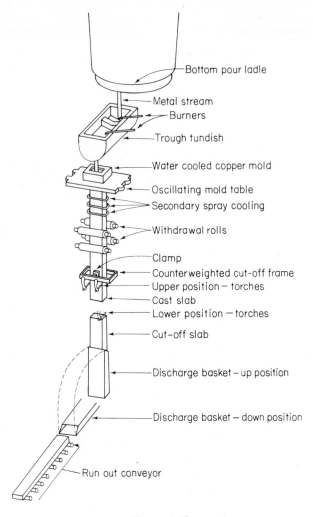

Bottom pour ladle
Metal stream
Burners
Trough tundish
Water cooled copper mold
Oscillating mold table
Secondary spray cooling
Withdrawal rolls
Clamp
Counterweighted cut-off frame
Upper position – torches
Cast slab
Lower position – torches
Cut-off slab
Discharge basket – up position
Discharge basket – down position
Run out conveyor

Figure 1-18

cutting torch to the billet. As torches and billet move down, the billet is cut into 16-ft. lengths. They are then transported to heat-treating furnaces for subsequent processing.

1-8   SOAKING PITS AND MILLS

Soaking pits are underground furnaces used to heat ingots once the molds have been stripped off. They heat the ingots to a uniform temperature of about 2200°F so that they may be rolled in the mills.

Blooming mills commonly used are two-high, electrically driven rollers that keep reversing their direction of rotation. The ingot passes back and forth through heavy grooves in the rollers as its size is being reduced.

Fingers, called manipulators, turn the steel so that its thickness and width may be controlled. The end results may be blooms, billets, or slabs.

A three-high mill uses three grooved rollers, one above the other. Thus no reversing is necessary. The steel passes between the bottom and the middle rollers, the table raises, and the steel is returned between the middle and top rollers to be caught by the platform on the starting side of the mill. This table lowers and the process repeats until the operation is completed, when the end product is sheared to length.

The flow chart (Fig. 1-19) shows many of the operations discussed so far.

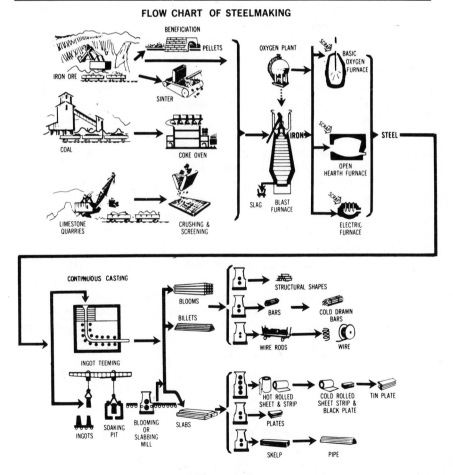

FLOW CHART OF STEELMAKING

Figure 1-19

1-9   PRODUCTION OF ALUMINUM AND MAGNESIUM

Aluminum is produced by electrolytic reduction of aluminum oxide. Bauxite, the only commercially important ore, is a hydrate of aluminum that is either $Al_2O_3 \cdot 3H_2O$, $Al_2O_3 \cdot H_2O$, or a mixture of both. The impurities inherent in bauxite are the oxides of iron, silicon, and titanium.

The *Bayer purification process* produces the largest amount of aluminum in the United States. The bauxite is pulverized, washed, and calcinated until it is a powder. It it then mixed into a hot solution of sodium hydroxide (NaOH) and pumped into large tanks containing caustic soda, which dissolves out the aluminum hydroxide. This forms a sodium aluminate solution. Crystals are formed when the sodium aluminate solution is seeded with crystals of aluminum hydroxide. After cleansing, the aluminum hydroxide crystals are calcinated to drive off the water. The resulting white powder, alumina (Al_2O_3), is the principal raw material used in the production of aluminum.

Other processes for producing alumina are the *line-sintering* process and the *Kalunite* process. The end product in both cases is ready for the Hall–Heroult direct reduction to aluminum.

The Hall–Heroult electrolytic process uses electricity to dissolve the alumina in a cryolite solution. Cryolite is an aluminum–sodium fluoride. As electrolysis progresses, oxygen is liberated and aluminum is deposited at the bottom of the container underneath the carbon cathode electrode. This aluminum is removed as a liquid. The process is continued since the alumina is constantly replenished.

The molten aluminum is poured into molds to produce 50-lb pigs. Remelting the pigs provides the opportunity for further refining.

It takes about 12 kilowatt-hours (kwh) of electricity, 2 lb of alumina, and $\frac{3}{4}$ lb of carbon electrode to produce 1 lb of approximately 95 per cent pure aluminum.

Another aluminum-producing process, the *Hoopes electrolytic* process, produces very pure (99.98 per cent) aluminum. This is not only a very expensive process, but it also produces smaller quantities of aluminum.

Magnesium is produced by either an electrolysis or chemical reduction process. The raw materials from which magnesium is extracted are the following:

1. *Magnesium chloride* (21 per cent magnesium) or carnallite (9 per cent magnesium) derived from salt water or sea deposits.

2. *Brucite* (42 per cent magnesium).

3. *Magnesite* (29 per cent magnesium) and dolomite (14 per cent magnesium).

The *Dow Chemical electrolytic process* was one of the first commercial processes to produce magnesium. This process drives the water off a magnesium brine, causing crystallization. The crystals are heated to yield magnesium chloride. The hydrated magnesium chloride is mixed with sodium and ammonium chloride, which allows the dehydration to proceed so that crystallization can proceed. The magnesium chloride and an alkali chloride are heated in an electrical cell. Chlorine is released at the positive electrode. Magnesium (99.9 per cent pure) collects at the bottom of the cell.

In the *Elektron process,* the first step is to grind magnesite or dolomite into a fine powder. This powder is concentrated by flotation and calcinated into magnesia. The magnesia is ground, briquetted with coke, and treated with chlorine to produce anhydrous magnesium chloride. The briquettes are converted to magnesium chloride, which is processed into magnesium and chlorine.

Another process, called the *Dow sea-water process,* produces flaked magnesium chloride, which is then electrolytically processed into 99.9 per cent pure magnesium.

Other than the processes indicated above, magnesium may also be produced by thermochemical processes such as the *ferrosilicon method.*

Zinc is also produced by distillation and electrolysis. The preliminary treatment (distillation) produces zinc sulfate, which is then processed to produce zinc.

1-10 PRODUCTION OF COPPER

There are many types of copper-bearing ores. The principal ores, found in the United States, contain about 1 per cent copper. Other copper-bearing ores are mined that contain copper and iron sulfides and small amounts of other elements, such as iron, silver, platinum, antimony, and zinc.

The first step in the refining process is accomplished by hearth, blast, or kiln *roasting.* Flat-hearth roasting is accomplished by raking the ore over a flat bed. In another hearth process, the copper is conveyed through a stack of hearths. The ore enters at the top and increases in temperature as it moves down through the stacks. In blast roasting, air is blown or drawn through a layer of ore that is on a moving grate. In the kiln process, the ore is charged into the top of a long revolving cylinder. The cylinder is inclined with the horizontal in such a way that the charge moves through the furnace at a predetermined speed. These processes either oxidize the sulfur, calcinate the carbon, or dry out the water by removing it in the form of vapor. Partial oxidizing of the sulfur or chlorine is also possible.

Oxide copper ore is melted in a vertical blast furnace after it has been charged with the copper ore and a carbon material. A *matte* and *slag* are the end products of this process. If the furnace is a reverberatory type, the melting process will separate the copper-containing matte from the slag.

The copper-containing mattes are melted again in either a horizontal or vertical converter. During this process, air is blown through the copper-bearing material so that the remaining sulfur is oxidized. The iron that is present is oxidized and removed by the slag. The gold and silver remaining are of such a poor grade that they are left in the copper. The copper remaining, *blister copper,* is about 99 per cent pure.

Blister copper is refined further by the *fire-refining* process. Green hardwood poles are repeatedly thrust into the melt. The wood decomposes, releasing carbon and hydrocarbon gases. The cuprous oxide formed is released, leaving a *tough-pitch.* This copper is ready to be cast into wire bars, billets, ingots, or cakes.

If the precious-metal content is high enough to warrant its extraction, the fire-refining process is not permitted to go to completion. The blister copper in this form is then further melted in a reverberatory or electric furnace so that anode casting may be done.

Lake copper is low in precious metal and is fire refined, cast, and used as low- or high-resistance lake copper.

1-11   PRODUCTION OF NICKEL, LEAD, AND TIN

Nickel is also found in very low percentages in its ore and must be concentrated before it is refined. The ore is briquetted with a sulfur (gypsum) material and smelted to produce an approximately 35 per cent nickel matte. Some sulfur and most of the iron are removed from the matte by blowing air through the melt in a converter. The end product is ground and the sulfur removed by roasting. The end product of roasting is pure nickel oxide, which is briquetted with a carbon material and heated to produce nickel that is better than 99 per cent pure.

If metallic nickel is to be made from the matte, sodium sulfate is added to the charge in a cupola or blast furnace. The sulfides of nickel and copper form layers; the top layer is nickel sulfide, and the bottom layer is copper sulfide. The nickel sulfide is converted to nickel oxide, reduced to metallic nickel, and refined electrolytically.

Lead is another metal that is first concentrated, roasted, and then smelted. The roasting, or sintering, controls the percentage of sulfur. This reduces the amount of lead remaining in the matte. In this instance, metal in the matte is not readily retrievable. If the lead sulfate is carefully controlled, the blast furnace will only reduce the lead oxide in its charge to metallic lead.

The lead bullion that results is then "softened." This process, known as *drossing,* consists of the removal of zinc, tin, copper, bismuth, and antimony as impurities. The metal is heated to its liquid state long enough to allow materials to float to the top so that they may be skimmed off. Other materials, such as sulfur or air, are added to the melt to form compounds or oxides, which float or settle out so that they may be removed.

The noble metals (gold and silver) are removed by causing them to react in the melt when in the presence of zinc. The intermetallic compound formed with zinc floats to the surface of the melt; it is removed and processed for recovery of zinc, gold, and silver. The remaining lead is processed for the removal of any remaining zinc, and then is ready for commercial use.

Tin is extracted from ore that contains impurities such as copper, iron, lead, zinc, bismuth, and antimony. They are difficult to remove. Washing and concentrating are the first steps. Roasting will remove sulfur, and form oxides and chlorides that may be removed with a dilute acid. Smelting further refines the tin. This may be done in a reverberatory furnace, which produces a very high purity tin and a high-tin-content slag. The high-tin-content slag is further refined, and the tin is poured into pigs for commercial use. Because of their higher melting point or lower specific gravity, impurities may be removed by a method called *liquating.* The process is carried out in a sloping furnace. The molten tin flows out the taphole; the residue remains behind. If desired, further refinement may take place to remove impurities such as bismuth and lead.

Tin may also be purified by electrolytic refinement. The end product is tin that is 99.9 per cent pure.

Problems

1-1. (a) What are tuyeres? (b) Describe their function. (c) What is the relationship of the tuyeres to the temperature of the region which they serve in the furnace?

1-2. List the various types of iron ore and indicate the percentage of iron that they contain.

1-3. Describe the constituents of a charge and the percentages of each present.

1-4. Describe the operation of a blast furnace. What is the function of each material that makes up the charge?

1-5. Describe the process, charge, and percentages of components of the end product of the cupola.

1-6. What is the function of the slag in the reduction process?

1-7. Describe the functions of the lining of a furnace in reducing the phosphorus content of pig iron.

1-8. What are the components of pig iron? What are the percentages of each?

1-9. Sketch a cupola. Label all components.

1-10. Describe the operation of a reverberating furnace when used to produce cast iron.

1-11. What is the duplexing process for making cast iron? When is it used?

1-12. What is the meaning of the term "direct process"? "Indirect process" when making steel?

1-13. Describe the kiln reduction process as a direct process for making cast iron.

1-14. Describe the retort reduction process as a direct process for making cast iron.

1-15. Describe one of the fluidized bed processes discussed as a direct process for making cast iron.

1-16. Describe the growth of graphite flakes in gray cast iron. What are "graphite formers"?

1-17. What is the effect of silicon, nickel, chromium, molybdenum, and vanadium upon the structure and physical characteristics of gray cast iron?

1-18. What effects do manganese, phosphorus, and sulfur have on gray cast iron?

1-19. What are some good qualities that gray cast iron has as a result of its structure?

1-20. Discuss the low impact resistance and low ductility qualities of gray cast iron.

1-21. Describe the procedure used to produce white cast iron. What is chilled iron?

1-22. Describe the procedure used to produce malleable iron and describe the structural changes that occur. How is mottled cast iron produced?

1-23. What is "temper carbon"?

1-24. What are the mechanical properties of ferritic malleable cast iron?

1-25. What are the mechanical properties of pearlitic malleable cast iron?

1-26. Describe the structure for nodular cast iron and list some of its physical properties.

1-27. Describe the process and resulting structure for making wrought iron.

1-28. What are the percentages of impurities that you might expect to find in wrought iron?

1-29. Discuss the ductility of wrought iron.

1-30. What are some of the mechanical properties of wrought iron?

1-31. Describe the fibrous structures in Fig. 1-9. Are they desirable?

1-32. In your own words, tell why each of the following are not desirable to the mechanical properties of steel: (a) large grains, (b) uneven distribution of grain, (c) uneven slag distribution.

1-33. Describe the use of acid and basic linings in the removal of impurities during the purification of steel.

1-34. Describe the acid Bessemer converter for making steel.

1-35. What is the purpose of blowing oxygen over the melt?

1-36. What is young blown steel? Full blown steel? Killed steel?

1-37. How does the basic Bessemer converter process for making steel differ from the acid Bessemer converter?

1-38. Describe the basic oxygen process for making steel.

1-39. What is the purpose of the double-slag process in the steel-making process?

1-40. Describe the Kaldo process for making steel.

1-41. Describe the rotor process for making steel.

1-42. Describe the basic open-hearth furnace.

1-43. Describe the five-step procedure for making steel in the open-hearth furnace.

1-44. Define duplexing and triplexing in making steel.

1-45. Describe the direct, indirect, and electrode-resistance arc furnaces.

1-46. List at least three methods that generate heat through resistance and are used for making steel.

1-47. Describe the double-slagging process used with the electric arc furnace for making steel.

1-48. Describe the induction process for making steel. What are some of the advantages that accrue because of the rapid heating cycle?

1-49. Describe the vacuum induction furnace operation and its advantages.

1-50. What is the purpose for the consumable electrode in the vacuum arc process?

1-51. What are "pipes" in ingots? How do they form?

1-52. Describe the mechanism called "segregation."

1-53. Describe the formation of killed steel; of rimmed steel; of semikilled steel; of capped steel.

1-54. Describe the continuous casting process discussed in this chapter.

1-55. Describe the operation of the two- and three-high rolling mills used to process ingots into blooms, billets, and slabs.

1-56. Describe the production of alumina from bauxite.

1-57. Describe the Hall–Heroult electrolytic purification of alumina.

1-58. List the three materials used in the production of magnesium.

1-59. Describe the Dow Chemical electrolytic process for making magnesium.

1-60. Describe the Elektron process for making magnesium.

1-61. May zinc be produced by distillation–electrolysis processes? Explain.

1-62. Define the term "roasting" as it relates to the production of copper. List three roasting processes. What is its purpose?

1-63. Describe the blast furnace process for producing mattes. What is a matte?

1-64. Describe the use of the converter to process copper mattes. What is the end product?

1-65. How is "blister" copper further refined? What is the end product if the process is permitted to go to completion? Why might the process not be permitted to go to completion? Explain.

1-66. Describe the production of nickel from its ore.

1-67. Describe the production of lead from its ore.

1-68. Describe the process of "softening" lead.

1-69. How are noble metals removed from lead?

1-70. Describe the production of tin from its ore.

1-71. Describe the liquating process for purifying tin.

2 | The Structure of Metals

2-1 THE ATOM

For many years scientists have been trying to develop an acceptable model of the micro- and the macro-universe. To this point in man's history, as instruments are refined and additional data are collected, the model of the macro-universe is being expanded. Much of this new information is deduced from energy that is continuously bombarding the earth's surface. If one reflects a moment, much of what man suspected about the moon was only a built-up model until it was confirmed by man's landing on its surface. It soon becomes apparent that much of what we actually know is still only a small part of that which actually exists.

In the same manner that man prefers to be able to assimilate his findings of the macro-universe into a model so that he may better understand it, so he has attempted to generate a model of the micro-universe.

For purposes of this text the model preferred is the one that views the atom as having almost all its mass concentrated in a positively charged *nucleus*. In orbit about this nucleus are one or more negatively charged particles called *electrons*. The net charge of the atom as a whole is zero. The nucleus is envisioned to have a positively charged particle called the *proton*. The number of protons in the nucleus remains unchanged for a particular element. A neutrally charged particle, called the *neutron,* may also be found in the nucleus. Together, the protons and neutrons are called *nucleons*. Thus the entire positive charge appears to be concentrated in the protons in the nucleus, which are balanced by the cumulative negative charges of the electrons.

The magnitude of the electron charge (and the proton charge) is 1.6×10^{-19} coulomb (C). Its mass is 9.108×10^{-28} gram (g). The mass of the proton is 1.673×10^{-24} g, and the mass of the neutron is 1.675×10^{-24} g. The mass is concentrated almost entirely in the nucleus.

Elements are designated according to their chemical symbols. Thus iron (Fe) has an *atomic number* of 26 and an *atomic mass number* of 56. It is designated as

$$^{56}_{26}\text{Fe}$$

This atom would consist of 26 protons and 26 electrons and 30 neutrons. If we let $A = 56$ and $Z = 26$, then the neutron number is

$$N = A{-}Z = 56{-}26 = 30$$

where N = neutron number
A = atomic mass number
Z = atomic number

Isotopes of any element have the same atomic number but differ in their atomic mass number because, even though they have the same number of electrons and protons (electrically neutral), they differ in the number of neutrons. Thus hydrogen and its isotopes are as follows:

Hydrogen($^1_1$H)	*Deuterium*($^2_1$H)	*Tritium*($^3_1$H)
1 electron	1 electron	1 electron
1 proton	1 proton	1 proton
0 neutrons	1 neutron	2 neutrons

Opposite charges attract each other, and the centrifugal forces would seem to cause the electrons to want to fly out of orbit. Under normal conditions this centrifugal force would be balanced by the *electrical forces* of attraction. However, a centripetal* force indicates an acceleration. Such an acceleration dictates that an electrically charged body must radiate energy. Such energy loss by the orbiting electron would cause it to fall into the nucleus. It is obvious that such a model is not acceptable.

To explain this unacceptable model, discrete energy orbits are assumed. If the electron is found in one of these orbits, it will have a fixed energy value. In these fixed energy levels, the electron does not radiate energy, and the energy level of the electron is balanced by the electrostatic attraction of the nucleus.

---

*A centripetal force is the reaction force to the centrifugal force. It is the force which acts toward the center of rotation.

Inside the nucleus the charges are positive. Since positive charges repel, it would seem that the nucleus should fly apart. However, this model of the atom dictates that there be very large forces of attraction at work inside the nucleus. These binding forces are called *nuclear forces*. Their magnitudes are much greater than the positive charges operating within the nucleus.

As already indicated, this model also envisions the orbital electrons as possessing discrete amounts of energy and therefore as being arranged in definite orbits dictated by the energy required to keep that electron in that particular orbit. These orbits are referred to as *shells*. The maximum number of electrons permitted in each shell is $2n^2$. The symbol n represents the shell number.

Example 1

Calculate the maximum number of electrons permitted in the first four shells of an atom.

solution:

The following shells close out with the numbers calculated. Thus,

The K shell closes with

$$2n^2 = 2(1)^2 = 2 \text{ electrons}$$

The L shell closes with

$$2n^2 = 2(2)^2 = 8 \text{ electrons}$$

The M shell closes with

$$2n^2 = 2(3)^2 = 18 \text{ electrons}$$

The N shell closes with

$$2n^2 = 2(4)^2 = 32 \text{ electrons}$$

Thus the hydrogen atom has one electron in the first shell and helium has two electrons in the first shell. According to Example 1, this closes the first (K) shell. Lithium has three electrons. The first two will fill the K shell and the one remaining electron will start a new shell L. Sodium, which requires 11 electrons, will have two electrons in the K shell, eight in the L shell, and one in the M shell. Starting with potassium, this scheme deviates somewhat as one progresses through the list of elements for the transition elements.

The capital letters K, L, M, N, O, P, and Q are used to designate the shells. Each shell is again divided so that no more than two electrons may occupy one subdivision. If two electrons occupy one energy level orbit, their spin must be in opposite directions. This introduces several ideas, which are identified as follows:

The *total energy* of an electron is designated by a *principal quantum number* n, where n is equal to 1, 2, 3, 4, . . .

The *angular momentum* of the electron is designated by the case letter l and is called the second quantum number. This angular momentum determines the subshell inhabited by the electrons; l has values from 0 to $(n-1)$ and the subshell is $2(2l+1)$. The subshells are lettered $s, p, d, f, g, h.$

Two additional quantum numbers are used to further define this model. M_l indicates the magnetic moment of the electron, has values of $+l$ to 0 to $-l$, and designates the orbit within a subshell corresponding to a particular energy level. M_s, which may have values of $+\frac{1}{2}$ or $-\frac{1}{2}$, indicates the direction of spin of the electron upon its own axis.

Thus no two elements may have all four quantum numbers alike.

Example 2

A bromine atom has an atomic number $Z = 35$. Find the number of electrons in the shells and subshells.

solution:

(a)   *K* shell: $n = 1; l = 0$

$$2n^2 = 2(1)^2 = 2 \text{ electrons in the first shell}$$

The subshell distribution is

$$2(2l + 1) = 2[2(0) + 1] = 2 \text{ electrons in the } 1s^2 \text{ subshell}$$

$$\overline{\text{Total} = 2 \text{ electrons closed}}$$

(b)   *L* shell: $n = 2; l = 0; l = 1$

$$2n^2 = 2(2)^2 = 8 \text{ electrons in the second shell}$$

The subshell distribution is

$$2(2l + 1) = 2[2(0) + 1] = 2 \text{ electrons in the } 2s^2 \text{ subshell}$$

$$2(2l + 1) = 2[2(1) + 1] = 6 \text{ electrons in the } 2p^6 \text{ subshell}$$

$$\overline{\text{Total} = 8 \text{ electrons closed}}$$

(c)   *M* shell: $n = 3; l = 0; l = 1; l = 2$

$$2n^2 = 2(3)^2 = 18 \text{ electrons in the third shell}$$

The subshell distribution is

$2(2l + 1) = 2[2(0) + 1] =$ 2 electrons in the $3s^2$ subshell

$2(2l + 1) = 2[2(1) + 1] =$ 6 electrons in the $3p^6$ subshell

$2(2l + 1) = 2[2(2) + 1] = 10$ electrons in the $3d^{10}$ subshell

Total = $\overline{18}$ electrons closed

(d) N shell: $n = 4$; $l = 0$; $l = 1$; $l = 2$; $l = 3$

$$2n^2 = 2(4)^2 = 32 \text{ electrons in the fourth shell}$$

The subshell distribution is

$2(2l + 1) = 2[2(0) + 1] = 2$ electrons in the $4s^2$ subshell

$2(2l + 1) = 2[2(1) + 1] = 6$ electrons in the $4p^6$ subshell

The total electrons so far allowed are

$2(K \text{ shell}) + 8(L \text{ shell}) + 18(M \text{ shell}) = 28$ electrons

Since $Z = 35$, the remaining seven electrons will populate the N shell. Of these, two electrons will fill the first subshell and five will enter the second subshell *but not close it*. The notation for bromine $Z = 35$ is

$$(1s)^2(2s)^2(2p)^6(3s)^2(3p)^6(3d)^{10}(4s)^2(4p)^5$$

The periodic table (Table 2-1) shows bromine (Br)35 as $4s^2 4p^5$. This is the N shell. A simple model is shown in Fig. 2-1.

The atomic model discussed so far was for electrically neutral atoms only and for atoms in their *ground state*. It was noted, however, that each of the n orbits has its own characteristic energy level. If for some reason (an electric discharge) the electron in its ground state is excited enough so that its energy level is increased, this increase of energy may be enough to knock it completely out of the atom. On the other hand, it may increase the energy of the atom by some discrete amount and cause the electron to enter one of the larger orbits. When an atom loses one or more of its electrons, it is said to be *ionized*. When an electron occupies a higher energy orbit, it is said to be in an *excited* state. An electron in the excited state will jump back to its ground state with one jump or several jumps. If it returns to its ground state in one jump [Fig. 2-2(a)], it will emit an amount of energy characteristic of that jump. This packet of energy is called a *quantum*. If it returns to its ground state through several jumps [Fig. 2-2(b)], each jump will require a release of a quantum of energy characteristic of that jump.

Table 2–1

Legend:

Electron configuration →

2	6 ← Atomic number
4	0.77 ← Covalent atomic radius (Å)
	C ← Symbol
	$2s^2 2p^2$ ← Electron distribution
	12.011 ← Atomic mass

TRANSITION METALS

1 1 0.37 H $1s^1$ 1.008								
2-1 3 1.52 Li $1s^2 2s$ 6.939	2-2 4 1.12 Be $1s^2 2s^2$ 9.012							
2-8-1 11 1.86 Na $2p^6 3s$ 22.99	2-8-2 12 1.60 Mg $2p^6 3s^2$ 24.31							
2-8-8-1 19 2.31 K $3p^6 4s$ 39.102	2-8-8-2 20 1.97 Ca $3p^6 4s^2$ 40.08	2-8-9-2 21 1.60 Sc $3d 4s^2$ 44.96	2-8-10-2 22 1.46 Ti $3d^2 4s^2$ 47.90	2-8-11-2 23 1.31 V $3d^3 4s^2$ 50.94	2-8-13-1 24 1.25 Cr $3d^5 4s$ 52.00	2-8-13-2 25 1.29 Mn $3d^5 4s^2$ 54.94	2-8-14-2 26 1.26 Fe $3d^6 4s^2$ 55.85	2-8-15-2 27 1.25 Co $3d^7 4s^2$ 58.93
2-8-18-8-1 37 2.44 Rb $4p^6 5s$ 85.47	2-8-18-8-2 38 2.15 Sr $4p^6 5s^2$ 87.62	2-8-18-9-2 39 1.80 Y $4d\ 5s^2$ 88.91	2-8-18-10-2 40 1.57 Zr $4d^2 5s^2$ 91.22	2-8-18-12-1 41 1.43 Nb $4d^4 5s^2$ 92.91	2-8-18-13-1 42 1.36 Mo $4d^5 5s$ 95.94	2-8-18-14-1 43 1.30 Tc $4d^6 5s$ (99)	2-8-18-15-1 44 1.33 Ru $4d^7 5s$ 101.1	2-8-18-16-1 45 1.34 Rh $4d^8 5s$ 102.90
2-8-18-18-8-1 55 2.62 Cs $4p^6 6s$ 132.91	2-8-18-18-8-2 56 2.17 Ba $5p^6 6s^2$ 137.34	2-8-18-18-18-8-2 57 1.87 See below 71	2-8-18-32-10-2 72 1.57 Hf $5d^2 6s^2$ 178.49	2-8-18-32-11-2 73 1.57 Ta $5d^3 6s^2$ 180.95	2-8-18-32-12-2 74 1.37 W $5d^4 6s^2$ 183.85	2-8-18-32-13-2 75 1.37 Re $5d^5 6s^2$ 186.2	2-8-18-32-14-2 76 1.34 Os $5d^6 6s^2$ 190.2	2-8-18-32-15-2 77 1.35 Ir $5d^9$ 192.2
2-8-18-32-18-8-1 87 2.70 Fr $6p^6 7s$	2-8-18-32-18-8-2 88 2.20 Ra $6p^6 7s^2$	2-8-18-32-18-9-2 89 2.00 See below						

RARE EARTHS

	57	58	59	60	61	62
Lanthanide series Fills $4f$ subshell $(4f)^n 5s^2 5p^6 6s^2$	2-8-18-18-9-2 1.87 La 138.91	2-8-18-20-8-2 1.65 Ce 140.12	2-8-18-21-8-2 1.65 Pr 140.91	2-8-18-22-8-2 1.64 Nd 144.24	2-8-18-23-8-2 1.63 Pm (147)	2-8-18-24-8-2 1.66 Sm 150.35
	89	90	91	92	93	94
Actinide series Fills $5f$ subshell $(5f)^n 6s^2 6p^2 7s^2$	2-8-18-32-18-9-2 2.0 Ac (227)	2-8-18-32-18-10-2 1.65 Th (232.04)	2-8-18-32-20-9-2 Pa (231)	2-8-18-32-21-9-2 1.40 U 238.03	2-8-18-32-22-9-2 Np (237)	2-8-18-32-24-8-2 Pu (242)

TRANSITION METALS

Shells	Z	(value)	Symbol	Config	Weight
2	2	0.93	He	$1s^2$	4.003
2 3	5	0.88	B	$2s^2 2p$	10.81
2 4	6	0.77	C	$2s^2 2p^2$	12.011
2 5	7	0.70	N	$2s^2 2p^3$	14.007
2 6	8	0.66	O	$2s^2 2p^4$	15.9994
2 7	9	0.64	F	$2s^2 2p^5$	18.998
2 8	10	1.12	Ne	$2s^2 2p^6$	201.83
2 8 3	13	1.43	Al	$3s^2 3p$	26.98
2 8 4	14	1.17	Si	$3s^2 3p^2$	28.09
2 8 5	15	1.10	P	$3s^2 3p^3$	30.97
2 8 6	16	1.04	S	$3s^2 3p^4$	32.064
2 8 7	17	0.99	Cl	$3s^2 3p^5$	35.453
2 8 8	18	1.54	Ar	$3s^2 3p^6$	83.948
2 8 16 2	28	1.24	Ni	$3d^8 4s^2$	58.71
2 8 18 1	29	1.28	Cu	$3d^{10} 4s$	63.54
2 8 18 2	30	1.33	Zn	$3d^{10} 4s^2$	65.37
2 8 18 3	31	1.22	Ga	$4s^2 4p$	69.72
2 8 18 4	32	1.22	Ge	$4s^2 4p^2$	72.59
2 8 18 5	33	1.21	As	$4s^2 4p^3$	74.92
2 8 18 6	34	1.17	Se	$4s^2 4p^4$	78.96
2 8 18 7	35	1.14	Br	$4s^2 4p^5$	79.91
2 8 18 8	36	1.69	Kr	$4s^2 4p^6$	83.80
2 8 18 18	46	1.38	Pd	$4d^{10}$	106.4
2 8 18 18 1	47	1.44	Ag	$4d^{10} 5s$	107.87
...2	48	1.49	Cd	$4d^{10} 5s^2$	112.40
...3	49	1.62	In	$5s^2 5p^1$	114.82
...4	50	1.40	Sn	$5s^2 5p^2$	118.69
...5	51	1.41	Sb	$5s^2 5p^1$	121.75
...6	52	1.37	Te	$5s^2 5p^4$	127.60
...7	53	1.33	I	$5s^2 5p^5$	126.90
...8	54	1.90	Xe	$5s^2 5p^6$	131.30
2 8 18 32 17 1	78	1.38	Pt	$5d^9 6s$	195.09
...1	79	1.44	Au	$5d^{10} 6s$	196.97
...2	80	1.55	Hg	$5d^{10} 6s^2$	200.59
...3	81	1.71	Tl	$6s^2 6p^1$	204.37
...4	82	1.75	Pb	$6s^2 6p^2$	207.19
...5	83	1.46	Bi	$6s^2 6p^3$	208.98
...6	84	1.40	Po	$6s^2 6p^4$	(210)
...7	85	1.40	At	$6s^2 6p^5$	(210)
...8	86	2.20	Rn	$6s^2 6p^6$	(222)

RARE EARTHS

Shells	Z	(value)	Symbol	Weight
2 8 18 25 8 2	63	1.65	Eu	151.96
2 8 18 25 9 2	64	1.61	Gd	157.25
2 8 18 27 8 2	65	1.59	Tb	158.92
2 8 18 28 8 2	66	1.59	Dy	162.50
2 8 18 29 8 2	67	1.58	Ho	164.93
2 8 18 30 8 2	68	1.57	Er	167.26
2 8 18 31 8 2	69	1.56	Tm	168.93
2 8 18 32 8 2	70	1.70	Yb	173.04
2 8 18 32 9 2	71	1.56	Lm	174.97
2 8 18 32 25 8 2	95		Am	(243)
2 8 18 32 25 9 2	96		Cm	(247)
2 8 18 32 26 9 2	97		Bk	(249)
2 8 18 32 28 8 2	98		Cf	(251)
2 8 18 32 29 8 2	99		Es	(254)
2 8 18 32 30 8 2	100		Fm	(253)
2 8 18 32 31 8 2	101		Md	(256)
2 8 18 32 32 8 2	102		No	(254)
2 8 18 32 32 9 2	103		Lw	(257)

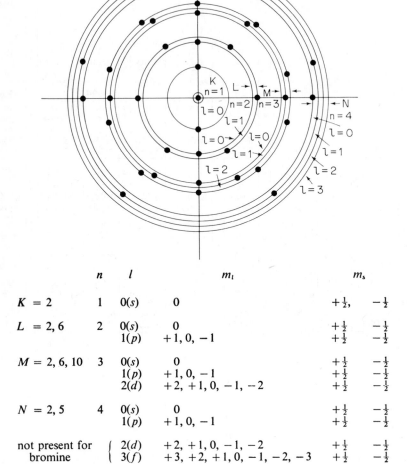

	n	l	m_l	m_s	
$K = 2$	1	0(s)	0	$+\frac{1}{2}$,	$-\frac{1}{2}$
$L = 2, 6$	2	0(s)	0	$+\frac{1}{2}$	$-\frac{1}{2}$
		1(p)	$+1, 0, -1$	$+\frac{1}{2}$	$-\frac{1}{2}$
$M = 2, 6, 10$	3	0(s)	0	$+\frac{1}{2}$	$-\frac{1}{2}$
		1(p)	$+1, 0, -1$	$+\frac{1}{2}$	$-\frac{1}{2}$
		2(d)	$+2, +1, 0, -1, -2$	$+\frac{1}{2}$	$-\frac{1}{2}$
$N = 2, 5$	4	0(s)	0	$+\frac{1}{2}$	$-\frac{1}{2}$
		1(p)	$+1, 0, -1$	$+\frac{1}{2}$	$-\frac{1}{2}$
not present for	{	2(d)	$+2, +1, 0, -1, -2$	$+\frac{1}{2}$	$-\frac{1}{2}$
bromine	{	3(f)	$+3, +2, +1, 0, -1, -2, -3$	$+\frac{1}{2}$	$-\frac{1}{2}$

Figure 2-1

2-2 LATTICE STRUCTURES

As we have seen, atoms are the building blocks of all materials. They are put together in a great variety of ways and bonded or "held together" by cohesive forces in a manner characteristic of a particular material.

In the *liquid state* the atoms of a metal are said to be in somewhat

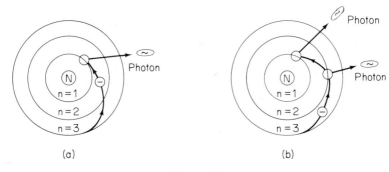

Figure 2-2

random arrangement, having short-range order. At times several unlike atoms will arrange themselves in the characteristic pattern of a particular metal. However, this is a probability event. Since the forces are weak and there is much activity taking place, they soon separate and re-form again. This phenomenon of random grouping, scattering, and regrouping for short periods of time is characteristic of the liquid state. As the random grouping mechanism becomes less frequent and the atomic movement of unlike atoms becomes more agitated, the material may become a gas.

As the energy input decreases, the random movement of the unlike atoms becomes less frequent, the bonding becomes stronger, and ordered arrays of atoms form *lattices*. These lattices form *crystals,* and many crystals* form a pattern, which we call the *solid* material.

Amorphous materials are those that retain their random disorder even when cooled into the solid state. Glass is such a material and is referred to as a supercooled liquid.

In all instances, all matter tends toward the equilibrium state, which is the lowest energy state. As is the case with sodium chloride, the neutral sodium atom has two electrons in its first shell, eight in its second shell, and one loosely bound electron in its third shell, whereas neutral chlorine has two electrons in its first shell, eight in its second shell, and seven electrons in its third shell. Since each atom attempts to complete its shells, to either 2, 8, 18, 32, etc., the neutral sodium gives up its one electron, which immediately migrates into the vacant seven-electron third shell of the chlorine atom. Thus the sodium atom by virtue of having lost one negative charge becomes a *positive ion,* because its nucleus still retains the same number of charges. The chlorine, on the other hand, has picked up this extra negative charge without changing the net positive configuration of its nucleus. Because of this extra electron, the atom becomes a *negative ion*.

*When one is speaking of many crystals, the structure is referred to as the *grain structure*.

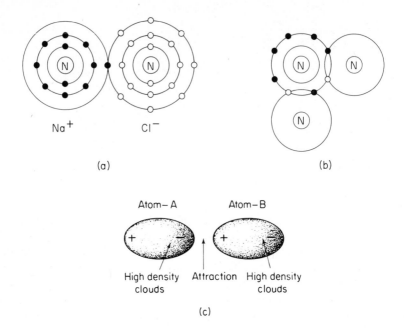

Na⁺ Cl⁻

(a)

(b)

Atom—A   Atom—B

High density   Attraction   High density
clouds   clouds

(c)

Figure 2-3

The positive and negative ions attract each other, as shown in Fig. 2-3(a). Six chlorine ions surround each sodium ion; likewise, the chlorine ion is surrounded by six sodium ions. The forces are, therefore, equal in all directions. The bonding mechanism, referred to as *ionic bonding,* takes place in many nonmetallic materials.

In Fig. 2-3(b), one oxygen atom is shown with six (black dots) of its outer electrons. Each electron (small circles) from the hydrogen atoms is shown completing the eight-electron shell and thus binding both hydrogen atoms to the oxygen to form a water molecule. This type of union is known as *covalent bonding.*

The *Van der Waal forces* are a third system of binding forces. These forces are generated at the instant that an electron cloud density occurs at one side of an atom during the electron flight about the nucleus. This creates a dipole wherein one side of the atom becomes electrically charged negative and the other side has a deficiency of electrons and is considerably charged positive. The atom is distorted as shown in Fig. 2-3(c).

If this atomic dipole approaches or is near another atomic dipole, they will attract each other if they are synchronized. These attracting forces may be considerable if the atoms are close together or if there are a greater number of electrons per molecule. One dipole atom may also cause a "spherical" atom to instantaneously become a dipole and generate a "chain" reaction of dipoles.

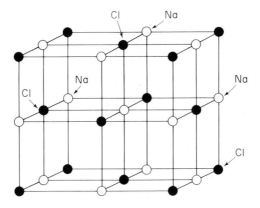

Figure 2-4

The fourth system of forces results in *metallic bonding*. If we consider the sodium chloride structure in Fig. 2-4, a careful examination shows that the center ion of sodium is surrounded by six oppositely charged chlorine ions. The forces between these six chlorine ions and the sodium ion are forces of attraction. They put the imaginary connecting "links" in tension. The other sodium ions repel the center sodium ion and therefore put these imaginary connecting links (not shown) into compression. That is, the Na^+ ion is pulling on all six Cl^- ions and pushing on all *other* Na^+ ions. The system is thus held together by these *ionic* forces.

Another system of forces operating in this metallic substance is *free-electron constraint*. This is due to the resonating of many covalent electrons shared by atoms. This comparatively free motion between atoms forms a negative cloud — almost a matrix — about the positive ions created by this movement of electrons. The vibration requires energy. Therefore, the energy level of the system is reduced, thus increasing the attraction forces between the atoms. The random vibration of the "free" electrons, constrained by the "cloud" within the structure, also makes the material electrically conducting. This is the case with metals.

In addition, the electron-filled outer shells of the metallic ions repel each other. This force results from the repulsion forces of the negative shells and is referred to as *electron interference*.

Many metals are bonded by combinations of the force systems just discussed.

The forces of attraction due to free-electron constraint and those of repulsion, which result from electron interference, together with the atomic repulsion–attraction make up the system of forces, referred to as metallic binding, which are called *rigidity*. A pattern of atomic structure such as that shown in Fig. 2-4 results.

The system, as indicated, is one in which the atoms vibrate about a central equilibrium point. They are not static, even though the system is

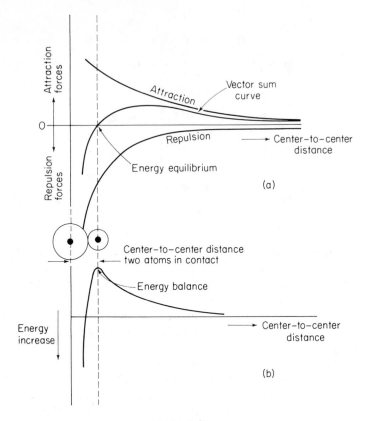

Figure 2-5

said to be rigid. They are said to be in dynamic equilibrium. As the energy of the system increases, the vibration of the atoms increases and the material expands, because the lattice parameters increase and the density of the crystal decreases.

If the atoms are considered solid spheres, then two atoms will approach each other until the forces of attraction balance those of repulsion. This point is taken to be the diameter of the atom [see Fig. 2-5(a)]. At this point, the net energy of the system will be a minimum [Fig. 2-5(b)]. The force required to push the atoms closer together or pull them farther apart increases.

The distance between atoms is different for different materials, since it is related to the number of shells populated by electrons and the number of valence electrons present. Thus the greater the number of shells, the greater the distance between the centers of the atoms that "touch." The greater the number of valence electrons present, the shorter the center-to-center distance between the atoms.

The imaginary lines that connect the centers of the atoms in a configuration are called *lattice structures*. Fortunately, the most common metals crystallize in one of three types of lattice systems: cubic, hexagonal, and tetragonal.

Figure 2-6(a) shows eight atoms "touching" each other to form a *simple cubic lattice* structure. The sides of the lattice are equal in length and form a cube. This unit cube may have another atom at its center [Fig. 2-6(b)], in which case it is called a *body-centered cubic lattice* (BCC). The unit cube may also have an atom in each of its faces as shown in Fig. 2-6(c). This lattice is called a *face-centered cubic lattice* (FCC).

Body-centered cubic lattice structures are found in the following metals: barium, cesium, potassium, lithium, molybdenum, sodium, rubidium, tantalum, vanadium, chromium, iron (alpha and delta), and tungsten.

Face-centered cubic lattice structures are found in the following metals: aluminum, copper, gold, iron (gamma), lead, nickel, platinum, and silver.

The *close-packed hexagonal* (CPH) lattice structure is shown in Fig. 2-6(d). Each of the three units that combine to form the close-packed hexagonal lattice is divided into two equilateral triangles. Perpendicular bisectors are dropped to the three sides as shown. An atom lies on this line entirely within the unit cell wedged in between its upper and lower planes.

Materials that have a CPH structure are beryllium, cadmium, cobalt, mangesium, osmium, tellurium, titanium, zinc, and zirconium.

Figure 2-6(e) shows a *body-centered tetragonal* (BCT) lattice. It is an elongated body-centered cubic lattice structure. Tin and manganese at elevated temperatures exhibit this lattice structure.

The atoms of all lattice structures are shared by other lattice structures that "touch." Therefore, atoms that share can only contribute a portion of their energies to a single lattice.

Example 3

Calculate the effective number of atoms contributing forces to the lattice structures of the following configurations: (a) BCC, (b) FCC, (c) CPH.

solution:

    (a) BCC

        Each corner atom is shared by 8 other lattices.

$$8 \times \tfrac{1}{8} = 1 \text{ atom}$$

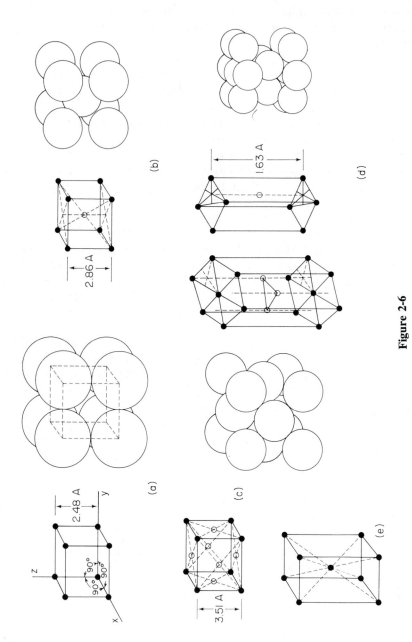

Figure 2-6

The center atom is entirely within the lattice.

The center atom = 1

Thus $1 + 1 = 2$ atoms per lattice.

(b) FCC

Each corner atom is shared by 8 other lattices.

$8 \times \frac{1}{8} = 1$ atom

Each face atom is shared by another lattice.

$6 \times \frac{1}{2} = 3$ atoms

Thus $1 + 3 = 4$ atoms per lattice.

(c) CPH

Each corner atom is shared by 6 other lattices in the upper plane. Since the lattice has both an upper and a lower plane, the atoms in one lattice contributed by both planes are

$2(6 \times \frac{1}{6}) = 2$ atoms

Each of the atoms in the upper and lower planes contributes half of its energies to the lattice.

$2 \times \frac{1}{2} = 1$ atom

There are 3 atoms entirely within the lattice. Thus

$2 + 1 + 3 = 6$ atoms per lattice

It is important to note that several of the metals in common use exhibit lattice structure changes as their temperatures increase.

Chromium: close-packed hexagonal, below 68°F
              body-centered cubic, 68 to 3270°F
Cobalt: close-packed hexagonal, below 788°F
              face-centered cubic, 788 to 2723°F
Iron: body-centered (α) cubic, below 1670°F
              face-centered (γ) cubic, 1670 to 2552°F
              body-centered (δ) cubic, 2552 to 2795°F
Manganese: cubic (α), below 1252°F
              cubic (β), 1252 to 2012°F
              tetragonal (γ), 2012 to 2080°F
              tetragonal (δ), 2080 to 2273°F
Tin: cubic (α), below 55.8°F
              tetragonal (β), 55.8 to 450°F
Titanium: close-packed hexagonal, below 1615°F
              body-centered cubic, 1615 to 3140°F

2-3   CRYSTAL FORMATION

Atomic motion in the liquid state of a metal is almost completely disordered. The movement of these atoms is without regard to the average of the ordered distances between atoms in the lattice structure so necessary in metals. As the energy in the liquid system decreases, the movement of the atoms decreases, the electrostatic forces become more effective, and the probability increases for the arrangement of a number of atoms into a lattice characteristic of that material. The energy level (temperature) at which these isolated lattices form is called the *freezing point*. If the gain and loss of heat energy of the system are in equilibrium, the production rate and melting rate of the crystals are the same, and the system exists as both a solid and a liquid.

The formation of lattices generates heat. As this heat is removed, the lattice growth continues about the few lattices that have already formed and about new lattices that are forming. This growth continues until stopped by some energy block, another lattice growth, or the wall of the container. Large crystals, if allowed to grow without any interference, will exhibit patterns characteristic of the material solidifying. Figure 2-7 shows the idealized representation of the mechanism of solidification. The growth stops at about one to two lattice spacings away from the next ordered growth. The space separating these lattice growths is called the *grain boundaries*.

Crystals are aggregates of space lattices that exhibit the same orientation. The terms *grains* and *crystals* are used interchangeably. The end result of the mechanism just described is the solid material.

It should be noted that a foreign body must be present to act as a nucleus for lattice growth to start. For very pure metals, which require time for the nucleus to grow, rapid cooling could drop the temperature of the system below the freezing temperature. As the nucleus forms, the temperature increases and then levels off as the lattice formation increases. This is called *supercooling* and is shown in Fig. 2-10(a).

It should be remembered that Fig. 2-7 shows growth in one plane. However, crystal growth takes place in three-dimensional space and in all directions. The growths *within* the crystals themselves are at right angles to each other, as shown in Fig. 2-8. The orientation of any adjacent crystal will be in a new direction. If a set of axes of adjacent crystals were aligned with its neighbor, it is probable that they would combine into one larger crystal.

The size of a grain is related to the nature of the metal, the temperature from which the liquid cools, and the cooling rate. Rapid removal of

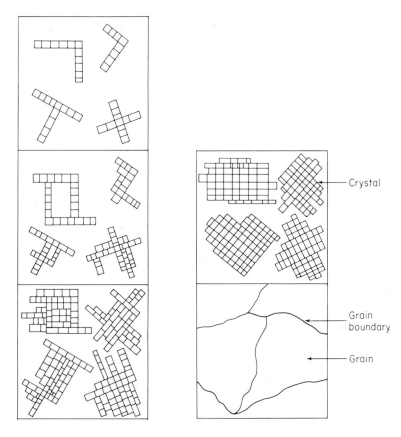

Figure 2-7

heat per unit time produces small grain structures. By slow cooling, some metals crystallize below the equilibrium melting point. If nuclei are formed at a rapid rate (nucleation) in comparison to the growth of the grains, the grains will not grow large. If the metal is cooled very rapidly and, as a result, the ratio of the nucleation to rate of growth increases, the grain will be very small. There is a break-even decrease in temperature rate below which the rate of growth increases with relationship to the rate of nucleation and, as a result, the grains grow large. In most metals, however, rapid cooling produces fine grain structures, and slow cooling produces large grains.

As the crystal grows, it releases heat energy, which increases the temperature in the direction and just ahead of growth. This increase in energy blocks the growth in that direction by blocking the addition of any

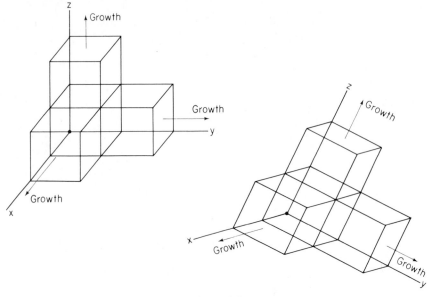

Figure 2-8

more electrons to the growing face. This blockage lasts until the heat energy is dissipated. The growth, which must continue, starts off in another direction, perpendicular to the former growth.

Metals that have a cubic lattice structure grow more rapidly in a direction perpendicular to each other. This type of growth is called *dendritic*. Dendrites are shown in Fig. 2-9(a) and (b).

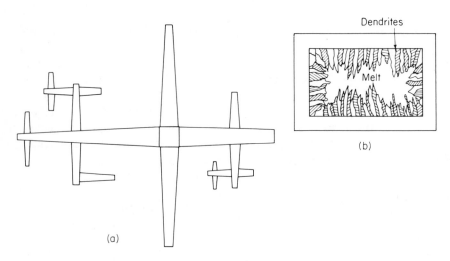

Figure 2-9

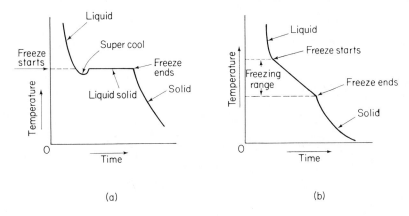

Figure 2-10

As growth continues, two adjacent dendrites touch each other to form *grain boundaries*. Since they solidify last, they give atoms a chance to diffuse and partially to equalize stressed regions in the boundaries. Some of the stresses remain and can be removed by subsequent heat treating.

The solidification process at this point is complete.

If water is permitted to freeze from the liquid to the solid state, the temperature at which solidification takes place is constant. As soon as solidification is complete, the temperature of the solid starts to decrease. A pure metal, when cooled from the liquid state, acts in a similar manner, and if the curve (temperature versus time) is plotted, Fig. 2-10(a) results. The flat curve (constant temperature) results from the giving off of the *latent heat of fusion*.

Figure 2-10(b) shows a metal that has two constituents. The cooling of the melt produces a curve somewhat like that of a pure metal. At the temperature at which freezing starts, the first solid that forms is richer in that constituent which has the higher freezing temperature. At the temperature at which the last freezing takes place, the metal is richer in that constituent which has the lower freezing point. This mechanism of multiple structure is called *coring*. The multiple structure results in some undesirable physical properties, which may be controlled by normalizing. The coring process will be discussed in a subsequent chapter.

2-4 MILLER INDEXES

At this point it becomes necessary to introduce a system for defining a plane in space. This is accomplished by a system known as the *Miller index*. The method is as follows:

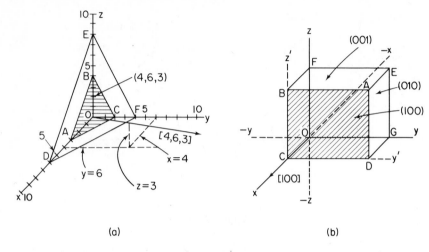

(a)   (b)

Figure 2-11

1. Coordinate axes are constructed so that one corner of the lattice is at the origin, as shown in Fig. 2-11(b). Note the position of the *x, y, z* axes.

2. The *intercept* of the plane with each axis is recorded.

3. The *reciprocal* of these intercepts constitutes the Miller indexes.

The cross-hatched plane *ABC* in Fig. 2-11(a) is recorded as follows:

1. The intercept of the plane with the *x*-axis is at 3; with the *y*-axis it is at 2; and with the *z*-axis it is at 4.

2. The reciprocal of these intercepts is *x* equals $\frac{1}{3}$, *y* equals $\frac{1}{2}$, and *z* equals $\frac{1}{4}$.

3. The least common denominator of these three fractions is 12. To clear the fractions, multiply through by 12. Thus

$$x = \tfrac{12}{3}, \qquad y = \tfrac{12}{2}, \qquad z = \tfrac{12}{4}$$

The Miller index is

$$x = 4, \qquad y = 6, \qquad z = 3$$

The notation is (463).

	x	*y*	*z*
intercept	3	2	4
reciprocal	$\frac{1}{3}$	$\frac{1}{2}$	$\frac{1}{4}$
index	4	6	3

4. Plane *DEF* intercepts the *x, y, z* axes at *x* = 6, *y* = 4, *z* = 8.

The reciprocals are

$$x = \tfrac{1}{6}, \qquad y = \tfrac{1}{4}, \qquad z = \tfrac{1}{8}$$

The multiplier is 24.

$$x = \tfrac{24}{6}, \qquad y = \tfrac{24}{4}, \qquad z = \tfrac{24}{8}$$

	x	*y*	*z*
intercept	6	4	8
reciprocal	$\frac{1}{6}$	$\frac{1}{4}$	$\frac{1}{8}$
index	4	6	3

The Miller index is

$$x = 4, \qquad y = 6, \qquad z = 3$$

and the notation is (463) and *represents the same plane orientation* as shown in item 3.

It is sometimes desirable to be able to draw a vector perpendicular to a family of planes that have the Miller index (463) as shown in Fig. 2-11(a). To do this, the index numbers are enclosed in brackets and traced from the origin through the x, y, z Miller numbers. This will yield a point on the arrow, as shown in Fig. 2-11(a). The other point is at the *origin*. This arrow will also be *perpendicular* to all planes parallel to (463).

In Fig. 2-11(b), plane $ABCD$ is the face of a *unit* cube. Arbitrarily, the atom at 0 is taken as the origin for the x, y, z axes. Since the plane $ABCD$ is parallel to the y-axis, it intersects that axis at infinity. This plane is also parallel to the z-axis and therefore intercepts the z-axis at infinity. It intercepts the x-axis at $+1$. Therefore, the reciprocals are

$$x\text{-axis} = \frac{1}{1} = 1$$

$$y\text{-axis} = \frac{1}{\infty} = 0 \qquad (100)$$

$$z\text{-axis} = \frac{1}{\infty} = 0$$

The Miller index is (100). The parentheses represent the specific plane described.

Since the origin was taken arbitrarily to be at 0 and since plane $EGOF$ goes through the origin and cannot be defined, it is permissible to move the origin to C. Under these circumstances the plane $EGOF$ intersects the

$$x\text{ -axis} = -1 = \bar{1}$$

$$y'\text{-axis} = \infty = 0 \qquad (\bar{1}00)$$

$$z'\text{-axis} = \infty = 0,$$

The family of (100) planes that have the same form is written with brackets [100]. Such a notation is the shorthand for

$$[100] = (100), (010), (001), (\bar{1}00), (0\bar{1}0), (00\bar{1})$$

In the cubic structure an arrow indicates the direction perpendicular to this family of planes. The reciprocal notation is not used. Therefore, in Fig. 2-11(b), starting at the origin and tracing the coordinates in the order 100, we determine the other point in space. The first point is at the origin. The coordinates of this arrow, which points in a direction perpendicular to the family of planes, are written in brackets [100].

Example 4

(a) Write the Miller index for the plane in Fig. 2-12(a). (b) Draw and identify the direction perpendicular to this plane.

solution:

(a)   The intercepts are
x-axis $= 4$

y-axis $= 3$

z-axis $= 2$

The reciprocals are
x-axis $= \frac{1}{4} = \frac{12}{4} = 3$

y-axis $= \frac{1}{3} = \frac{12}{3} = 4$

z-axis $= \frac{1}{2} = \frac{12}{2} = 6$

The Miller index is (346).

(b)   The direction perpendicular to this plane is shown in Fig. 2-12(b) and is designated as [346].

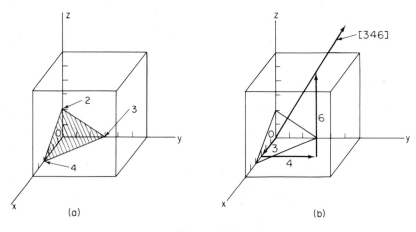

Figure 2-12

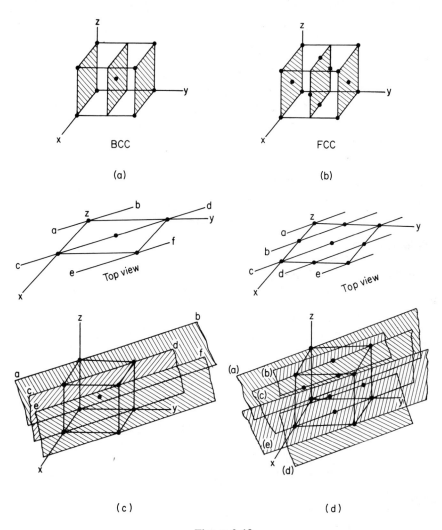

Figure 2-13

The population of several planes for the BCC and FCC cubic lattice is shown in Fig. 2-13(a, b, c, d). The top views show the planes as they cut through the various atoms that lie in that plane.

Example 5

Examine the following planes and indicate the packing of atoms in one lattice for (111), (100), (110), (430), (220), ($\bar{1}$11): (a) in the BCC and (b) in the FCC.

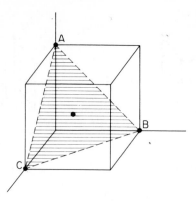

Figure 2-14

solution:

(a)   For the BCC plane:

Plane	No. atoms		Plane	No. atoms
111	3		430	0
100	4		220	0
110	5		$\bar{1}$11	3

(b)   For the FCC plane

Plane	No. atoms		Plane	No. atoms
111	6		430	1
100	5		220	2
110	6		$\bar{1}$11	6

The plane (111) is shown to pack three atoms, A, B, C, in Fig. 2-14. The center atom is not cut by the plane.

Problems

2-1.   (a) Describe the atomic model as explained in this chapter. (b) State the magnitude of the charge on one electron, one proton, and one neutron. (c) Which of these three particles are referred to as nucleons? (d) State the mass of each of the three particles.

2-2.   How many protons, neutrons, and electrons are there in: (a) $^{234}_{90}$Th; (b) $^{10}_{5}$B; (c) $^{27}_{13}$Al; (d) $^{9}_{4}$Be; (e) $^{65}_{29}$Cu; (f) $^{52}_{24}$Cr; (g) $^{7}_{3}$Li; (h) $^{56}_{26}$Fe; (i) $^{210}_{83}$Bi; (j) $^{17}_{9}$F; (k) $^{38}_{18}$Ar; (l) $^{30}_{14}$Si.

2-3.   (a) How do isotopes differ from the most abundant configuration for any one element? Check a table of isotopes, select an element and at least one isotope for each element, and show how it differs from the base element. (b) When is an atom said to be ionized?

2-4. Hydrogen has two isotopes. Name them. How many neutrons, protons, and electrons are there in hydrogen and its isotopes?

2-5. (a) Indicate the forces operating in an atom and show how they affect the stability of the structure. (b) How does the orbital electron model differ from the orbital earth–sun model?

2-6. Calculate the maximum number of electrons in each of the shells for $_{14}^{28}$Si.

2-7. Calculate the maximum number of electrons in each of the shells for $_{18}^{40}$A.

2-8. Explain the four quantum numbers used when subdividing the shells into subshells.

2-9. A selenium (Se) atom has an atomic number of $Z = 34$. Calculate the number of electrons in the shells and subshells.

2-10. A krypton atom has an atomic number of $Z = 36$. Calculate the number of electrons in the shells and subshells.

2-11. A lead atom has an atomic number of $Z = 46$. Calculate the number of electrons in the shells and subshells.

2-12. List the shells and subshells in a chart similar to that shown in Fig. 2-1 for the elements in Problems 2–9 through 2–11.

2-13. Explain the mechanism of energy levels and the results of electrons (a) leaving their shells, and (b) returning to ground state.

2-14. Explain the mechanism (a) of random grouping in the liquid state of a material, and (b) of lattice formation.

2-15. Describe ionic bonding. Support your explanation with a sketch of the model.

2-16. Describe covalent bonding and support your explanation with a sketch.

2-17. Describe metallic bonding, using a model to support your description.

2-18. Explain the mechanisms of "free electron constraint" and "electron interference" as forces acting to bind a metallic structure together.

2-19. In your own words, describe what you understand dynamic equilibrium to mean when related to atoms and lattice structures.

2-20. How are the "sizes" of atoms determined? Explain fully.

2-21. Draw the lattice structures of (a) the simple cube, (b) BCC, (c) FCC, (d) CPH, and (e) BCT.

2-22. Calculate the whole number of atoms contributed to a single lattice structure for (a) BCC, (b) FCC, and (c) CPH.

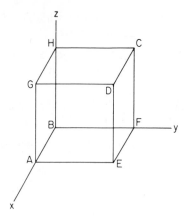

Figure 2-15

2-23. Describe the mechanism by which metals freeze to form crystals.

2-24. Describe dendritic growth and how it progresses.

2-25. Draw and describe the cooling curves (a) when a pure metal freezes and (b) when an alloy of two metals freezes.

2-26. In Fig. 2-15 calculate the Miller index for each of the planes *AGDE, CDEF, CDGH, EFGH,* and *AGCF.*

2-27. Calculate the Miller indexes for Fig. 2-16(a), (b), and (c).

2-28. Draw the direction arrow perpendicular to each of the planes in Problem 26.

2-29. Draw the direction arrow perpendicular to each of the planes in Problem 27.

2-30. Determine the atom packing in the following planes for the BCC and the FCC: (a) 100, (b) (011), (c) (0$\bar{1}$2), (d) (122), (e) (121).

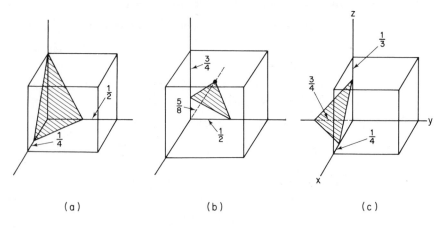

(a)   (b)   (c)

Figure 2-16

3 | Inspection and Testing of Materials

3-1 PREPARATION OF PHOTOMICROGRAPHIC SPECIMENS

Two very useful tools employed in the study of the internal structure of metals are the metallurgical microscope [Fig. 3-1(a)] and the metallograph [Fig. 3-1(b)]. Both instruments use reflected light to study the surface structure of a prepared specimen.

The first step is to *choose a sample* of the material to be tested. This sample is then cut with great care from the section of the material that is

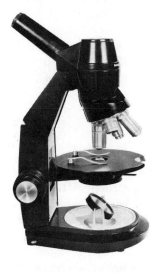

(a)

(b)

Figure 3-1

to be examined. Care must be taken not to do anything to the material that will change its structure. Overheating the sample will change its structure and must be avoided.

Once cut, the sample piece is either ground by hand, or, if the sample is small, *mounted* in a Lucite or Bakelite mold before grinding. The *rough grinding* should remove all evidence of the cutting operation and leave a flat smooth surface. Movement of the specimen at right angles to the previous grind (or polish) will make it easier to detect the removal of marks left from previous processing.

After rough grinding, *preliminary polishing* with successively finer and finer emery paper is the next step. These polishing operations are done dry; No. 1, 1/0 through $\frac{1}{4}$ polishing papers are used.

Fine polishing is accomplished with wet rotating discs. These discs are covered with polishing cloths that are charged with an abrasive powder suspended in water. The type of polishing cloth and the abrasive used depend upon the material being polished. The particles removed from the samples are generally washed away by having a few drops per second of water fall on the rotating wheel. Again, the purpose is to get a flat, scratch-free surface.

After polishing, samples should show only impurities. *Etching* makes the microstructure visible to the microscope by highlighting composition differences, differences in orientation of grain structure, and selectively attached grain boundaries.

Figure 3-2(a) shows a sample immediately after it has been polished. The metal, as a result of plastic flow, has covered the grain orientation so that it is not visible. Figure 3-2(b) shows the grain orientation after the etchant has done its work. Those grain surfaces oriented perpendicular to the light source reflect the greatest percentage of light along a line parallel to the incoming light and therefore appear almost white. Those that reflect at an angle to the incoming light cause the surface to appear dark. The greater the angle of reflection, the darker the appearance of the grain. Figure 3-2(c) shows grains, grain boundaries, etc., in a photomicrograph of steel.

Etchants and the materials upon which they are used are listed in Table 3-1.

Once the specimens have been prepared, they may be viewed through a microscope. The optical system of a metallurgical microscope is shown in Fig. 3-3(a). It has a capability of magnifying structures by as much as 2000 times. High magnifications of this order generally require the use of oil-immersion lenses. A drop of specially prepared oil (cedar oil) is placed on the polished specimen. The objective lens is brought into contact with the oil. The lens is withdrawn, leaving a layer of oil on the sample and the lens. The effect is to increase resolution, or the light-gathering power, of

Table 3-1 Etching Reagents

Alloy material base	Etching reagent	Composition	Remark
Lead	Nitric acid	5% in water	Dip 5–15 seconds (s)
Aluminum	Hydrofluoric acid	0.5% in water	Swab 5–15s
	Sodium hydroxide	1% in water	Swab 5–15s
	Keller's etch	1% HF, 1.5% HCL, 2.5% HNO$_3$, 95% H$_2$O	Dip 30–60s
Copper	Hydroxide–peroxide	50% H$_2$O$_2$–50% NH$_4$OH	Swab 5–15s
	Ferric chloride	5g FeCL$_3$–10 cm^3 HCl–100 cm^3 H$_2$O	Swab 5–15s
Nickel	Mixed acid	50% HNO$_3$ (conc.)–50% glacial acetic acid	Swab 10–15s
Iron			
Low carbon	Nital	2% HNO$_3$ in ethyl alcohol	Dip 10–30s
Welds	Nital		
Wrought iron	Nital		
Malleable iron	Nital		
Martensite	Nital		
Pearlitic	Picral	5% Picric acid in ethyl alcohol	Dip 15–30s
Sorbite	Picral		
Cast iron	Picral		
Med. st.	Picral		
High carbon	Picral		
Stainless	Mixed acid (or)	75% HCL, 25% HNO$_3$	Swab 10–30s
Austenitic	Ferric chloride	5g FeCL$_3$–50 cm^3 HCL–100 cm^3 H$_2$O	Swab 10–30s

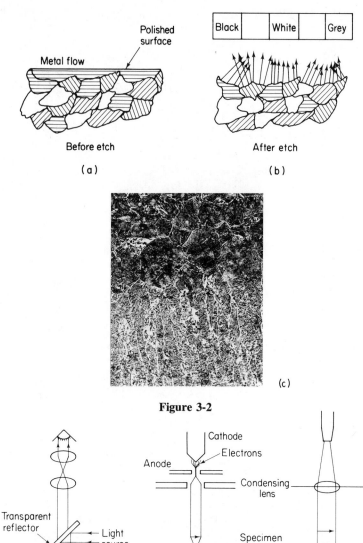

Figure 3-2

Figure 3-3

the lens so that clarity of image may be retained as the magnification increases.

The electron microscope in Fig. 3-3(b) is an instrument that uses a high-velocity electron beam as an energy source. Whereas optical microscopes use lenses as light-gathering and focusing mechanisms, the electron microscope uses a magnetic field to control and focus an electron beam. Because of the very much shorter wavelengths of electron waves as contrasted with light waves, the resolution of an electron microscope is very much higher than that of the optical microscope. Magnifications are of the order of 35,000 diameters, and may be increased when accessories are used to 200,000 diameters. Figure 3-3(b) shows the comparison between an electron microscope and an optical system.

3-2 ELASTIC AND INELASTIC DISLOCATION DEFINITIONS

Stress σ is defined mathematically as the ratio between a load F applied to a sample and the original cross-sectional area A_o.

$$\sigma = \frac{F}{A_o} = \frac{\text{lb}}{\text{in.}^2} = \text{psi}$$

$F = \text{force, lb}$
$A_o = \text{original area, in.}^2$
$\sigma = \text{stress, psi}$

Strain ϵ is defined mathematically as the ratio between the change in length Δl of a stressed sample to the original length l_o of that sample.

$$\epsilon = \frac{\Delta l}{l_o} = \frac{\text{in.}}{\text{in.}} = \text{no. units}$$

$l_o = \text{original length, in.}$
$\Delta l = \text{change in length, in.}$
$\epsilon = \text{strain, in./in.}$

Once a loaded sample starts to neck down, another concept called *true stress* and *true strain* is used. True stress is obtained from the cross-sectional area of the necked-down portion at the instant the load is applied. The true strain is obtained from the summation of all the $\Delta l/l$ segments from zero to rupture load. Δl is the change in length corresponding to an increment loading, and l is the actual length at this increment load. We shall restrict our discussion for the most part to actual stress–strain concepts rather than true stress–strain.

A series of events takes place when a specimen is at room temperature (68–70°F) and is loaded in various ways. *Elastic strain* occurs when a load is applied to a sample that deforms it. Upon removal of the load, the sample returns to its original length. *Inelastic strain* occurs when the sample *does not* return to its original shape after the deforming load is removed. *Rupture* is the limiting state where separation takes place.

The *modulus of elasticity E* is the ratio of stress to strain:

$$E = \frac{\sigma}{\epsilon} = \text{psi}$$

σ = stress, psi
ϵ = strain, in./in.
E = mod. of elast., psi

This ratio E is valid only for elastic strain and is measured from the stress–strain curves shown in Fig. 3-4(a). The graph in Fig. 3-4(a) is the first part of the graph in Fig. 3-4(b) (note the abscissa increments on both graphs). The value E for a material shows the elastic resistance to an applied load that causes deformation. It is a measure of the *stiffness* of a material. Thus,

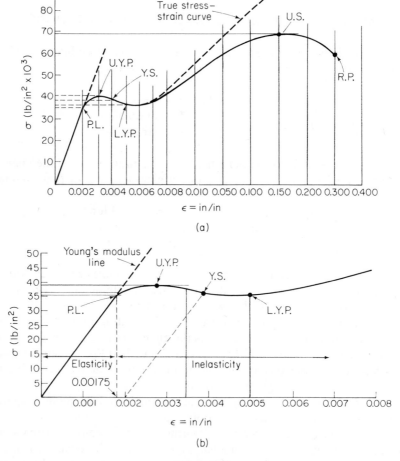

(a)

(b)

Figure 3-4

if the modulus of elasticity of carbon steel is 30×10^6 psi and aluminum is 10×10^6 psi, then a steel beam will support a three times greater load than an aluminum beam of the same dimensions and configuration, and with the same deflection. Some representative values for E for various materials are shown in Table 3-2.

Figure 3-4(a) shows the plot of certain increments of elongation when certain loads are applied.

Proportional Limit (P.L.). On the curve of Fig. 3-4(a) from 0 to the P.L., the stress applied to a sample is directly proportional to the strain, and the curve is a straight line. At that point the curve is marked P.L. It is the place on the curve where it deviates for the first time from a straight line. The material at this load is still elastic.

Elastic Limit (E.L.). A very short distance farther along on the curve beyond P.L. is the last point of elasticity for this sample. The elastic limit is the maximum load that can be applied to the sample *without* permanently deforming it. This value is determined by applying and then removing a load followed by a precise measurement. The load is increased slightly and the process repeated. The last measurement that does not produce a permanent set is called the elastic limit.

Yield Strength (Y.S.). If a slight increase in loading is applied to the elastic limit, the material will deform for the first time without any increase in loading. The method for finding the yield strength of the sample is to draw a line parallel to the curve [Fig. 3-4(a)] from some value for ϵ (usually between 0.001 and 0.003, indicated as 0.1 to 0.3 per cent) on the abscissa. The value of σ is taken from the point of intersection of this line and the curve.

Referring to the curve will show that its slope is horizontal just beyond the Y.S. point. At this point, elongation takes place without the need of applying any additional load. The elongation continues for a time (lower yield point) until the slope of the curve is zero again.

Ultimate Strength (U.S.). This is the point on the curve which represents the highest stress that can be applied to a ductile material before the sample begins to rupture. For brittle materials the rupture point is essentially the same as the ultimate strength. It is shown as the highest point on the curve in Fig. 3-4(b). Thus, increasing the stress will increase the strain along the graph until the point marked U.S. is reached. This is the maximum strength developed by the material in resisting the applied stress based on the original cross-sectional area. It is also referred to as the *tensile strength* of the material.

Table 3-2 Properties of Metals

Material	Density lb/in.³	Sp. gr.	Hardness BHN	Thermal expansion 10^{-6} in./in./°F	Tensile elastic 10^6 psi	Torsion mod. 10^6 psi	Yield st. 0.2% offset $\times 10^3$ psi	Tensile st. 10^3 psi	Elongation in 2 in. %
Carbon st. (1020)									
Annealed	0.284	7.86	130	6.7	30	16	38	65	30
Hot rolled			135				42	68	32
Hardened			179				62	90	25
Gray cast iron	0.260	7.20	180	6.7	13		6	25	0.5
Malleable iron	0.264	7.32	130	6.6	25		33	52	12
Steel (stainless)									
302 annealed	0.281	7.80	155	10.0	28		35	90	60
304 annealed	0.281	7.78	150	10.0	28		30	80	45
431 annealed	0.280	7.75	250	6.5	29	10.6	85	120	25
Wrought iron	0.278	7.70	100	6.35	29		30	48	30
Aluminum									
Annealed	0.103	2.71	23	13	10.4		5	13	45
Cold rolled	0.098		32		10	3.8	17	18	20
Copper									
Annealed	0.322	8.9	42	9.3	17	6.4	10	32	45
Cold drawn			90				40	45	15
Magnesium									
Sand cast	0.066	1.83	50	14.5	6.5	2.4	14	24	6
Manganese–bronze									
Cold drawn	0.302	8.36	180	11.2	15	5.6	50	80	20
Nickel									
Annealed	0.322	8.91		7.8	30	11	8.5	46	30
Red brass									
Cold drawn	0.316	8.75	120	9.8	17	64	55	70	15
Titanium									
Annealed	0.163	4.54	200	5.0	16.5	6.6	70	90	23

Rupture Strength (R.S.). This value is determined by dividing the load at the point of rupture by the cross-sectional area of the original sample. The curve in Fig. 3-4(b) shows that, once the ultimate strength of the material has been exceeded, a localized reduction in cross-sectional area of the sample takes place. This is referred to as "necking down." The separation that results takes place rapidly.

If the stresses are divided by the actual cross-sectional areas for the actual load at that time, the *true rupture strength* of the material can be found. See the dashed curve in Fig. 3-4(b).

Ductility is a matter of inelastic deformation prior to rupture. It is a measure of *toughness*. The ability of a material to absorb energy without rupture is a function of ductility and a measure of toughness.

Ductility is indicated by the tensile property of *percentage of elongation*. Mathematically, it is the percentage of change in length after application of a load compared to the original length of the specimen before the load was applied. Thus

$$\text{Percentage of elongation} = \frac{l_f - l_o}{l_o} \times 100 \qquad \begin{array}{l} l_o = \text{initial gage length} \\ l_f = \text{final gage length} \end{array}$$

Percentage of reduction in cross-sectional area of a specimen is another way to indicate the tensile property of ductility. Thus

$$\text{Percentage of reduction in area} = \frac{A_f - A_o}{A_o} \times 100 \qquad \begin{array}{l} A_f = \text{final area} \\ A_o = \text{initial area} \end{array}$$

If the percentage of elongation and reduction of cross-sectional area are large, the material is said to be *ductile*. When they are low, the material is said to be *brittle*.

Figure 3-5(a) shows standard ASTM specimens used to check tensile properties of materials. The gage length is taken as 2.000 in. A pair of dividers may be adjusted to match these gage marks. The specimen is pulled by applying a desired load, after which the changes in length and area are measured. The percentages of each may then be calculated.

In some instances, it is difficult to detect changes in length with dividers while the elongation is taking place. In such cases, extensometers, or *strain gages,* are used. These instruments magnify slight increases in length so that they can easily be detected, read, and measured. The magnification is accomplished with electrical, optical, or mechanical devices.

Figure 3-5(b) shows a ductile specimen in which the reduction of area and the elongation are uniform. Figure 3-5(c) shows a ductile specimen in which the reduction of area and the elongation are nonuniform. This results from a localized weakness in the material. It results in a

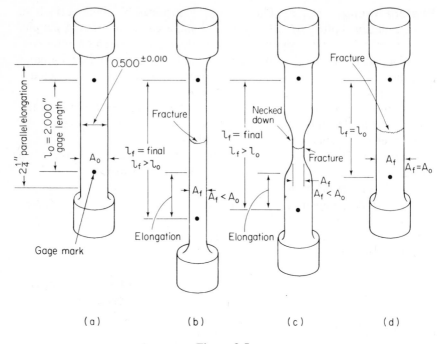

Figure 3-5

marked reduction in area and less elongation before rupture than shown in Fig. 3-5(b). Brittleness in a material is indicated when there is negligible elongation and no reduction in area of the specimen after it ruptures. The latter is shown in Fig. 3-5(d).

Toughness is measured by calculating the area under the stress–strain curve. This is in effect the product of the abscissa times the ordinate. Toughness is a combination of strength and ductility absorbed by a specimen up to the point of rupture.

Example 1

You are given the curve in Fig. 3-4(a) and (b) and the following data: Test bar diameter prior to test is 0.502 in., and after rupture the diameter of the necked-down cross-sectional area is 0.430 in. The gaging length was 2.000 in. before the test and 2.250 in. after rupture. Calculate the following: (a) the modulus of elasticity, (b) the proportional limit, (c) the ultimate strength, (d) the rupture strength for the original area, (e) the rupture strength based on the actual area, (f) the yield strength for 0.02 per cent offset, (g) the percentage of elongation, and (h) the percentage of reduction in area.

solution:

(a) The modulus of elasticity E is equal to the slope of the straight-line portion of the curve. From the curve of Fig. 3-4(b),

$$E = \frac{\sigma}{\epsilon} = \frac{35,000 \text{ psi}}{0.00175} = 20 \times 10^6 \text{ psi}$$

$\sigma = $ stress, psi

$\epsilon = $ strain, psi

(b) The proportional limit (P.L.) is taken from the graph where the stress deviates and is no longer proportional to the strain.

P.L. = prop. limit, psi

P.L. = 35,000 psi

(c) The ultimate strength (U.S.) is taken from the graph at the maximum point of stress, Fig. 3-4(b).

U.S. = ultimate strength, psi

U.S. = 68,000 psi

(d) The rupture strength (R.S.) based on the original area is taken from the graph, Fig. 3-4(b).

R.S. = rupture strength, psi

R.S. = 60,000 psi

(e) The true rupture strength is based on the actual area at rupture. The load at failure is

$$F = (\text{R.S.}_o)A_o$$

$A_o = $ original area

$d_o = $ original diameter

$= 0.502$ in.

$$= 60,000 \left[\frac{\pi}{4}(0.502)^2 \right]$$

$$= 11,900 \text{ lb}$$

The true rupture strength is

$$\text{R.S.} = \frac{F}{A_f} = \frac{11,900}{\frac{\pi}{4}(0.430)^2}$$

$d_f = $ necked-down diameter

$= 0.430$ in.

$$= 82,000 \text{ psi}$$

(f) The yield strength (0.02% offset from the graph) is obtained by constructing a parallel line to the Young's modulus line, Fig. 3-4(a).

Y.S. = 38,000 psi

(g) The percentage of elongation is

$$\text{Percentage of elongation} = \frac{l_f - l_o}{l_o} \times 100 \qquad \begin{array}{l} l_f = 2.250 \text{ in.} \\ l_o = 2.000 \text{ in.} \end{array}$$

$$= \frac{2.250 - 2.000}{2.000} \times 100$$

$$= 12.5\%$$

(h) The percentage of reduction in area is

$$\text{Percentage of reduction in area} = \frac{A_o - A_f}{A_o} \times 100$$

$$= \frac{\left[\frac{\pi}{4}(0.502)^2 - \frac{\pi}{4}(0.430)^2 \right] 100}{\frac{\pi}{4}(0.502)^2} = 26.6\%$$

3-3 STATIC FAILURE: SLIP AND TWINNING

In Chapter 2 we saw that there are forces operating to hold atoms in an equilibrium position within the structure of a solid material. This equilibrium position is created by a balance between the forces of attraction and repulsion. Figure 3-6 shows two curves plotted which represent the forces of attraction and repulsion that exist as the atoms come closer together. Attraction results from binding valence electron clouds and charged positive ions attracting negative electrons. Repulsion results from positive ions repelling. Other positive ions and negative nonvalence electron shells repel other nonvalence electron shells.

These forces are shown in Fig. 3-6. They are not equal. If the curves are added, a resultant (dashed) curve results. It should be noted that there is an equilibrium point where the summation curve crosses the ordinate, and it is at this interatomic spacing that the forces are in equilibrium. If the spacing is reduced, the forces become repulsive very rapidly. If the spacing is increased beyond this point, the forces become strongly attractive. In other words, the effort is directed toward bringing the atoms back to their equilibrium spacing. This mechanism is the basis for elasticity.

If metals are subjected to high enough forces, the equilibrium position for the two atoms discussed becomes ineffective. Since these two atoms are surrounded by other atoms, the forces start to operate for these atoms and with one of its neighbors. The new set of forces does not take over before there has been some dislocation (see Fig. 3-7).

One mechanism of dislocation is called *slip*. This mechanism takes place when a layer of atoms (ions) moves in a plane parallel to its neighbor.

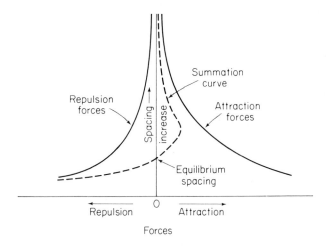

Figure 3-6

It is also seen that the plane, parallel to which the atoms move, separates two densely populated planes of atoms. This is so because the forces binding one atom to another are stronger in a heavily populated plane than in a less densely populated plane. Slip planes are shown in Fig. 3-7(a). They show up as *slip lines*. An aggregate of slip lines adjacent to each other is called a *slip band*. Figure 3-7(b) shows one of the more densely populated planes of a body-centered cubic lattice. The direction of slip is shown by the arrow. Figure 3-7(c) shows a BCC structure that has slipped two atomic layers. Actually, this mechanism takes place through several thousand atomic layers.

Slip, as a mechanism, appears to take place as a movement of an entire plane over another plane of atoms. However, the movement is more like the movement of an energy pulse, where the *dislocation* moves forward a step at a time, as shown in Fig. 3-7(d). Slip lines are shown in the photomicrograph Fig. 3-7(e).

Body-centered cubic lattice structures have many planes through which slip may occur—of the order of 50. Face-centered cubic lattice structures have the fewest number of planes—about 3—through which slip may occur.

Twins are dislocations wherein the atomic planes move over each other some fractional part of the interatomic spacing, depending on the planes' distance from the twin plane. This is in contrast to slip, which occurs in whole-number increments of interatomic spacing. Twinning, however, also takes place along select planes of atoms. The difference is that, whereas after slip the atoms come to rest without changing their orienta-

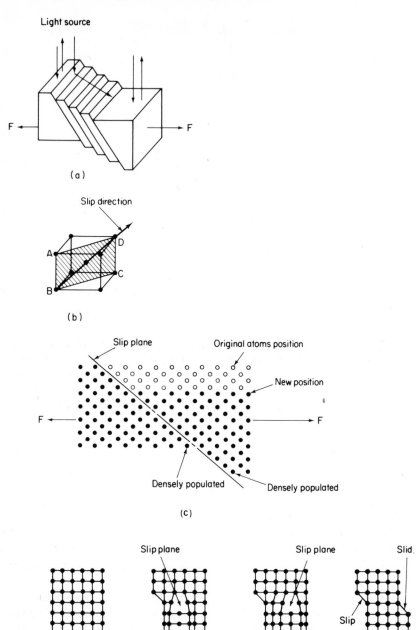

Figure 3-7

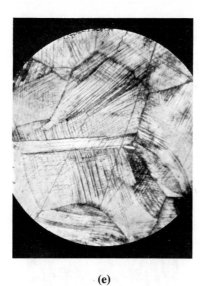

(e)

Figure 3-7 (continued)

tion, twinning requires that the dislocated plane come to rest so that it appears to be a mirror image of the plane on the opposite side of the twinning plane. This is shown in Fig. 3-8(a).

The twinned layer, because of its changed crystal orientation, may be readily visible upon etching, even after it has been polished. Twins appear as broad bands, as shown in Fig. 3-8(b).

The mechanisms just discussed take place when metals are *cold worked*. When cold-worked metals deform, the ductility and malleability are lowered, and the tensile strength, yield point, and hardness increase. These conditions are referred to as *work hardening* or *strain hardening* and are currently explained by the dislocation theory just discussed.

There are three hypotheses that attempt to explain work hardening.

1. *Amorphous metal hypothesis:* this hypothesis presumes that atoms are torn from their lattices, deposit between the slip planes, and impede the sliding of one plane over another.

2. *Fragmentation:* the assumption is that, instead of atoms tearing loose, large segments of lattices tear loose and deposit in slip planes. These impede the sliding of one plane over another.

3. The *lattice-distortion* hypothesis dictates that the distortion of lattices when stressed interferes with the smooth gliding of one plane on another.

Each of these explanations fails to explain work hardening completely. One oversimplified explanation for *work hardening* and *recovery*

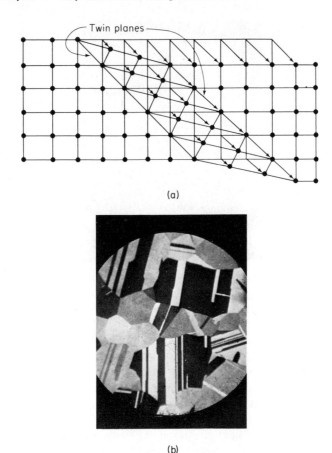

(a)

(b)

Figure 3-8

states that the yield stress of a crystal in a perfect metal is much higher than in a common variety metal. The reason seems to be internal imperfections such as foreign atoms, grain boundaries, and other dislocations, which reduce the yield strength of a metal structure up to a point. Beyond an optimum density of imperfections, the yield strength increases again. As new imperfections are created at *low temperature,* the stress required to cause further deformation increases.

This mechanism is called *work or strain hardening* and is characteristic of cold working. During hot working, even though the imperfections increase, the thermal energy assists the movement of dislocation. In fact, it aids the destruction of imperfections already created. This process is known as *recovery.* The dislocations are shown in the electron micrograph in Fig. 3-9(a), (b), (c), and (d).

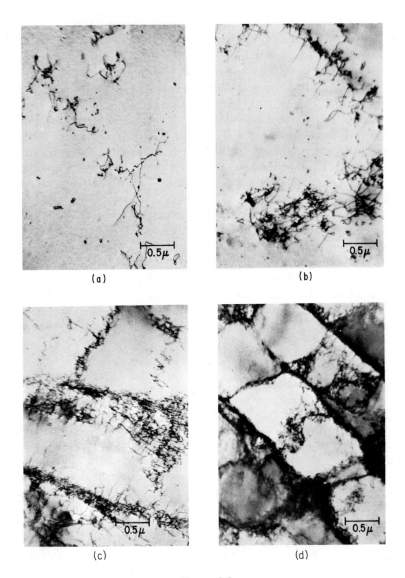

(a)

(b)

(c)

(d)

Figure 3-9

3-4 CREEP

Another type of inelastic action is *creep*. We have seen that slip is dislocation that takes place for a short period of time after the stress is applied but not increased, after which strain stops. If the mechanism of creep is operating, the strain or elongation continues as long as the load is present.

Figure 3-10(a) is a plot of an increasing load being applied to a specimen that causes strain. If the stress is removed, the specimen returns to its original shape. Under conditions of slip [Fig. 3-10(b)], as the stress increases, the strain increases. The strain is shown to increase slightly after the stress becomes constant and then levels out. After the stress is removed, the material retracts, but a *permanent set* remains. Under the conditions of Fig. 3-10(c), the strain continues to increase even though the stress has become constant. This is called *creep*. Once the stress has been removed, the strain decreases. However, the permanent set is greater than that for slip. Furthermore, since the creep strain is a function of time, the specimen will rupture if the stress is allowed to remain. This is the dashed line in Fig. 3-10(c).

The *elastic limit* of a material is the limiting stress below which no slip will occur. It appears as though metals that creep have no limit below which creep will not occur. Increasing the temperature of any metal increases its creep potential. This appears to be because at elevated temperatures the grain boundaries are weaker than the crystals, and rotation of the crystal takes place followed by separation at the grain boundaries (intercrystalline). As temperatures are lowered, the grain boundary strength increases, and the tendency for separation to take place within the crystals increases in the form of slip.

3-5   FATIGUE

When a load below the ultimate strength of some materials is applied repeatedly to a metal specimen, localized hardening occurs. Then a small crack appears. This crack is a line of stress concentration, which causes it to grow. As the crack grows, the cross-sectional area of the sound metal gets smaller until it can no longer support the load. Fracture takes place. The loading is called *fatigue loading*. The fracture is called *fatigue failure*.

Cracks generally start at the surface of a metal part. As the crack grows, the two surfaces rub against each other, polishing both faces to a dull metallic finish, whereas the fractured surfaces show signs of plastic deformation and a crystalline finish.

Figure 3-11(a) shows a schematic drawing of the test equipment used to load a specimen while it is rotating, thus generating fatigue. The stress is then plotted as the ordinate, and the number of cycles is plotted as the abscissa of an $S-N$ fatigue graph. Nonferrous materials when stressed for a given number of cycles eventually rupture. Such a plot is shown in Fig. 3-11(b).

Some materials do not rupture below a given limiting stress, no mat-

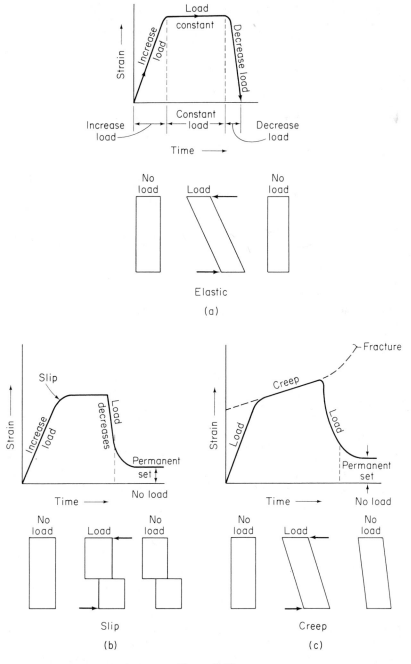

Figure 3-10

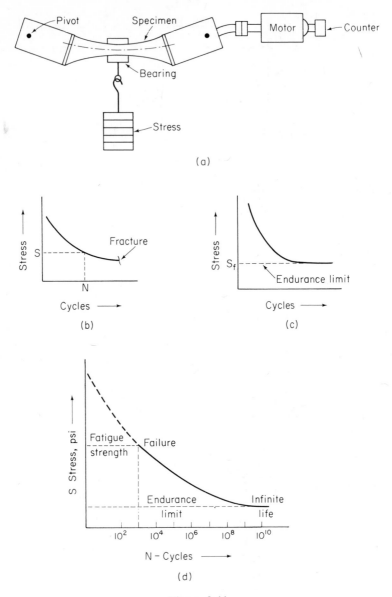

Figure 3-11

ter how many fatigue cycles they are subjected to. Such a stress is plotted in the curve of Fig. 3-11(c) and is called the *endurance limit.*

The *fatigue strength* refers to the stress above the endurance limit at which failure is likely to occur after a given number of cycles [Fig. 3-11(d)].

In certain materials, failure as a result of repeatedly cycled stress generates localized slip patterns. Each slip segment work hardens so that very small cracks form in the material. The notch effect causes the cracks to multiply until a network develops to cause fracture. If these cracks are reversible — sealed — with each cycle, the material is said to be ductile. If not, it will fracture.

The mechanism just described occurs readily in the grain boundaries at existing imperfections inside the structure and on the surface of materials. It is, therefore, important that when a structure is to be cycled (i.e., vibrated) sharp corners, surface scratches, or notches must be avoided by the tool designer. These types of imperfections plastically deform more readily and more severely than smooth, flowing contours or polished surfaces.

3-6 IMPACT TESTING

Impact testing provides for testing materials under conditions of shock loading at fixed temperatures. The procedure is to use the notched specimen shown in Fig. 3-12(b) and (c). The hammer is released from an elevated position [Fig. 3-12(a)]. At this position the hammer has potential energy based on the height above the specimen multiplied by its mass. Thus, if the mass is released when there is no specimen in the vise, the hammer should swing to essentially the same height on the opposite side of the machine. There are some losses due to friction, but they are negligible. The kinetic energy at the vise will be approximately equal to the potential energy at the start.

If the specimen is placed in the vise, it will absorb some of the kinetic energy, and the hammer will not swing as high; that is, the potential energy after impact will be less than the potential energy before impact. The change in potential energy times the mass will be the energy that was required to fracture the specimen. The difference in height of the swinging hammer is recorded on a dial.

The notch in the specimen serves the purpose of concentrating the stresses so that plastic flow is reduced to a minimum. Thus all the stress goes into fracturing the specimen instead of some of the stress being lost to strain.

In the Izod test the specimen is cantilevered [Fig. 3-12(b)] so that the notch faces the hammer. In the Charpy test the notch may be either a V-notch or a keyhole, as shown in Fig. 3-12(c).

Another test that may be done with this equipment is the tension impact test. The specimen and holder are shown in Fig. 3-12(d).

It is also possible to test many specimens under varying temperature

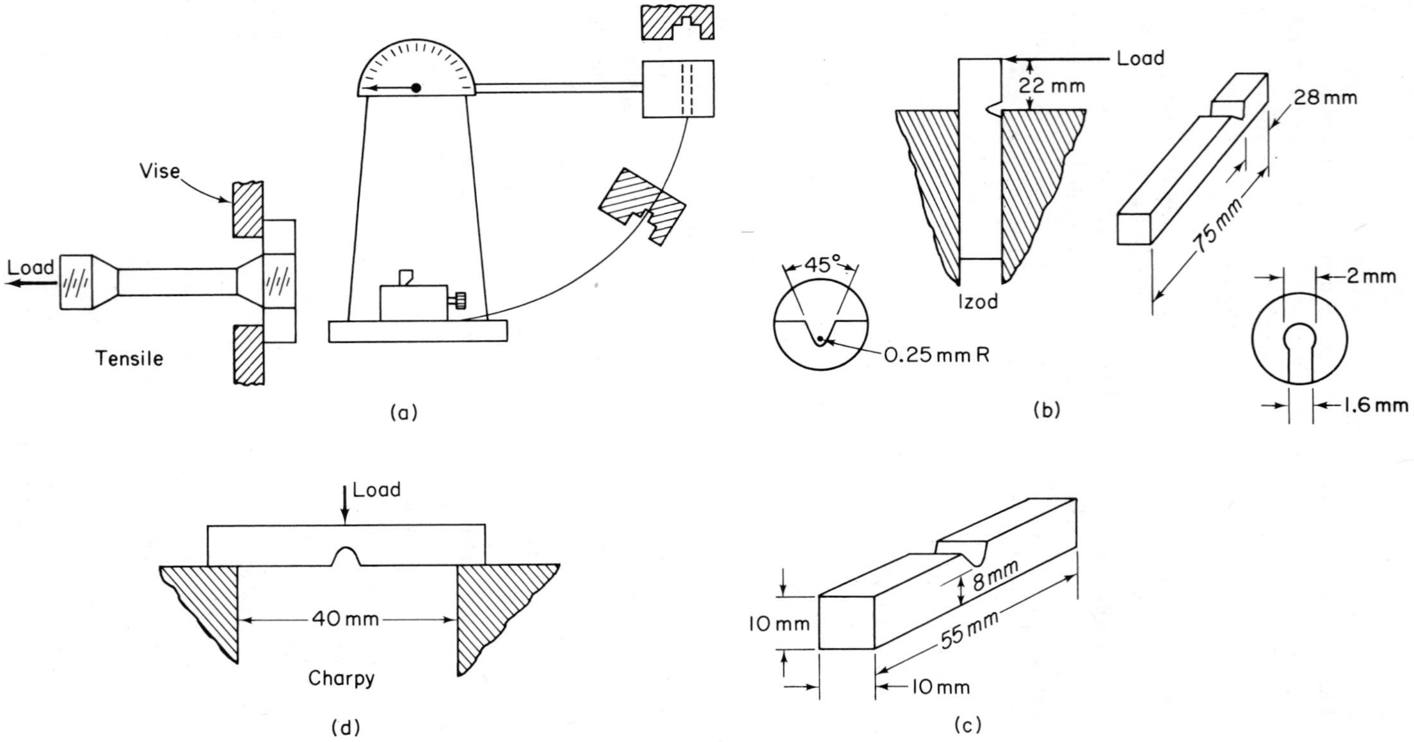

Figure 3-12

82

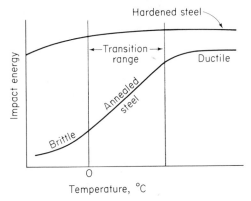

Figure 3-13

conditions. A curve of impact versus temperature is shown in Fig. 3-13. This curve shows that *above* a certain temperature range, called the *transition temperature,* the metal is ductile; *below* this temperature range the metal is brittle. A low transition temperature may be interpreted as being desirable, whereas a high transition temperature may indicate that the metal is subject to brittle failure.

3-7 HARDNESS TESTING

There are several methods used to check the hardness of materials. These test values may be related to values of tensile and yield strength of the material. See Table 3-3. The most commonly used hardness tests are those that measure the resistance to penetration of the material. The procedure, in general, is to force an indenter into the surface of the material and then to evaluate the penetration. Other types of hardness tests rely upon the elasticity of the material. Still others evaluate the effect of the resistance offered to abrasion, scratching, or cutting of materials.

The tests described in this section leave some type of impression in the surface of the work. If this is objectionable, the tests should be performed in a place that will not interfere with the operation of the workpiece.

If a diamond penetrator is used during Rockwell testing, the impression will be less noticeable than if a 10-mm steel ball is used with the Brinell machine. The Superficial Rockwell penetrator is smaller than both the Rockwell and the Brinell testers. Microtesters leave even smaller indentations. Under no conditions should the penetrator affect the underside of the workpiece.

The relationship of size of penetrator and load used to the thickness of the material is also important. Brinell and Rockwell tests are usually

Table 3-3 Hardness Numbers

| Indentation (mm) | Brinell 10 mm 3000 kg | | Rockwell number | | | | Superficial Rockwell | | | Vickers | Scleroscope | Tensile strength | MOH 10 |
	Standard	Tungsten carbide	A 60 kg, Brale	B 100 kg, 1/16 in. Brale	C 150 kg, Brale	D 100 kg, Brale	15N 15 kg	30N 30 kg	45N 45 kg	Pyramid diamond	Number	×10³ psi	Scale
2.25		745	84.1		65.3	74.8	92.3	82.2	72.2	840	91		
2.30		710	83.0		63.3	73.3	91.5	80.4	70.2	780	87		
2.35		682	82.2		61.7	72.0	91.0	79.0	68.5	737	84		
2.40		653	81.2		60.0	70.7	90.2	77.5	66.5	697	81		
2.45		627	80.5		58.7	69.7	89.6	76.3	65.1	667	79	323	8.0
2.50		601	79.8		57.3	68.7	89.0	75.1	63.5	677	77	309	
2.55		578	79.1		56.0	67.7	88.4	73.9	62.1	640	75	297	
2.60		555	78.4		54.7	66.7	87.8	72.7	60.6	591	73	285	7.5
2.65		534	77.8		53.5	65.8	87.2	71.6	59.2	569	71	274	
2.70		514	76.9		52.1	65.0	86.7	70.7	58.0	547	70	263	
2.75	495	495	76.3		51.0	63.8	85.9	69.4	56.1	528	68	253	
2.80	477	477	75.6		49.6	62.7	85.3	68.2	54.5	508	66	243	
2.85	461	461	74.9		48.5	61.7	84.7	67.2	53.2	491	65	235	7.0
2.90	444	444	74.2		47.1	60.8	84.0	65.8	51.5	472	63	225	
2.95	429	429	73.4		45.7	59.7	83.4	64.6	49.9	455	61	217	
3.00	415	415	72.8		44.5	58.8	82.8	63.5	48.4	440	59	210	
3.05	401	401	72.0		43.1	57.8	82.0	62.3	46.9	425	58	202	
3.10	388	388	71.4		41.8	56.8	81.4	61.1	45.3	410	56	195	
3.15	375	375	70.6		40.4	55.7	80.6	59.9	43.6	396	54	188	6.5
3.20	363	363	70.0		39.1	54.6	80.0	58.7	42.0	383	52	182	

Table 3-3 Hardness Numbers (Continued)

Brinell 10 mm 3000 kg			Rockwell number				Superficial Rockwell			Vickers	Scleroscope	Tensile strength	MOH 10
Indentation (mm)	Standard	Tungsten carbide	A 60 kg, Brale	B 100 kg, 1/16 in. Brale	C 150 kg, Brale	D 100 kg, Brale	15N 15 kg	30N 30 kg	45N 45 kg	Pyramid diamond	Number	×10³ psi	Scale
3.25	352	352	69.3	110.0	37.9	53.8	79.3	57.6	40.5	372	51	176	
3.30	341	341	68.7	109.0	36.6	52.8	78.6	56.4	39.1	360	50	170	
3.35	331	331	68.1	108.5	35.5	51.9	78.0	55.4	37.8	350	48	166	
3.40	321	321	67.5	108.0	34.3	51.0	77.3	54.3	36.4	339	47	160	
3.45	311	311	66.9	107.5	33.1	50.0	76.7	53.3	34.4	328	46	155	
3.50	302	302	66.3	107.0	32.1	49.3	76.1	52.2	33.8	319	45	150	6.0
3.55	293	293	65.7	106.0	30.8	48.3	75.5	51.2	32.4	309	43	145	
3.60	285	285	65.3	105.5	29.9	47.6	75.0	50.3	31.2	301	42	141	
3.65	277	277	64.6	104.5	28.8	46.7	74.4	49.3	29.9	292	41	137	
3.70	269	269	64.1	104.0	27.6	45.9	73.7	48.3	28.5	284	40	133	
3.75	262	262	63.6	103.0	26.6	45.0	73.1	47.3	27.3	276	39	129	
3.80	255	255	63.0	102.0	25.4	44.2	72.5	46.2	26.0	269	38	126	5.5
3.85	248	248	62.5	101.0	24.2	43.2	71.7	45.1	24.5	261	37	122	
3.90	241	241	61.8	100.0	22.8	42.0	70.9	43.9	22.8	253	36	118	
3.95	235	235	61.4	99.0	21.7	41.4	70.3	42.9	21.5	247	35	115	
4.00	229	229	60.8	98.2	20.5	40.5	69.7	41.9	20.1	241	34	111	
4.05	223	223		97.3	18.8					234			
4.10	217	217		96.4	17.5					228	33	105	
4.15	212	212		95.5	16.0					222		102	
4.20	207	207		94.6	15.2					218	32	100	

Table 3-3 Hardness Numbers (Continued)

Brinell 10 mm 3000 kg			Rockwell number				Superficial Rockwell			Vickers	Sclero-scope	Tensile strength	MOH 10
Indentation (mm)	Standard	Tungsten carbide	A 60 kg, Brale	B 100 kg, 1/16 in. Brale	C 150 kg, Brale	D 100 kg, Brale	15N 15 kg	30N 30 kg	45N 45 kg	Pyramid diamond	Number	×10³ psi	Scale
4.25	201	201		93.8	13.8					212	31	98	
4.30	197	197		92.8	12.7					207	30	95	
4.35	192	192		91.9	11.5					202	29	93	
4.40	187	187		90.7	10.0					196		90	
4.45	183	183		90.0	9.0					192	28	89	5.0
4.50	179	179		89.0	8.0					188	27	87	
4.55	174	174		87.8	6.4					182		85	
4.60	170	170		86.8	5.4					178	26	83	
4.65	167	167		86.0	4.4					175		81	
4.70	163	163		85.0	3.3					171	25	79	
4.80	156	156		82.9	0.9					163		76	4.5
4.90	149	149		80.8						156	23	73	
5.00	143	143		78.7						150	22	71	
5.10	137	137		76.4						143	21	67	
5.20	131	131		74.0						137		65	
5.30	126	126		72.0						132	20	63	
5.40	121	121		69.8						127	19	60	
5.50	116	116		67.6						122	18	58	
5.60	111	111		65.7						117	15	56	

Source: U.S. Steel Carilloy Steel, pp. 164–165–166. Carnegie-Illinois Steel Corp., Pittsburgh, Pa. (1948).

reserved for thick sections. Nonhomogeneous materials are best tested with a steel ball penetrator.

Brinell Hardness Test. This test (Fig. 3-14) relies on the resistance to penetration of the material being tested. A 10-mm hardened steel ball, or tungsten carbide ball, is pressed into the surface of the specimen with a 500- or 3000-kg load. The 500-kg load is used when one is testing soft materials, such as copper or aluminum. The 3000-kg load is used to test steels and other hard materials. The penetrator is applied for a period of 10 seconds (s) for 500-kg and 30 s for 3000-kg loads.

The Brinell hardness number (BHN) is the ratio between the load and the surface area of the indentation. Thus

$$\text{BHN} = \frac{\text{load}}{\text{surface area of indentation}}$$

$$= \frac{P}{\frac{\pi D}{2}(D - \sqrt{D^2 - d^2})}$$

$P = $ load, kg
$D = $ dia. of indenter, mm
$d = $ dia. of indentation, mm
BHN = Brinell hardness number

Example 2

Assume that a Brinell test is made using a 10-mm steel ball and a 3000-kg load. What is the BHN if the diameter of the spherical impression is 4.00 mm?

Figure 3-14

solution:

$$\text{BHN} = \frac{P}{\dfrac{\pi D}{2}(D - \sqrt{D^2 - d^2})}$$

$$= \frac{3000}{\dfrac{\pi 10}{2}(10 - \sqrt{10^2 - 4^2})} = \frac{3000}{13.116}$$

$$= 229$$

$P = 3000$ kg
$D = 10$ mm
$d = 4.00$ mm

The value for d is read with a calibrated microscope. Then either BHN is calculated, a conversion chart is consulted, or in some instances the dial of the machine is calibrated to read BHN directly.

Some limitations of the Brinell test are the following:

1. It should not be used on workpieces less than $\frac{1}{8}$ in. thick.

2. If several readings must be taken on the same specimen, they should be spaced away from each other and away from the edges of the workpiece.

3. These indentations may be objectionable if the surface of the workpiece is finished.

4. The test may be unreliable for very hard or very soft materials.

Rockwell Hardness Test.   This test [Fig. 3-15(a)] is calibrated on the basis of the depth of penetration of the indenter [Fig. 3-15(b)]. The workpiece is elevated until the indenter touches the work. A 10-kg minor load is applied. As the load is applied, an indicator needle registers zero. Then,

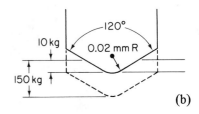

(b)

(a)

Figure 3-15

a load lever being released, the major load is applied uniformly. This creates the impression in the workpiece. The major load is removed, and the Rockwell number is read while the minor load is still applied. The Rockwell number reflects the penetration of the major load only. This eliminates the effect of the elastic recovery of the metal in the Rockwell hardness number.

An inverse relationship exists between depth of penetration and hardness. That is, the softer the material tested, the deeper the penetration of the indenter; the higher the Rockwell number, the harder the material. Because of the wide range of materials to be tested, several Rockwell scales are available. In all cases the *depth* of penetration is measured and calibrated. The scales available are lettered A through G. The scales, indenter, and load used are shown in Table 3-4. A diamond brale is generally used on hard material. A $\frac{1}{16}$-in.-diameter steel ball is used for soft material.

Figure 3-15(b) shows a minor load of 10 kg and a major load of 140 kg as required by the Rockwell C scale. The brale is a 120-degree diamond indenter, with a radius of 0.02 mm. The C scale, most commonly used, is represented by the notation

$$60R_c$$

This symbol reads 60 Rockwell hardness on the C scale.

One Rockwell number represents 0.000080 in. (0.002 mm) of penetration. Thus the difference in penetration between a $60R_c$ reading and a $65R_c$ reading is $5 \times 0.000080 = 0.0004$ in.

Irregular surfaces, scales, and imperfect flatness will yield false readings. Very thin pieces will also give false readings because they will reflect the hardness of the supporting anvil as well as the hardness of the workpiece. Repeated readings should not be taken at the same spot on the workpiece, and several readings should be averaged.

Table 3-4

Scale	Indenter	Load kg
A	Brale	60
B	$\frac{1}{16}$-in. ball	100
C	Brale	150
D	Brale	100
F	$\frac{1}{16}$-in. ball	60
G	$\frac{1}{16}$-in. ball	150

For checking the hardness of thin workpieces, a *Superficial Rockwell* tester may be used. This machine uses a 3-kg minor load, and a 15-, 30-, or 45-kg major load. It is calibrated for incremental depths of 0.000040 in. for each dial reading.

Vickers Hardness Tester.   Another instrument used to test hardness is the Vickers hardness tester (Fig. 3-16). It operates very much like the Brinell machine in that the diagonal of the impression made by a pyramidal diamond penetrater is viewed through a calibrated microscope. The pyramid has a 136-degree inclined angle between its faces, and uses loads of 5 to 100 kg applied for periods of up to 20 s. The measurement of the diagonal is then referred to a conversion chart, or used in the equation

$$\text{DPH} = \frac{2P \sin \alpha/2}{d^2}$$

$$= \frac{1.854P}{d^2}$$

$\alpha = 136$ degrees
DPH = diamond pyramid number
d = diagonal, mm
P = load, kg

Tukon Hardness Tester.   This tester (Fig. 3-17) uses a knoop pyramidal-shaped diamond indenter that has a longitudinal angle of 172 degrees, 30 minutes and a transverse angle of 130 degrees. A square-based diamond indenter with a 136-degree included angle may also be used. Once the indentation is made, the workpiece is viewed with a calibrated microscope to measure the long diagonal. This long diagonal is about seven times longer than the short diagonal. The numerical measurement viewed through the microscope may be compared with a chart to determine the knoop hardness number, or it may be obtained from the formula

Figure 3-16

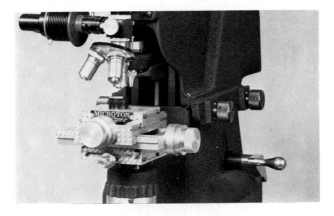

Figure 3-17

$$I = \frac{P}{A} = \frac{P}{l^2 C}$$

I = knoop hardness number
P = load, kg
A = projected area of indentation, mm^2
l = length of long diagonal, mm
C = constant

If the indenter used is a square-base pyramid type, then the measurement of the diagonal is inserted into the equation used for the Vickers test. This yields the DPH number.

Since the indentation is very small, this test is suitable for testing the hardness of thin sections, small-diameter wire, the case of hardened surfaces, brittle materials (glass), foil, microconstituents of metals, etc.

Shore Scleroscope.   This instrument (Fig. 3-18) has a diamond-tipped hammer that drops onto the surface to be tested. The harder the material to be tested, the higher the rebound, and the higher the hardness number. The height of rebound is a function of the elasticity of the material.

Ultrasonic Hardness Tester.   A diamond-tipped rod vibrating at ultrasonic frequencies is brought into contact with the part to be tested at a load of about 2 lb. The tip penetrates the surface of the material to be tested and records this depth on a dial in terms of standard numbers. The shallow indentation is about 50-millionths of an inch.

Durometer Hardness Tester.   This test is used to test the hardness of rubber and is a measure of resistance to elastic deformation.

Scratch Hardness Tests.   This test is a rough measure of the hardness

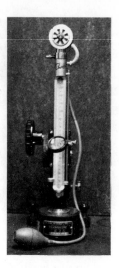

Figure 3-18

of materials. It is based on the concept that hard materials will scratch softer material.

The *file test* simply indicates that the material, if capable of being scratched by a file, is softer than the file.

The MOH scale compares the hardness of the material tested with 10 materials that have been graded, number 1 being very soft and number 10 being very hard. They are (1) talc, (2) gypsum, (3) calcite, (4) fluorite, (5) apatite, (6) feldspar, (7) quartz, (8) topaz, (9) ruby, and (10) diamond. Thus, if the material tested is scratched by 8 and not by 7, the MOH number is taken as 7.5. Like the file test, this test is only an approximation.

3-8   NONDESTRUCTIVE TESTING

Magnaflux.   The principle of the magnaflux test [Fig. 3-19(a)] can be used only with magnetic materials. A current flow will generate magnetic lines of flux that leak wherever there is a crack in the surface of the material being tested. Magnetic flux lines are induced at right angles to the direction of the current flow producing them. If fine iron filings are applied to the surface of the piece either as a liquid suspension or dusted as a powder, these filings will be held in the cracks by the concentration of the flux at that point. If the filings are made fluorescent, they can be viewed by black light (ultraviolet) very clearly. This latter process is called *magnaglo*.

Surface cracks become visible when the magnaflux process is used.

(a)

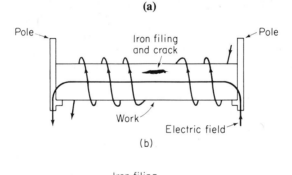

(b)

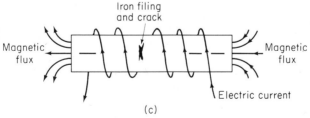

(c)

Figure 3-19

Visible evidence of defects up to about $\frac{1}{4}$ in. below the surface or deeper can be detected by a trained operator using this process. Cracks that are perpendicular to the direction of the lines of magnetic flux are more visible than those that are parallel to the flux lines. Two methods of generating flux lines are shown in Fig. 3-19(b) and (c). If the configuration of the workpiece permits, it should be rotated and checked in several positions. *Duovec,* a process developed by the Magnaflux Corporation, generates a changing current and thus makes possible a one-step inspection.

All parts magnafluxed must be demagnetized. This is accomplished by passing the part through a coil that is oppositely polarized from the part.

Visual Inspection.   Parts are sprayed with a fluorescent liquid. This penetrates the small cracks, so that when the excess material is washed away a substantial amount remains in the cracks. In some instances, after the excess liquid has been removed, a dry powder developer is dusted on the part. This powder will cling to the cracks if the rest of the workpiece has previously been dried. When viewed by black light, the cracks fluoresce and become visible.

The Magnaflux Corporation has developed a similar powder that may be used on nonmagnetic materials. The parts are immersed in a fluorescent liquid, drained, washed, and dusted with an absorbent powder, which brings the penetrant to the surface of the defect, where it can be viewed with ultraviolet light. It is called the *Zyglo** process. The Dy-Chek† process is similar, except that a developing solution is used as a final step instead of the ultraviolet light. The developing solution reacts with the penetrant, which turns red and becomes visible.

Ultrasonic tests use a piezoelectric material and the fact that such materials change size if a voltage is applied to them. It is also possible to reverse the process and generate a voltage when the size of the crystal is changed by an applied force. This applied force could be an ultrasonic wave that impinges on the crystal.

Figure 3-20(a) shows a crystal (usually quartz) placed on the workpiece to be tested. In the *reflecting type* of instrument, an electric current sets the crystal vibrating; this sets up sound vibrations of the order of 10^6 hertz (Hz) in the workpiece. Contact between the crystal and the workpiece is assured by a film of oil or by immersing the workpiece in a liquid bath.

Thus, in the reflecting type of instrument, the transmitted waves, if interfered with, will return to the crystal and generate a voltage, which is made visible on an oscilloscope screen. The samples that interfere are the upper and lower (reflecting) surfaces and the surface of the defect. Since the sound in a material travels at a constant speed and in a straight line, it is time related; therefore, the location of the flow can be calculated from the image on the screen.

In the *transmission type instrument* [Fig. 3-20(b)], a flaw blocks some of the sound. Therefore, the energy collected by the receiving crystal is not as great as that which left the transmitting crystal. This type of in-

*Trade name of the Magnaflux Corporation.
†Trade name of the Dy-Chek Co., division of Northrup Aircraft, Inc.

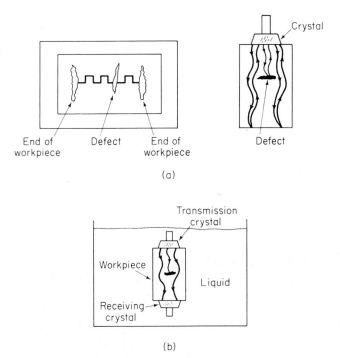

Figure 3-20

strument is suitable for the inspection of parts up to about 5 in. thick. The reflecting instrument can be used on parts five or six times thicker.

The *liquid crystal method* was recently developed by the Boeing Company for inspecting metals and plastics. The liquid crystals used are cholesterol derivatives. They exhibit many of the optical characteristics of solid crystals. The liquid mixture is brushed on the workpiece to be inspected, heated slightly, and then cooled to 85°F. Any flaws in the workpiece material will cause its surface to change color. A slight change in temperature of the surface causes a change in spacing of the crystal layers, which in turn causes the color to change from purple at 86°F through the visible spectrum to red at 84°F. Cooling in a flawless workpiece changes the colors uniformly, whereas any flaw will cause that part of the workpiece to change colors either faster or slower than normal.

Radiography is an inspection method that uses penetrating X-rays or gamma rays. The process is to expose the work piece to penetrating X-ray radiation. Dense materials absorb X-rays, whereas less dense materials permit a greater number of X-rays to be transmitted through the material. The transmitted rays are caused to fall on a film negative and to produce a radiograph. The greater the number of rays per unit area transmitted, the darker the appearance of the radiograph. Thus, voids appear

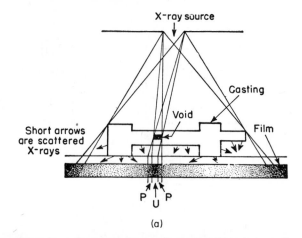

(a)

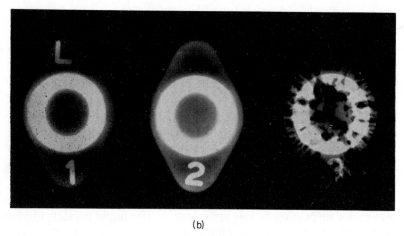

(b)

Figure 3-21

excessively dark on the negative. Inclusions that are more dense than the base metal will appear excessively light. Figure 3-21(a) shows the X-rays impinging on a casting. Figure 3-21(b) shows three views of castings, one of which shows porosity.

The limitations that are encountered when one is using X-rays for inspection result primarily because of scattering and the lack of parallelism of the rays. These characteristics result in the umbra-penumbra effect and therefore in fuzziness of the outline of the work piece. These limitations may be avoided by applying existing radiographic techniques.

Gamma radiography uses high-energy gamma radiation from a radioactive source for inspection of metals. The high-energy rays, their deep penetration characteristics, the expense of the source (ridium-192), and the exposure time required are some of the conditions that make it

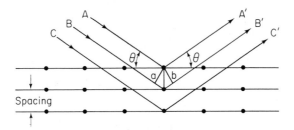

Figure 3-22

necessary to exercise precaution and maintain a high degree of control of the process when it is used.

X-ray diffraction is the most useful method for investigating crystal structure. X-rays that impinge upon parallel atomic layers of a material cause the atoms to vibrate and emit energy of the same wavelength as that of the incident X-ray. If the X-ray beam is directed at an angle of incidence that is the same as the angle of reflection, the waves will constructively reinforce each other as shown in Fig. 3-22. Thus, if $a+b$ equals a whole number of wavelengths, BB' must be in phase with AA', and reinforcement occurs.

Since the X-rays used are of the order of the atomic planar spacing, the lattice spacing and structure, phase determination, orientation of crystals, etc. may be studied.

Electron diffraction may also be used, since lattices that are present in metals cause electrons to be diffracted. This process is used for checking thin films and surface conditions only.

3-9 TEMPERATURE MEASUREMENT

Disappearing-Filament Optical Pyrometer. A hot wire will glow with a color characteristic of the material in the wire and the temperature of the wire. If this hot wire is placed in the operator's field of vision as he views a hot object, the wire will visually disappear when the hot object is at the same temperature as the hot wire. If the wire is cooler than the object, it will appear black. If it is hotter than the object, it will glow more brightly. The procedure is to sight through the wire at the heated area to be checked. The temperature of the wire is increased until it can no longer be seen. The temperature reading is then taken from a calibrated dial. This method is very effective for checking temperatures from about 1100 to 2000°F.

Radiation Pyrometer. This instrument uses a reflector to collect and focus radiant heat energy on a thermocouple. The thermocouple transduces the heat energy into electrical energy, which is recorded and calibrated to read temperature.

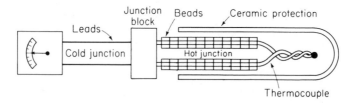

Figure 3-23

Thermocouples (Fig. 3-23). If two wires made from dissimilar materials are joined intimately at the end and the end is heated, a flow of electrons across the junction will exist. This happens because every material has its own characteristic flow of electrons for every degree of increase in temperature. This current flow will be increased or decreased, depending upon an increase or decrease in temperature. If the cold ends of the wires are attached to a voltmeter and the dial is calibrated to read degrees, it will then be possible to read temperature directly when the fused end (hot) is inserted into a furnace.

Thermocouples are made from various materials, which determines the temperature range at which they operate. They must be corrosion resistant and temperature resistant over that range. Table 3-5 shows these materials and the range through which they operate.

Thermocouples may be calibrated by keeping the cold end at the freezing point of water and the hot end at a particular elevated temperature.

Table 3-5   Thermocouples

Materials	Temp., °F
Chromel–alumel	325–2200
Copper–constantan*	325– 800
Iron–constantan	325–1400
Platinum–platinum rhodium	32–2880

* Constantan is 55 per cent Cu, 45 per cent Ni.

Resistance Thermometer. This instrument operates on the principle that changes of temperature will cause a change of resistance in the structure of a material. Small changes in resistance can be detected by an electrical Wheatstone bridge. If one resistor of a bridge is made of a material that has a high temperature coefficient of resistance and is capable of being matched against resistors in the bridge that have normal coefficients, a satisfactory method for checking temperature is available.

Problems

3-1. Describe the procedure for preparing a photomicrographic specimen.

3-2. Show by diagram, and explain, the various shades of gray seen when viewing an etched sample with a microscope.

3-3. Compare the optical magnification of the metallurgical microscope and the electron microscope.

3-4. Compare the magnification capabilities of an electron microscope with an optical microscope.

3-5. Define the terms (a) stress, (b) strain, and (c) modulus of elasticity.

3-6. Define (a) elastic strain, (b) inelastic strain, and (c) rupture. Relate each to the stress–strain diagram in this chapter.

3-7. Describe the difference between (a) stress and true stress and (b) strain and true strain.

3-8. Explain each of the following terms as they relate to the stressing of a sample of ductile material: (a) stiffness, (b) proportional limit, (c) elastic limit, (d) yield strength, (e) ultimate strength, (f) tensile strength, (g) rupture strength, and (h) toughness.

3-9. Describe the concepts of (a) percentage of elongation and (b) percentage of reduction in area.

3-10. Explain the following statement: "Percentage of elongation is related to the tensile properties of a material, and the percentage of reduction in area is related to its ductile properties."

3-11. What is the basic principle upon which an extensometer operates?

3-12. Referring to the tensile specimens as drawn in this chapter, explain the reasons for the structures as they evolve after being loaded.

3-13. Given the following data for Fig. 3-24: original diameter, 0.502 in.; necked-down diameter, 0.470 in.; gage length, 2.000 in.; gage length at fracture, 2.460 in. Calculate: (a) modulus of elasticity, (b) proportional limit, (c) ultimate strength, (d) rupture strength, (e) true rupture strength, (f) yield strength, (g) percentage of elongation, and (h) percentage of reduction in area.
Ans. (a) 15×10^6 psi, (b) 30,000 psi, (c) 58,000 psi, (d) 50,000 psi, (e) 57,040 psi, (f) 32,000 psi, (g) 23 per cent, (h) 12.3 per cent.

3-14. Given the following data for Fig. 3-24: original diameter, 0.504 in.; necked-down diameter, 0.420 in.; gage length, 2.000 in.; gage length at fracture, 2.625 in. Calculate: (a) modulus of elasticity, (b) proportional limit, (c) ultimate strength, (d) rupture strength, (e) true rupture strength, (f) yield

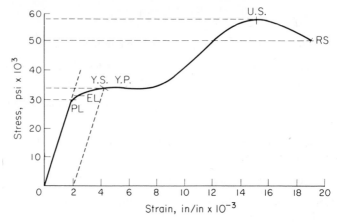

Figure 3-24

strength, (g) percentage of elongation, and (h) percentage of reduction of area.

Ans.   (a) 15 × 10⁶ psi, (b) 30,000 psi, (c) 58,000 psi, (d) 50,000 psi, (e) 71,986 psi, (f) 32,000 psi, (g) 31.25 per cent, (h) 30.5 per cent.

3-15.   Given a test sample having an original diameter of 0.500 in., a final diameter at rupture of 0.455 in., an original gage length of 2.000 in., and a final gage length of 2.270 in. The data collected are shown in Table 3-6. Calculate: (a) stress, (b) strain, (c) the plot of the stress–strain curve, (d) modulus of elasticity, (e) proportional limit, (f) ultimate strength, (g) rupture strength, (h) true rupture strength, (i) yield strength, (j) percentage of elongation, and (k) percentage of reduction in area.

Ans.   (d) 32 × 10⁶ psi, (e) 60,000 psi, (f) 76,000 psi, (g) 62,000 psi, (h) 74,900 psi, (i) 65,000 psi, (j) 13.5 per cent, (k) 17.3 per cent.

Table 3-6

Load, lb	Δl, in.
1,000	0.00032
2,000	0.00064
4,000	0.00127
5,000	0.00159
6,000	0.00191
8,000	0.00255
10,000	0.00318
11,930	0.00380
11,970	0.00390
12,170	0.00400
12,560	0.00450
12,760	0.00500
12,950	0.00550
13,150	0.00600

Table 3-6 (continued)

Load, lb	Δl, in.
12,760	0.00700
12,560	0.00800
12,950	0.00900
14,720	0.01200
14,920	0.01400
14,325	0.01600
12,170	0.01800

3-16. Refer to Chapter 2 and review the binding forces that hold a material together. Then, using Fig. 3-6, explain elasticity.

3-17. Describe the mechanism of slip.

3-18. Describe the mechanism of twinning. How does it differ from slip?

3-19. Explain the concept of work hardening and the hypotheses of dislocation.

3-20. Describe the mechanism and the progress of fatigue failure.

3-21. How does recovery differ from work hardening?

3-22. What is creep? Explain the three diagrams in Fig. 3-10.

3-23. Describe the S–N diagrams for (a) ferrous metals and (b) nonferrous metals.

3-24. Describe the terms (a) endurance limit and (b) fatigue strength.

3-25. Explain the three graphs in Fig. 3-11, which show the differences between elasticity, slip, and creep.

3-26. What effect does an increase in temperature have upon the mechanism of creep?

3-27. Explain the notch effect on ductility.

3-28. Explain inter- and intracrystalline failure at various temperatures.

3-29. Describe the Izod and the Charpy tests.

3-30. Explain the value of knowing the transition temperature range of a material.

3-31. Describe the Brinell hardness test method.

3-32. Calculate the Brinell hardness number, given the following data: load, 3000 kg; ball diameter, 10 mm; diameter of impression, 4.10 mm.
Ans. BHN = 217.

3-33. Calculate the Brinell hardness number, given the following data: load, 3000 kg; diameter of ball, 10 mm; diameter of impression, 3.50.
Ans. BHN = 302.

3-34. Calculate the BHN, given the following data: load, 3000 kg; diameter of

ball, 10 mm; diameter of indentation, 4.60 mm.
Ans.   BHN = 171.

3-35.   List the limitations of the Brinell hardness test.

3-36.   Describe the Rockwell hardness test and its calibration. Explain the purpose of the minor load.

3-37.   Why is it necessary to have so many different Rockwell hardness scales?

3-38.   Calculate the difference in the depth of penetration between a reading of $50R_c$ and one of $30R_c$.
Ans.   0.0016 in.

3-39.   Repeat Problem 3-38 if the readings are $62R_c$ and $55R_c$.
Ans.   0.0056 in.

3-40.   List some advantages and disadvantages of the Rockwell machine as a hardness tester.

3-41.   What is a Superficial Rockwell tester? How does it operate?

3-42.   Describe the operation of the Vickers hardness tester. Make sure you discuss the characteristics of the knoop penetrater and the square penetrater.

3-43.   How does the Shore scleroscope operate?

3-44.   Describe the principle behind all scratch tests. What is the MOH scale?

3-45.   Describe the magnaflux process. How does it differ from the magnaglo process?

3-46.   Describe the Zyglo process of visual inspection of nonmagnetic materials.

3-47.   Describe the *reflection* and *transmission* processes of supersonic wave inspection of workpieces.

3-48.   Describe the liquid crystal method of inspecting materials.

3-49.   Describe the use of X-rays in producing radiographs of materials to be inspected.

3-50.   Explain the use of the X-ray diffraction technique for examining structures of materials.

3-51.   Describe the disappearing-filament optical pyrometer as a method for determining the temperature of a furnace.

3-52.   Describe the operation of a thermocouple when used to check temperature.

3-53.   How does a resistance thermometer work?

4 | Equilibrium Diagrams

4-1 COOLING CURVES

A cooling curve [Fig. 4-2(a)] is a graph of the structure of a pure metal or a combination of that metal with another metal. The latter is called an *alloy*. Thus studying a particular cooling curve yields data related to a particular combination of two or more metals of the entire temperature range through which an alloy cools. If the characteristics of another alloy of the same two metals are desired, the cooling curve for the new combination is needed. An *equilibrium diagram* is a composite of all cooling curves of all the possible combinations of two or more metals. It should be stressed that equilibrium diagrams are related to the conditions of cooling that occur slowly enough to be considered stable — hence the term equilibrium.

Assume a closed system. If the ice at −10°F is heated, it will absorb heat until the ice reaches 32°F, at which point it will begin to melt. The mixture of ice and water will remain at 32°F, because all the heat is being used to melt whatever ice is being transformed to water at 32°F. Once all the ice has been transformed to water at 32°F, the heat added to the system will be used to raise the temperature of the water to 212°F. At this *point* the heat will all go toward transforming the water into steam at 212°F. The mixture of water and steam remains at 212°F until all the water converts to steam, at which time the heat input raises the temperature of the steam. The plot of this heating system is shown in Fig. 4-1(a).

The reverse mechanism takes place [Fig. 4-1(b)] if the steam is in a closed system and if it is allowed to cool to −10°F. The plateaus at 32°F

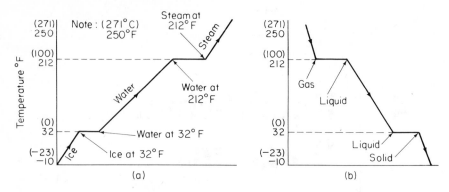

Figure 4-1

and 212°F are the temperatures on the curve at which *phase* changes occur.

Several definitions are in order at this time: (1) The elements that make up an alloy or the chemical compounds that may be formed from these elements are called *components*. They are considered components if chemical means are needed to separate them. (2) A *phase* is a part of an alloy that does not require chemical methods to make that part distinguishable from other parts. That is, physical means can be used to distinguish the parts of the alloy.

If a *pure metal* is heated and then cooled very slowly, its cooling curve may be plotted as shown in Fig. 4-2(a). The curve *ab* represents the cooling of the melt. At temperature *T* and at point *b*, the pure metal starts to precipitate out of solution. At point *c* the entire melt has transformed to the solid pure metal. Note that curve *bc* is level and that the phase change occurs at one temperature *T*. From *c* to *d* the solid metal undergoes cooling.

In Fig. 4-2(b), *two metals* have been heated to point *a*, where both are liquid and are dissolved in each other. On slow cooling they remain liquid until they reach T_1, at which temperature solidification starts. At T_2 the solidification has been completed, and the entire mixture is a solid solution. The cooling of the solid continues to *d*. Point *b* is the point

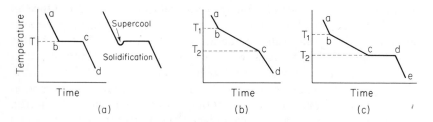

Figure 4-2

where crystallization begins and is referred to as the *liquidus* point. Point *c* is the point on the curve where final solidification takes place. It is referred to as the *solidus* point.

Another type of cooling curve is one that exhibits characteristics as shown in Fig. 4-2(c). This occurs when two metals *A* and *B* are soluble in the liquid state but insoluble in the solid state.

In this mechanism the liquid solution starts to cool at *a*. At *b*, pure metal *A* starts to form and precipitate out of the liquid phase. As the temperature drops from T_1 to T_2, pure metal *A* continues to precipitate. As metal *A* precipitates out, the ratio of *B* to *A* increases, and we say that the liquid remaining becomes richer in *B*. This mechanism of freezing results in the slope of the curve *bc*.

At *c* the freezing is that of the remaining metals of *A* and *B*. The temperature remains constant until all solidification is complete at *d*. Line *de* represents the normal cooling of the solid phases.

Figure 4-3(a) shows several cooling curves of combinations of metals *A* and *B*. These curves have been rotated through 90 degrees, and the points *a* and *b* have been projected to the back plane as shown. That is, all points *a* are on the liquidus line and all points *b* are on the solidus line. When all the liquidus points are connected by a smooth line, and all the solidus points are connected by another smooth line, the equilibrium diagram results as shown in Fig. 4-3(b).

4-2 THE ONE-TWO-ONE AND INVERSE LEVER RULES

The *one-two-one* rule provides a simple method for determining the phases that exist in a particular region. The procedure is to draw a horizontal line [*xy* in Fig. 4-3(b)] that starts at the solid boundary *x* and terminates at the liquid boundary *y*. The *xy* line itself lies in a two-phase region. The two phases are solid and liquid.

It should also be noted that in our example the overall composition of the alloy *P* is always 75 per cent metal *A* and 25 per cent metal *B*. Since there is only one phase (liquid) for temperatures above T_1, the composition of the alloy is always liquid. The same is true for the single-phase solid that exists below the temperature T_3 for alloy *P*.

However, at the intersection *z* of the temperature line *xy* and the composition line *P*, two phases exist, namely liquid and solid. The composition of the liquid phase and the solid phase varies as the alloy cools along line *P*.

The line *xy*, through point *z*, intersects the solidus line at *x*. If a perpendicular is dropped from point *x* to the abscissa of the graph, it will be seen that the line shows the composition of the solid phase as being

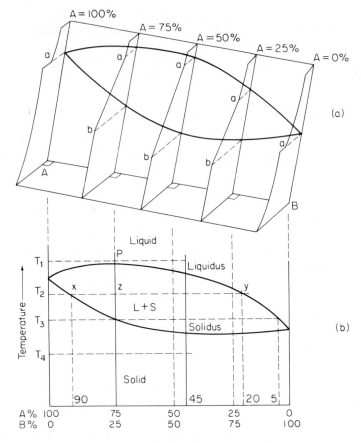

Figure 4-3

composed of 90 per cent element A and 10 per cent element B ($A + B$ must total 100 per cent).

The line xy, through point z, also intersects the liquidus line at y. If a perpendicular is dropped from y to the abscissa, the composition of the liquid will be seen to be 20 per cent element A alloyed with 80 per cent element B.

Inverse Lever Rule.   It now becomes necessary to determine the percentages of each phase that is present at various temperatures. Let us look at alloy P and determine the percentages of L and S present at T_2.

The length of the line xy represents the sum of the two phases as 100 per cent. The inverse rule states that the *liquid phase* can be calculated by taking the length of the line xz and dividing this by xy. The *solid phase* may be calculated by taking the length of the line *zy* and dividing this by *xy*. To get the percentage the values are multiplied by 100. Thus

$$\text{Liquid phase} = \frac{xz}{xy} \times 100 = \frac{90 - 75}{90 - 20} \times 100 = 21.4\%$$

$$\text{Solid phase} = \frac{zy}{xy} \times 100 = \frac{75 - 20}{90 - 20} \times 100 = 78.6\%$$

Example 1 illustrates the composite calculation of another combination, 45 and 55 per cent.

Example 1

Given an alloy Q, which has a composition of 45 per cent A and 55 per cent B, calculate and tabulate the requirements in Table 4-1 for the various temperatures in Fig. 4-3(b).

solution: See Table 4-1.

4-3 TYPES OF EQUILIBRIUM DIAGRAMS

There are several combinations of liquid solubility and solid solubility states that yield characteristic equilibrium diagrams. Equilibrium diagrams reflect combinations of these characteristic diagrams and may be classified with reference to their solid–liquid state combinations. That is, the components that enter into an alloy may be insoluble, partially soluble, or completely soluble in each other in the liquid or solid state. Several of these combinations are discussed with the use of hypothetical diagrams.

Certain elements may exist as two separate phases in the liquid state and two distinct phases in the solid state. This indicates a system completely insoluble in the liquid state and completely insoluble in the solid state. The idealized equilibrium diagram, characteristic of such a system, is shown in Fig. 4-4(a). The liquid solubility of element A in element B is practically zero. As the system cools below T_1, solid A_s forms and precipitates out, and two phases exist. At T_2, solid B_s forms and exists as a separate phase with A_s.

Figure 4-4(b) shows the left portion of such a system. At the top of the diagram two liquids are shown existing as separate phases. The nonvertical boundary lines would indicate a very slight solubility. Parallel boundaries (idealized) would indicate no solubility of the two liquids.

Table 4-1

TEMP.	PHASES	% OF EACH PHASE	% CHEMICAL COMP EACH PHASE	WEIGHT PER 100 lb.
T_1	Liquid	Single	A = 45 B = 55	A = 45 lb B = 55 lb. 00.0 lb

T_2	L + S	$L = \dfrac{90-45}{90-20} \times 100 = 64.3$
		
		$S = \dfrac{45-20}{90-20} \times 100 = 35.7$
			L < A = 20
			    B = 80
			
			S < A = 90
			    B = 10
				L < A = 64.3 × 0.20 = 12.9 lb
				    B = 64.3 × 0.80 = 5.14 lb
				
				S < A = 35.7 × 0.90 = 32.1 lb
				    B = 35.7 × 0.10 = 3.6 lb

T_3	L + S	$L = \dfrac{75-45}{75-5} \times 100 = 42.9$
		
		$S = \dfrac{45-5}{75-5} \times 100 = 57.1$
			L < A = 5
			    B = 95
			
			S < A = 75
			    B = 15
				L < A = 42.9 × 0.05 = 2.1 lb
				    B = 42.9 × 0.95 = 40.8 lb
				
				L < A = 57.1 × 0.75 = 42.8 lb
				    B = 57.1 × 0.25 = 14.3 lb

T_4	Solid	Single	A = 45 B = 55	A = 45 lb. B = 55 lb.

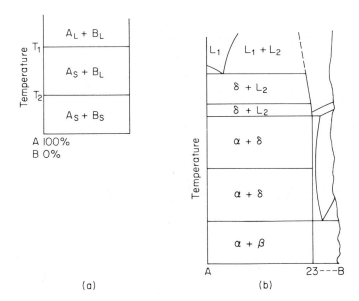

Figure 4-4

Systems that show this type of solubility combinations in the liquid and solid states are iron–oxygen, silver–iron, silver–tungsten, and copper–tungsten.

Another system is that in which the components are partially soluble in the liquid state. They may be partially soluble in the solid state, or they may be completely or insoluble in the solid state.

$$\text{Partially soluble in liquid state} \begin{cases} \text{insoluble in solid state} \\ \text{partially soluble in the solid state} \\ \text{completely soluble in solid state} \end{cases}$$

Figure 4-5 shows a diagram in which the elements are partially soluble in the liquid state and insoluble in the solid state. For alloy X the single-phase liquid state exists above T. Below the upper transformation line there exist two distinct liquid phases: one rich in element A and the other rich in element B. The partial solubility in the liquid phases exists in this region of the diagram. Below T_1, the A-rich phase solidifies, and a liquid–solid combination exists. As the temperature drops below T_2, B solidifies independently of A, so that B coexists with the A phase.

Figure 4-6 shows a diagram in which the two liquid phases are partially soluble ($L_1 + L_2$) and the solid phases are partially soluble

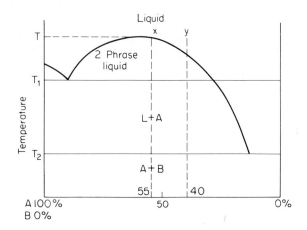

Figure 4-5

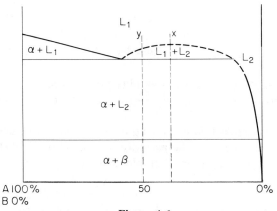

Figure 4-6

($\alpha + \beta$). Thus, as the two partially soluble liquids cool, A and B exist as partially soluble in each other as an alpha (α) phase. The remaining liquid L_2 solidifies as a partially soluble beta (β) phase. Systems that show partial liquid solubility and partial solid solubility are zinc–bismuth, aluminum–lead, copper–iron, and zinc–lead.

Figure 4-7 shows a diagram in which the two components are partially soluble in the liquid state and completely soluble in the solid state. That is, elements A and B combine to form a single phase on cooling.

Another series of equilibrium diagrams is that in which the constituents are completely soluble in the liquid state in combination with varying degrees of solubility in the solid state.

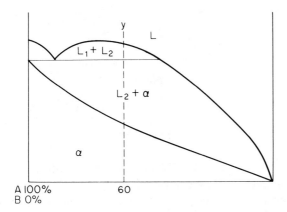

Figure 4-7

Completely soluble in the liquid state — completely soluble in the solid state / completely insoluble in the solid state / partially soluble in the solid state

    Systems of complete solubility of the components in the liquid state and complete solubility in the solid state are those that exhibit the characteristic shape shown in Fig. 4-8. In any combination of A and B (except for the pure metals A or B), the components are soluble above the liquidus line. Between the two solubility lines, the metals A and B are soluble in the liquid phase and also in the solid phase. Below the solidus line, the solid solution is that of the composition of the solid being considered, where one component is completely soluble in the other. Systems

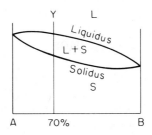

Figure 4-8

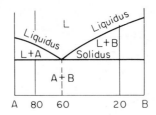

Figure 4-9

such as these are nickel–copper, antimony–bismuth, chromium–molybdenum, gold–silver, and tungsten–molybdenum.

In some instances components may be completely soluble in the liquid state and completely insoluble in the solid state. One such equilibrium diagram is shown in Fig. 4-9. Bismuth–cadmium forms such a diagram. In this example, the metals *A* and *B* in the alloy *X* are completely soluble above the liquidus line. As the temperature drops below the liquidus line, *A* precipitates out as heat is removed. At *T* the two metals *A* and *B* are evidently completely insoluble in each other, since the cooling curves exhibit a flat, or hold. Below *T*, the metals *A* and *B* are insoluble in each other.

A special invariant combination of *A* and *B* exists in this diagram. Whenever a combination of 60 per cent *A* with 40 per cent *B* exists, the two solidify as a *eutectic*. However, the eutectic also exhibits the fact that *A* and *B* are completely insoluble in the solid state.

The most common solubility combination of components is the case where the components are completely soluble in the liquid state and partially soluble in the solid state. Such systems are formed by combining copper–silver, copper–zinc, copper–beryllium, copper–tin, copper–aluminum, magnesium–aluminum, lead–antimony, lead–tin, and aluminum–silicon. Figure 4-13 shows that the metals *A* and *B* are completely soluble in the liquid state and that in *no* case is pure *A* or pure *B* formed, except at the terminal sides of the diagram.

4-4   INTERMETALLIC COMPOUNDS

The addition of alloys to pure metals may alter the dimensions or the configuration of their lattice structures. The addition of alloying elements (solute) to a pure metal (solvent) may be accomplished in several ways.

In some instances, solute atoms may displace solvent atoms in a solid lattice. When this occurs, it is called a *substitutional solid solution*. This is shown in Fig. 4-10(a). Since it is rare that the solute atoms are the

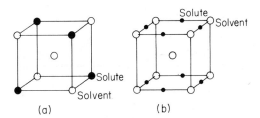

Figure 4-10

same size as the solvent atoms that they displace, the result is a distortion of the lattice.

If elements are to be completely soluble in one another, they must have the same type of lattice structure. The greater the size difference between the atoms of the two elements, the less their solid solubility in substitutional solutions. That is, if atom A is larger than atom B, it will be less likely to substitute for a B atom in the lattice than would an atom C whose size is more nearly the size of B. It is also true that the tendency toward the formation of intermetallic compounds takes precedent generally over the formation of a substitutional solid solution. As the concentration of the solute increases, so does the probability of formation of a substitutional solid solution increase.

When the solute atoms lodge in the space between the solvent atoms [Fig. 4-10(b)], the system is called an *interstitial solid solution*. This happens when the solute atoms are small, about half the diameter of the solvent atoms, and are able to fit in between the larger atoms in a lattice.

Obviously, the size of the space between the solvent atoms, as well as the size of the solute atoms, is important to the formation of an interstitial solid solution. Iron, nickel, chromium, manganese, molybdenum, tungsten, and vanadium are elements that lend themselves to being the solvent in the formation of interstitial solid solutions. Carbon, hydrogen, boron, nitrogen, and oxygen are atoms that have diameters small enough to act as the solute atoms in interstitial solid solutions. Iron, at room temperature, dissolves very small amounts of carbon interstitially. Above 1333°F the lattice structure changes from body-centered cubic lattice to face-centered cubic lattice, and the interatomic spacing increases, which makes it possible for carbon to form interstitial solid solutions.

If the atoms are insoluble in each other, a *mechanical mixture* results.

The *intermetallic compound* results when the two, solute and solvent, atoms join each other. These space lattices are very complex. They transform from the liquid phase to the solid compound phase at a fixed temperature and at a fixed composition. They differ from chemical compounds in that the rules of valence bonding, described earlier, do not

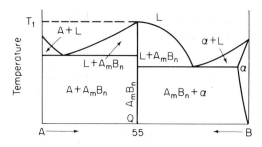

Figure 4-11

apply. The mechanical properties associated with intermetallic compounds are high strength and hardness, low ductility, and low conductivity.

As stated, intermetallic compounds form at one temperature. In Fig. 4-11, T_1, the alloy of A and B that has a composition of 55 per cent A and 45 per cent B freezes from the liquid directly to our intermetallic compound. The region characteristic of such a compound is a straight line that divides the diagram into two distinct sections. Figure 4-11 shows an intermetallic compound marked A_mB_n.

4-5 INVARIANT SYSTEMS

Several additional combinations of a two-component system form when two elements are in the required fixed percentages for a particular change in phase to take place. In the binary system there are five such important reversible and invariant transformations. They are as follows (the diagrams are hypothetical):

1. The *monotectic* system is a transformation that takes place when a single-phase liquid transforms directly into a liquid and solid two-phase structure. This is shown in Fig. 4-12(a) at point P.

2. The *peritectic* system is a transformation that takes place when a liquid and a two-phase solid transform directly to a single-phase solid structure. This is shown in Fig. 4-12(b) at point P.

3. The *peritectoid* system is shown in Fig. 4-12(c) at point P. In this system a two-phase solid transforms directly to a single-phase solid.

4. The *eutectic* system is a transformation that takes place when a single-phase liquid transforms directly to a two-phase solid, as shown in Fig. 4-12(c) at q.

5. The *eutectoid* system is shown in Fig. 4-12(d) and occurs when a single-phase solid transforms directly to a two-phase solid, as shown at point P.

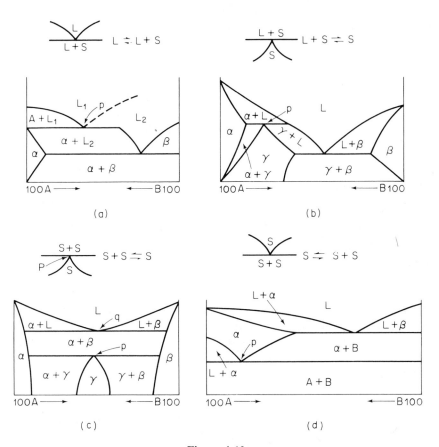

Figure 4-12

These are invariant systems because they occur whenever an alloy has a fixed composition. For example, in Fig. 4-13, at a temperature T_E, anytime the composition of the liquid phase is 40 per cent A and 60 per cent B, that liquid will transform directly into a two-phase $\alpha + \beta$. It will behave as though it is a pure metal. As indicated, such a transformation is called a eutectic transformation.

The inverse lever rule was discussed in Sec. 4-2. If this rule is applied to alloy R, Fig. 4-13, it will be seen that the liquid phase at e has the necessary 40 per cent A, 60 per cent B composition to be classified as a eutectic. At T_1 and q, solidification starts in a manner such that its composition is z. As the alloy cools further to T_2, the composition of the solid phase changes from z to x, and the liquid phase changes from q to y. It should be remembered that the composition of each phase is determined by dropping a perpendicular to the abscissa and reading the percentages of A and B. Thus the percentage of B in the solid will increase

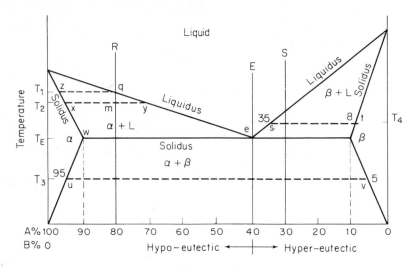

Figure 4-13

from about $z = 4$ per cent to $x = 6$ per cent, while the percentage of B in the liquid will be enriched from about $q = 20$ per cent to $y = 30$ per cent. The percentages of A in the solid and liquid will decrease accordingly. Also note in Fig. 4-13 that the α-region represents a solid solution in which metal A is the solvent.

At T_E, the composition of the solid has moved from $x = 6$ per cent B to $w = 10$ per cent B, while the composition of the *liquid* has moved from $y = 30$ per cent B to $e = 60$ per cent B. Thus the liquid has a eutectic composition at e of 40 per cent A and 60 per cent B. As alloy R cools below T_E, this eutectic liquid solidifies as if it were a pure metal [see Fig. 4-2(a)]. The result is solid A and B in the form of an alpha (α) phase, and A and B in the form of an eutectic phase.

Also, any alpha that solidifies *above* T_E, between compositions $A = 90$ per cent and $A = 40$ per cent, is called *primary alpha* (α_p). If beta solidifies above T_E, between compositions $A = 40$ per cent and $A = 10$ per cent, it is called *primary beta* (β_p).

There is still another movement that takes place. Below T_E, the α-phase is rich in component B. It cannot hold all the A component as it cools and forces it out of the lattice structure. In a like manner, some of the B component is forced out of the β-phase, which is A rich.

The composite analysis just discussed is shown in Table 4-2. The same analysis can be made for alloy S, Fig. 4-13.

It should be emphasized that a eutectic alloy has a fixed percentage of A to B. Since the eutectic liquid in Fig. 4-13 solidifies along the *wu*

Table 4-2

TEMP	PHASES	% OF EACH PHASE × 100	% COMPOSITION	WEIGHT PER 100 lb.
T_1	L	$\alpha = \dfrac{0}{96-80} = 0$ $L = \dfrac{96-80}{96-80} = 100$	$\alpha < \begin{cases} A = 94 \\ B = 4 \end{cases}$ $L < \begin{cases} A = 80 \\ B = 20 \end{cases}$	$\alpha < \begin{cases} A = 0 \times 0.94 = 0 \\ B = 0 \times 0.04 = 0 \end{cases}$ $L < \begin{cases} A = 100 \times 0.80 = 80.0 \\ B = 100 \times 0.20 = 20.0 \end{cases}$
T_2	$\alpha + L$	$\alpha = \dfrac{80-70}{94-70} = 41.7$ $L = \dfrac{94-80}{94-70} = 58.3$	$\alpha < \begin{cases} A = 94 \\ B = 6 \end{cases}$ $L < \begin{cases} A = 70 \\ B = 30 \end{cases}$	$\alpha < \begin{cases} A = 41.7 \times 0.94 = 39.2 \\ B = 41.7 \times 0.06 = 2.5 \end{cases}$ $L < \begin{cases} A = 58.3 \times 0.70 = 40.8 \\ B = 58.3 \times 0.30 = 17.5 \end{cases}$
T_E	overall $\alpha_p + \beta$	$\alpha_p = \dfrac{80-10}{90-10} = 87.5$ $\beta = \dfrac{90-80}{90-10} = 12.5$	$\alpha_p < \begin{cases} A = 90 \\ B = 10 \end{cases}$ $\beta < \begin{cases} A = 10 \\ B = 90 \end{cases}$	$\alpha_p < \begin{cases} A = 87.5 \times 0.90 = 78.7 \\ B = 87.5 \times 0.10 = 8.7 \end{cases}$ $\beta < \begin{cases} A = 12.5 \times 0.10 = 1.3 \\ B = 12.5 \times 0.90 = 11.3 \end{cases}$
T_E	$\alpha_p + E$	$\alpha_p = \dfrac{80-40}{90-40} = 80.0$ $E = \dfrac{90-80}{90-40} = 20.0$	$\alpha_p < \begin{cases} A = 90 \\ B = 10 \end{cases}$ $E < \begin{cases} A = 40 \\ B = 60 \end{cases}$	$\alpha_p < \begin{cases} A = 80.0 \times 0.90 = 72.0 \\ B = 80.0 \times 0.10 = 8.0 \end{cases}$ $E < \begin{cases} A = 20.0 \times 0.40 = 8.0 \\ B = 20.0 \times 0.60 = 12.0 \end{cases}$
T_3	overall $\alpha + \beta$	$\alpha = \dfrac{80-5}{95-5} = 83.3$ $\beta = \dfrac{95-80}{95-5} = 16.7$	$\alpha < \begin{cases} A = 95 \\ B = 5 \end{cases}$ $\beta < \begin{cases} A = 5 \\ B = 95 \end{cases}$	$\alpha < \begin{cases} A = 83.3 \times 0.95 = 79.1 \\ B = 83.3 \times 0.05 = 4.2 \end{cases}$ $\beta < \begin{cases} A = 16.7 \times 0.05 = 0.8 \\ B = 16.7 \times 0.95 = 15.9 \end{cases}$
T_3	$\alpha_p + E$	$\alpha_p = \dfrac{80-40}{90-40} = 80.0$ $E = \dfrac{90-80}{90-40} = 20.0$	$\alpha_p < \begin{cases} A = 90 \\ B = 10 \end{cases}$ $E < \begin{cases} A = 40 \\ B = 60 \end{cases}$	$\alpha_p < \begin{cases} A = 80.0 \times 0.90 = 72.0 \\ B = 80.0 \times 0.10 = 8.0 \end{cases}$ $E < \begin{cases} A = 20.0 \times 0.40 = 8.0 \\ B = 20.0 \times 0.60 = 12.0 \end{cases}$
T_3	E	$\alpha_E = \dfrac{40-10}{90-10} = 37.5$ $\beta_E = \dfrac{90-40}{90-10} = 62.5$	$\alpha_E < \begin{cases} A = 90 \\ B = 10 \end{cases}$ $\beta_E < \begin{cases} A = 10 \\ B = 90 \end{cases}$	$\alpha_E < \begin{cases} A = 37.5 \times 0.90 = 33.7 \\ B = 37.5 \times 0.10 = 3.7 \end{cases}$ $\beta_E < \begin{cases} A = 62.5 \times 0.10 = 6.3 \\ B = 62.5 \times 0.90 = 56.3 \end{cases}$

horizontal line, there can be only eutectic composition between $A = 90$ per cent and $A = 10$ per cent.

It should also be noted that only at composition E does the eutectic, and only the eutectic, form. To the *left* of E, $A = 90$ per cent to $A = 40$ per cent, the eutectic and α crystals form. This is called a *hypoeutectic* structure. To the right of E, $A = 40$ per cent to $A = 10$ per cent, the eutectic and β crystals form. This is called a *hypereutectic* structure.

Problems

4-1. Describe the relationship that exists between cooling curves and equilibrium diagrams.

4-2. Define the terms (a) component and (b) phase.

4-3. Discuss the cooling of (a) a pure metal and (b) an alloy.

4-4. Discuss the cooling curve characteristics when two metals are soluble in each other in the liquid state and insoluble in the solid state.

4-5. In relation to Fig. 4-12(a)–(d), discuss the one-two-one rule.

4-6. Label the missing regions in Fig. 4-14 and also label the liquidus and solidus lines.

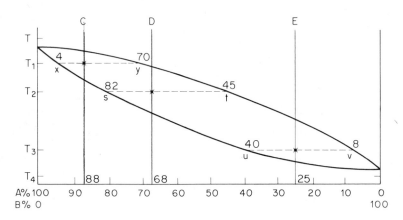

Figure 4-14

4-7. Indicate the percentages of element A and B in the alloys in Fig. 4-14:
(a) C, (b) D, (c) E.

4-8. Construct a table similar to Table 4-1 and calculate the percentages for alloy C in Fig. 4-14 for the temperatures T, T_1, T_2.

4-9. Repeat Problem 4-8 for alloy D for temperatures T_1, T_2, T_3.

4-10. Repeat Problem 4-8 for alloy E for temperatures T_2, T_3, T_4.

4-11. Draw the equilibrium diagram for two elements insoluble in the liquid state and insoluble in the solid state. Explain the phase changes as the alloy cools from the liquid state to room temperature.

4-12. Trace the cooling of an alloy y in Fig. 4-5 ($A = 40$ per cent) when the elements are partially soluble in the liquid state and insoluble in the solid state.

4-13. Trace the cooling of an alloy y in Fig. 4-6 ($A = 50$ per cent) when the elements are partially soluble in the liquid state and partially soluble in the solid state.

4-14. Trace the cooling of an alloy y in Fig. 4-7 ($A = 60$ per cent) when the elements are partially soluble in the liquid state and completely soluble in the solid state.

4-15. Explain the solubility of A and B in Fig. 4-8 for a 70 per cent alloy as the alloy y cools.

4-16. In Fig. 4-9 allow the alloy, 20 per cent A, to cool from the liquid state to room temperature. Describe the solubility changes that occur.

4-17. Trace the cooling of the alloys R, S, and E, in that order, from the liquid state to room temperature in Fig. 4-13. Describe the solubility changes that take place.

4-18. In your own words, describe the difference between the term *solute* and *solvent*.

4-19. Explain the mechanisms of substitutional and interstitial solid solutions.

4-20. Describe an intermetallic compound.

4-21. List the phase changes that take place during the following transformations: (a) eutectic, (b) peritectic, (c) eutectoid, (d) peritectoid, (e) monotectic.

4-22. Describe the phase changes that take place when a 40–60 per cent alloy E, Fig. 4-13, cools to room temperature from the liquid state.

4-23. Repeat Problem 4-22 for alloy S in Fig. 4-13.

4-24. Set up a table such as Table 4-2 and calculate the percentages for alloy E in Fig. 4-13 for temperatures T_2, T_E, T_3.

4-25. Repeat Problem 4-24 for alloy S in Fig. 4-13 for temperatures T_1, T_2, T_4, T_E, T_3.

4-26. Define a hypo- and a hypereutectic transformation from Fig. 4-13.

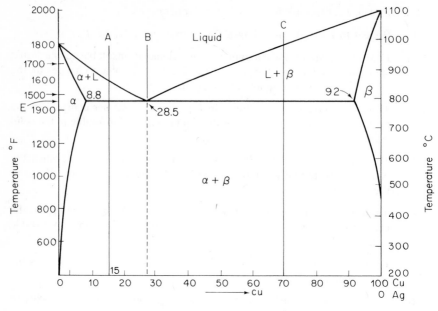

Figure 4-15

4-27. Set up a table such as Table 4-2 and calculate the percentages for alloy *A* in Fig. 4-15 for temperatures of 1700°F, 1500°F, *E*, 1000°F.

4-28. Repeat Problem 4-27 for alloy *B*.

4-29. Repeat Problem 4-27 for alloy *C*.

5 | The Iron–Iron Carbide Diagram: Steel

5-1 IRON–IRON CARBIDE DIAGRAM

Figure 5-1(a) shows the equilibrium diagram for combinations of carbon in a solid solution of iron. This diagram may be used to evaluate the physical state of the "plain" carbon steels. Equilibrium means very slow cooling. The diagram shows iron and carbons combined to form Fe_3C, an intermetallic compound at the 6.67 per cent carbon end of the diagram. The left side of the diagram is pure iron. Iron and carbon are the two dominant constituents of steel. Alloys are used to control the effects of each on the properties of steel. To the right of the pure iron line is carbon in combination with various forms of iron, which are called alpha (α), iron, or *ferrite*; gamma (γ) iron, or austenite; and delta (δ) iron.

Figure 5-1(b) is the cooling curve for pure iron. It shows the melt above 2795°F, where the first allotropic change takes place. Allotropic changes take place when there is a change in lattice structure and consequently in crystalline form. From 2795 to 2535°F the delta iron has a body-centered-cubic lattice structure. At 2535°F a second allotropic change occurs. The lattice structure changes from a body-centered-cubic to a face-centered-cubic lattice structure. It is called *austenite*. At 1670°F another allotropic change occurs and the lattice again changes from face-centered back to body-centered-cubic lattice. This ferrite structure, however, is nonmagnetic from 1670 to 1420°F. At about 1420°F, and below, the ferrite retains its body-centered-cubic lattice structure but once again becomes magnetic. It should be noted that at 1420°F, even

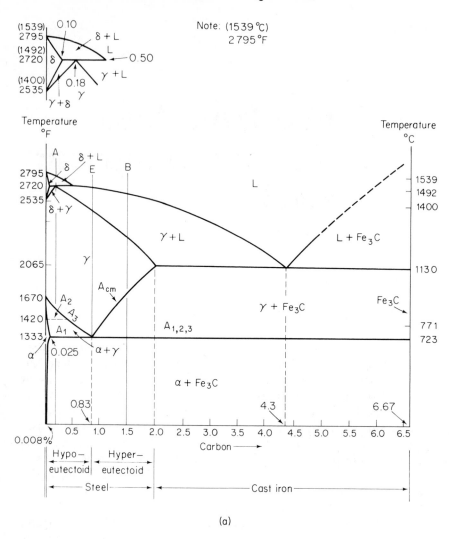

(a)

Figure 5-1

though the curve shows a plateau (dashed lines), this does not signify an allotropic change. This is the curve temperature where the metal changes its magnetic properties.

Two very important phase changes take place at 0.83 per cent carbon and at 4.3 per cent carbon in Fig. 5-1(a). The former is the *eutectoid* composition, called *pearlite*. The latter is the *eutectic* composition, called *ledeburite*.

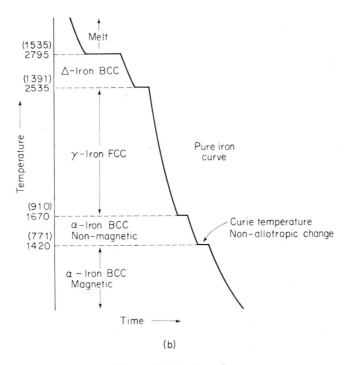

(b)

Figure 5-1 (continued)

5-2 ALLOTROPIC FORMS

Delta (δ) Iron. As has been noted, delta iron exists between 2795 and 2535°F. Figure 5-1(a) shows that it may exist in combination with the melt to about 0.50 per cent carbon, in combination with austenite to about 0.18 per cent carbon, and in a single-phase state out to about 0.10 per cent carbon. Delta iron has the BCC lattice structure and is magnetic. Upon heating, the allotropic change that takes place when gamma iron (FCC) changes to delta iron (BCC) is accompanied by an expansion of the lattice.

Gamma (γ) Iron. Delta iron, upon cooling into the gamma range, is accompanied by a contraction of the lattice. This crystal structure has the capability of absorbing greater amounts of carbon. *Austenite* is the solid solution of carbon in gamma iron. Under normal conditions austenite transforms below 1333°F.

Austenite. Austenite [Fig. 5-2(a)] is soft and ductile. Its maximum

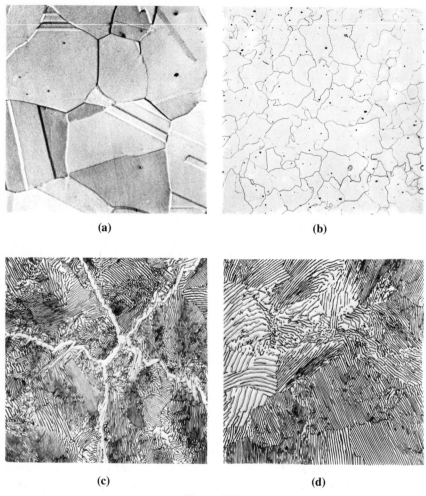

(a)   (b)

(c)   (d)

Figure 5-2

solubility of carbon is seen to be 2 per cent carbon at 2065°F. This is considered to be the steel portion of the diagram. Even though austenite can exist only at elevated temperatures, it is possible to have austenite at room temperature by adding special alloys to the metal. Some stainless steels, for instance, show austenite at room temperature.

Alpha (α) Iron. This iron exists at temperatures below 1670°F. The body-centered structure exists over 1333°F, even though the iron is nonmagnetic between 1420 and 1670°F. *Ferrite* is the solid solution of

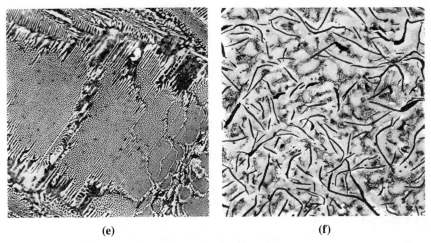

(e)   (f)

Figure 5-2 (continued)

carbon in alpha iron. It is the softest component in steel and is very ductile. It is shown in Fig. 5-2(b) as the white constituent.

Cementite. Cementite (Fe_3C) is the hardest and most brittle component of steel. Since its carbon content is 6.67 per cent, it is found in various combinations with ferrite and pearlite. Figure 5-2(c) shows cementite in the pearlite and also in the grain boundaries.

Pearlite. Pearlite is shown in Fig. 5-2(d). It forms a lamellar structure, as shown in Fig. 5-3. The carbon diffuses toward the cementite plates that have grown out from the austenite. Since cementite requires more carbon than austenite, the carbon diffuses to the cementite plates and depletes the austenite of carbon to form ferrite on cooling. It forms when 0.83 per cent carbon exists in combination with iron. It possesses phy-

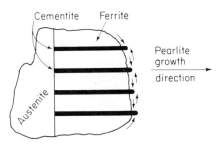

Figure 5-3

sical properties that are between those of the very hard cementite constituent and the very soft ferrite constituent.

A fixed amount of carbon and a fixed amount of iron are needed to form cementite (Fe_3C). Also, pearlite needs fixed amounts of cementite and ferrite. The equilibrium diagram [Fig. 5-1(a)] and the photomicrograph [Fig. 5-2(d)] show only pearlite forming when the carbon (0.83 per cent) percentage exactly fills the cementite requirements.

If there is not enough carbon present, that is, less than 0.83 per cent, the carbon and the iron will combine to form Fe_3C until all the carbon is consumed. This cementite will combine with the required amount of ferrite to form pearlite. The ferrite remaining will stay in the structure. This "free" ferrite is called *proeutectoid ferrite*. The structure, pearlite and proeutectoid ferrite, is referred to as *hypoeutectoid* and is shown in Fig. 5-2(b).

If, however, there is an excess of carbon above 0.83 per cent in the austenite, pearlite will form and the excess carbon above 0.83 per cent will form cementite. This "excess" cementite deposits in the grain boundaries. It is called *proeutectoid cementite*. The structure, pearlite and proeutectoid cementite, is called *hypereutectoid* and is shown in Fig. 5-2(c).

Ledeburite.   Ledeburite [Fig. 5-2(e)] is the eutectic of cast iron. It exists when the carbon content is greater than 2 per cent, which represents the dividing line on the equilibrium diagram, Fig. 5-1(a), between steel and cast iron. It contains 4.3 per cent carbon in combination with iron.

Graphite.   Graphite is the other major constituent of cast iron. It and iron are the end products of slowly cooled cementite above 2 per cent on the equilibrium diagram. It is shown as long slivers in the photomicrograph of gray cast iron [Fig. 5-2(f)].

In Fig. 5-1(a) are also shown three solubility lines, A_1, A_3, and A_{cm}. The A_1 line is called the *lower-critical-temperature* line. It is the lowest temperature at which FCC iron can exist. It is also the temperature at which the eutectoid forms. The A_3 line is the *upper-critical-temperature* line. As noted, it is the temperature at which the FCC lattice changes to BCC lattice.

To the right of the eutectoid (0.83 per cent) line the A_1 and A_3 lines coexist at 1333°F. At the eutectoid and to the right of the eutectoid the *upper critical temperature* follows the A_{cm} line. The A_3 line represents an allotropic change. The A_{cm} line represents a change in carbon solubility.

5-3   EQUILIBRIUM CALCULATION: STEEL

Three significant calculations can be made relative to the steel portion of the iron–iron carbide diagram [Fig. 5-1(a)]. They are E, the eutectoid, A, the hypoeutectoid, and B, the hypereutectoid composition. All are shown in Fig. 5-4.

Example 1

Calculate the phases at the eutectoid composition line E, Fig. 5-4, at the temperature (a) T_1, (b) T_3, (c) T_4, (d) T_E, and (e) T_6.

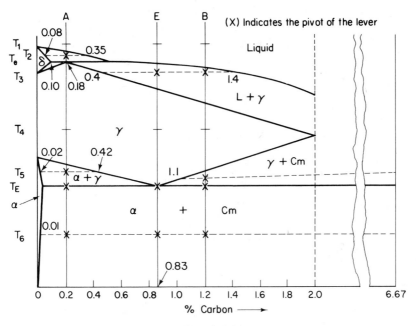

Figure 5-4

solution:

The composition E is eutectoid, and the carbon content is 0.83.

(a)   At T_1 there exists a single-phase liquid.

(b) At T_3 there are two phases. The boundaries of the level are equal to solid 0.4 C and the liquid at 1.4 C. The percentages of each are

$$\text{Liquid} \quad = \frac{0.83 - 0.40}{1.4 - 0.40} \times 100 = 43\%$$

$$\text{Austenite} = \frac{1.4 - 0.83}{1.4 - 0.40} \times 100 = 57\%$$

(c) At T_4 there exists a single-phase austenite.

(d) At T_E the eutectoid composition forms. The percentages of ferrite and cementite that form the eutectoid are

$$\text{Eutectoid ferrite} \quad = \frac{6.67 - 0.83}{6.67 - 0.025} \times 100 = 88\%$$

$$\text{Eutectoid cementite} = \frac{0.83 - 0.025}{6.67 - 0.025} \times 100 = 12\%$$

(e) At T_6 the overall percentages of ferrite and cementite are

$$\text{Ferrite} \quad = \frac{6.67 - 0.83}{6.67 - 0.01} \times 100 = 87.7\%$$

$$\text{Cementite} = \frac{0.83 - 0.01}{6.67 - 0.01} \times 100 = 12.3\%$$

The composition of the pearlite does not change. It is

$$\text{Eutectoid ferrite} \quad = 88\%$$
$$\text{Eutectoid cementite} = 12\%$$

Example 2

Calculate the phases present at composition A, Fig. 5-4, for (a) T_1, (b) T_2, (c) T_e, (d) T_4, (e) T_5, (f) T_E, and (g) T_6.

solution:

(a) At T_1 there exists a single-phase hypoeutectoid liquid that has 0.2 per cent carbon.

(b) At T_2 solid delta iron has precipitated out of the liquid. Two phases exist at this temperature. The ends of the lever terminate at 0.08 per cent C and 0.35 per cent C. The percentages of delta and liquid phases are

$$\text{Liquid} \quad = \frac{0.20 - 0.08}{0.35 - 0.08} \times 100 = 44\%$$

$$\text{Delta iron} = \frac{0.35 - 0.20}{0.35 - 0.08} \times 100 = 56\%$$

(c) At T_e a peritectic forms and transforms directly into austenite. Thus, just prior to the change into the FCC lattice from the BCC lattice, the percentages of liquid to delta iron are

$$\text{Liquid} \quad = \frac{0.18 - 0.1}{0.5 - 0.1} \times 100 = 20\%$$

$$\text{Delta iron} = \frac{0.5 - 0.18}{0.5 - 0.1} \times 100 = 80\%$$

(d) At T_4 austenite exists as a single-phase solid.

(e) At T_5 two phases exist, ferrite and austenite. The composition of each is

$$\text{Ferrite} \quad = \frac{0.42 - 0.20}{0.42 - 0.02} \times 100 = 55\%$$

$$\text{Austenite} = \frac{0.20 - 0.02}{0.42 - 0.02} \times 100 = 45\%$$

(f) At T_E and at the *beginning* of the transformation from FCC to BCC lattice structure, two phases exist, ferrite and austenite. The percentages of each are

$$\text{Proeutectoid ferrite} = \frac{0.83 - 0.20}{0.83 - 0.025} \times 100 = 78\%$$

$$\text{Austenite} \quad = \frac{0.20 - 0.025}{0.83 - 0.025} \times 100 = 22\%$$

The composition of the austenite is eutectoid. The percentages of ferrite to cementite are

$$\text{Eutectoid ferrite} \quad = \frac{6.67 - 0.83}{6.67 - 0.025} \times 100 = 88\%$$

$$\text{Eutectoid cementite} = \frac{0.83 - 0.025}{6.67 - 0.025} \times 100 = 12\%$$

Thus it is seen that the percentages of ferrite and cementite in the eutectoid are the same as they were for the eutectoid composition for alloy E.

The overall ferrite and cementite at T_E are

$$\text{Eutectoid and proeutectoid ferrite} = \frac{6.67 - 0.20}{6.67 - 0.025} = 100 = 97.4\%$$

$$\text{Eutectoid cementite} = \frac{0.20 - 0.025}{6.67 - 0.025} \times 100 = 2.6\%$$

The latter calculations may be confirmed in the following manner:

$$\begin{aligned}
\text{Proeutectoid ferrite} \quad &= 78.0\% \\
\text{Eutectoid ferrite} \quad &= 22 \times 88\% = \underline{19.4\%} \\
\text{Total ferrite} &= 97.4\% \; check
\end{aligned}$$

Eutectoid cementite $= 22 \times 12\% = 2.6\%$ *check*

(g) At T_6 a very small amount of cementite will precipitate and continue to precipitate following the solubility line at the left from 0.025 per cent at 1333°F to 0.008 per cent C at room temperature. See Fig. 5-1(a). The overall percentages of ferrite and cementite are

$$\text{Ferrite} \quad = \frac{6.67 - 0.2}{6.67 - 0.01} \times 100 = 97.1\%$$

$$\text{Cementite} = \frac{0.2 - 0.01}{6.67 - 0.01} \times 100 = 2.9\%$$

The percentages of ferrite and pearlite are

$$\text{Pearlite} \quad = \frac{0.2 - 0.01}{0.83 - 0.01} \times 100 = 23.2\%$$

$$\text{Proeutectoid ferrite} = \frac{0.83 - 0.2}{0.83 - 0.01} \times 100 = 76.8\%$$

The percentages of the ferrite and cementite phases in the pearlite remain unchanged

Eutectoid ferrite $\quad = 88\%$
Eutectoid cementite $= 12\%$

Example 3

Calculate the percentages of the phases for the composition at B in Fig. 5-4 at temperatures (a) T_1, (b) T_3, (c) T_4, (d) T_5, (e) T_E, and (f) T_6.

solution:

(a) At T_1 the physical state of the composition is a hypereutectoid liquid.

(b) At T_3 two phases exist. They are

$$\text{Austenite} = \frac{1.4 - 1.2}{1.4 - 0.4} \times 100 = 20\%$$

$$\text{Liquid} \quad = \frac{1.2 - 0.4}{1.4 - 0.4} \times 100 = 80\%$$

(c) At T_4 the single-phase austenite exists.

(d) At T_5 two phases exist. They are

$$\text{Austenite} \quad = \frac{6.67 - 1.2}{6.67 - 1.1} \times 100 = 98.2\%$$

$$\text{Cementite} = \frac{1.2 - 1.1}{6.67 - 1.1} \times 100 = 1.8\%$$

(e) At T_E two phases exist. They are

$$\text{Proeutectoid cementite} = \frac{1.2 - 0.83}{6.67 - 0.83} \times 100 = 6.3\%$$

$$\text{Austenite} = \frac{6.67 - 1.2}{6.67 - 0.83} \times 100 = 93.7\%$$

The eutectoid phases are

$$\text{Eutectoid cementite} = \frac{0.83 - 0.025}{6.67 - 0.025} \times 100 = 12\%$$

$$\text{Eutectoid ferrite} = \frac{6.67 - 0.83}{6.67 - 0.025} \times 100 = 88\%$$

The overall ferrite and cementite are

$$\text{Eutectoid and proeutectoid cementite} = \frac{1.2 - 0.025}{6.67 - 0.025} \times 100 = 17.7\%$$

$$\text{Eutectoid ferrite} = \frac{6.67 - 1.2}{6.67 - 0.025} \times 100 = 82.3\%$$

The last calculation may be checked as follows:

Cementite in the austenite $= 93.7 \times 12\% = 11.2\%$
Proeutectoid cementite $\qquad = \; 6.3\%$
Total cementite $\qquad\qquad = 17.5\% \; check$
Eutectoid ferrite $\qquad = 93.7 \times 88\% = 82.5\% \; check$

(f) At T_6 the overall percentages of cementite and ferrite are

$$\text{Cementite} = \frac{1.2 - 0.01}{6.67 - 0.01} \times 100 = 17.9\%$$

$$\text{Ferrite} = \frac{6.67 - 1.2}{6.67 - 0.01} \times 100 = 82.1\%$$

The percentages of pearlite and cementite are

$$\text{Pearlite} = \frac{6.67 - 1.2}{6.67 - 0.83} \times 100 = 93.7\%$$

$$\text{Proeutectoid cementite} = \frac{1.2 - 0.83}{6.67 - 0.83} \times 100 = 6.3\%$$

The percentages of ferrite and cementite in the pearlite do not change.
They are

Eutectoid ferrite $\quad = 88\%$
Eutectoid cementite $= 12\%$

5-4   CARBON AND THE PHYSICAL PROPERTIES OF STEEL

In general, the Brinell hardness increases as the carbon content increases up to about 2.0 per cent. The tensile strength and yield strength also increase to about 0.83 per cent carbon. Thereafter, they level out [Fig. 5-5(a)]. The elongation in 2 in. and the reduction in area drop sharply with increase in carbon content, going almost to zero at about 1.5 per cent carbon. The Charpy impact also decreases very sharply up to about 0.83 per cent carbon and then levels out. These characteristics are shown in Fig. 5-5(b).

As indicated earlier, ferrite is a solid solution of carbon in alpha iron. On slow cooling it is the soft constituent in iron. It has a tensile strength of about 40,000 psi. Cementite, which is a compound of iron and carbon has a very low tensile strength of 40,000 to 60,000 psi. It is

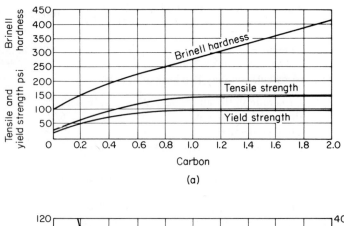

(a)

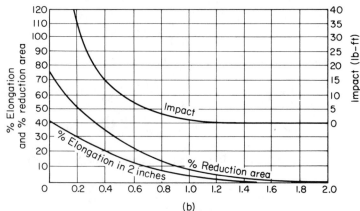

(b)

Figure 5-5

the hard brittle component of steel. Pearlite, which is a combination of ferrite and cementite, has a high tensile strength in the neighborhood of 100,000 psi. It is much harder than ferrite but much less hard than cementite.

It would appear to be reasonable to expect the tensile strength and hardness to be affected as the ratio of ferrite to cementite in the structure of the steel changes. As the percentage of pearlite increases in the hypoeutectoid steels, the tensile strength increases. The hardness does not increase dramatically. The hypereutectoid steels show only a slight increase in strength as the cementite-to-ferrite ratio increases. This is probably due to the fact that the excess cementite acts as an envelope about the softer pearlite phase.

5-5   ALLOYING ELEMENTS AND THE CRITICAL TEMPERATURE

Alloying elements also have a profound effect upon the iron–iron carbide equilibrium diagram. The addition of alloying elements will form solid solutions, intermetallic compounds, or a combination of the two. The effect will be to widen or constrict the temperature range through which gamma iron is stable. That is, it will either raise or lower the critical temperature.

Alloys such as cobalt, manganese, and nickel increase the temperature range through which gamma iron is stable or lower the critical temperature [see Fig. 5-6(a)]. They are very soluble in the gamma iron. Carbon, copper, zinc, and nitrogen form intermetallic compounds or solid solutions that are rich in these elements. Again the result is to widen the temperature range through which the gamma iron is stable, thus dropping the critical temperature.

Alloys such as aluminum, beryllium, chromium, molybdenum, phosphorus, silicon, tin, tungsten, or vanadium all tend to form solid solutions with alpha iron. This constricts the temperature region through which gamma iron is stable. Sulfur, tantalum, and boron, besides forming solid solutions with alpha iron, also form intermetallic compounds. This also causes the constriction of the temperature range through which gamma iron is stable. Figure 5-6(b) shows the very marked reduction of the region through which austenite is stable.

5-6   ALLOYING ELEMENTS AND THEIR EFFECT ON HARDNESS AND TENSILE STRENGTH

The elements shown in Fig. 5-7(a) have the greatest solubility in ferrite and also influence the hardenability of iron when in the presence of car-

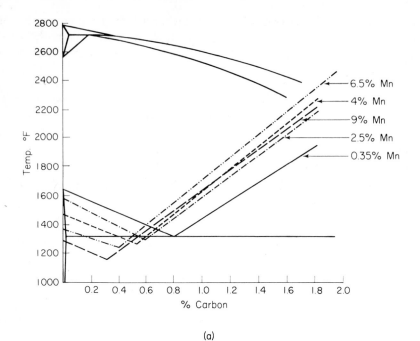

(a)

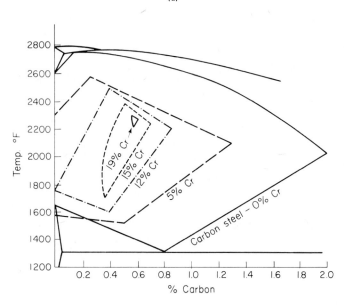

(b)

Figure 5-6

bon. With but a slight increase in the carbon content, they respond markedly to heat treating, because the carbon acts as a ferrite strengthener.

Example 4

Refer to Fig. 5-7(a); compute the increase in the Brinell hardness of a sample of steel when the following alloy percentage increase is combined with alpha iron: (a) chromium, 8 to 16; (b) vanadium, 6 to 18; (c) manganese, 2 to 4.

solution:

    (a) Chromium, 8 to 16 per cent:

$$\text{Chromium,} \quad 16\% = 125 \text{ BHN}$$
$$\text{Chromium,} \quad \ 8\% = \underline{100} \text{ BHN}$$
$$\text{Decrease of} \ \ 8\% = \ \ 25 \text{ BHN}$$

    (b) Vanadium, 6 to 18 per cent:

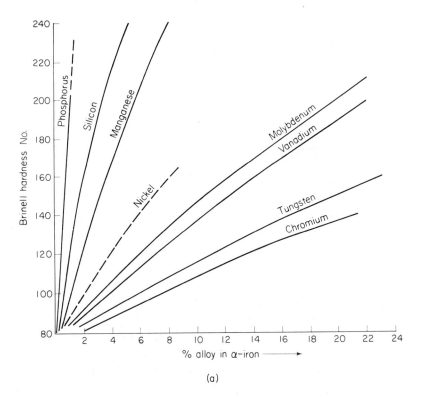

(a)

Figure 5-7

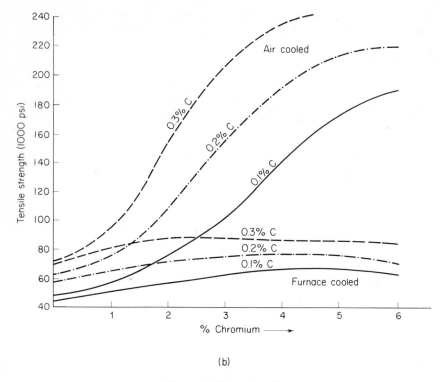

(b)

Figure 5-7 (continued)

Vanadium, 18% = 180 BHN
Vanadium,   6% = 110 BHN
Decrease of 12% = 70 BHN

(c) Manganese, 2 to 4 per cent:

Manganese, 4% = 165 BHN
Manganese, 2% = 125 BHN
Decrease of 2% = 40 BHN

Figure 5-7(b) shows the effect upon the tensile strength of three percentages of carbon in the presence of chromium. Furnace cooling has very little effect on the tensile strength of the material. The addition of chromium to the steel does not change the slope of the curves appreciably. If these same steels are air cooled at the same rate, the slope of the curves increases very drastically with but a slight increase in the chromium content.

Example 5

In Fig. 5-7(b), if the chromium content is 3.5 per cent, calculate the percentage of increase of tensile strength for furnace versus air-cooled steel with a carbon content of (a) 0.1 per cent, (b) 0.2 per cent, and (c) 0.3 per cent.

solution:

(a) The percentage of increase in tensile strength for a 3.5 per cent Cr, 0.1 per cent C steel when it is furnace versus air cooled is

Air cooled = 120,000 psi
Furnace cooled = 65,000 psi
Change in tensile = 55,000 psi
  strength

Increase $= \dfrac{55,000}{120,000} \times 100 = 45.8\%$

(b) The percentage of tensile strength increase for a 3.5 per cent Cr, 0.2 per cent C steel is

Air cooled = 180,000 psi
Furnace cooled = 75,000 psi
Change in tensile = 105,000 psi
  strength

Increase $= \dfrac{105,000}{180,000} \times 100 = 58.3\%$

(c) The percentage of tensile strength increase for a 3.5 per cent Cr, 0.3 per cent C steel is

Air cooled = 220,000 psi
Furnace cooled = 85,000 psi
Change in tensile = 145,000 psi
  strength

Increase $= \dfrac{145,000}{220,000} \times 100 = 65.9\%$

Example 6

Calculate the increase in tensile strength from the graphs of Fig. 5-7(b) when the chromium content increases from 2 to 3.5 per cent for steels with 0.1C, 0.2C, 0.3 per cent C when the steels are (a) furnace cooled and (b) air cooled.

solution:

(a) Furnace-cooled steel:

$$3.5\% \text{ Cr} - 0.1\% \text{ C} = \quad 65,000 \text{ psi}$$
$$2.0\% \text{ Cr} - 0.1\% \text{ C} = \quad \underline{55,000} \text{ psi}$$
$$\Delta\text{T.S.} = \quad 10,000 \text{ psi}$$

$$3.5\% \text{ Cr} - 0.2\% \text{ C} = \quad 75,000 \text{ psi}$$
$$2.0\% \text{ Cr} - 0.2\% \text{ C} = \quad \underline{70,000} \text{ psi}$$
$$\Delta\text{T.S.} = \quad 5,000 \text{ psi}$$

$$3.5\% \text{ Cr} - 0.3\% \text{ C} = \quad 85,000 \text{ psi}$$
$$2.0\% \text{ Cr} - 0.3\% \text{ C} = \quad \underline{90,000} \text{ psi}$$
$$\Delta\text{T.S.} = \quad -5,000 \text{ psi}$$

(b) Air-cooled steel:

$$3.5\% \text{ Cr} - 0.1\% \text{ C} = 120,000 \text{ psi}$$
$$2.0\% \text{ Cr} - 0.1\% \text{ C} = \underline{75,000} \text{ psi}$$
$$\Delta\text{T.S.} = 45,000 \text{ psi}$$

$$3.5\% \text{ Cr} - 0.2\% \text{ C} = 175,000 \text{ psi}$$
$$2.0\% \text{ Cr} - 0.2\% \text{ C} = \underline{110,000} \text{ psi}$$
$$\Delta\text{T.S.} = 65,000 \text{ psi}$$

$$3.5\% \text{ Cr} - 0.3\% \text{ C} = 220,000 \text{ psi}$$
$$2.0\% \text{ Cr} - 0.3\% \text{ C} = \underline{150,000} \text{ psi}$$
$$\Delta\text{T.S.} = 70,000 \text{ psi}$$

Problems

5-1. Draw the iron–iron carbide diagram from memory, inserting (a) the eutectoid composition, (b) the solubility limit for steel, (c) the eutectic, and (d) the intermetallic compound. (e) Label all regions, and (f) label the allotropic transformations.

5-2. Relate the lattice structures to the allotropic transformations.

5-3. Explain what is meant by a hypo- and hypereutectoid transformation as related to the iron–iron carbide diagram, Fig. 5-1(a).

5-4. Explain the changes that occur as pure iron cools from the melt. See Fig. 5-1(b).

5-5. Describe the two invariant phase changes that take place on the equilibrium diagram [Fig. 5-1(a)] at 0.83 and 4.3 per cent carbon.

5-6. (a) What is pearlite? (b) Ledeburite?

5-7. (a) What is austenite? (b) How does it differ from gamma iron? (c) Is it ever stable at room temperature for plain carbon steel? (d) What is its lattice structure?

5-8. (a) What is alpha iron? (b) How does it differ from ferrite? (c) Describe its solubility, magnetic properties, lattice structure, and mechanical properties.

5-9. What is cementite?

5-10. Describe the growth of pearlite and list some of its physical properties.

5-11. Describe the relationship between the carbon available in the structure of steel and the formation of (a) eutectoid, (b) hypoeutectoid, (c) hyper-eutectoid, (d) proeutectoid cementite, and (e) proeutectoid ferrite.

5-12. Discuss (a) ledeburite and (b) graphite.

5-13. Discuss the A_1, A_2, A_3, and A_{cm} lines in Fig. 5-1(a).

5-14. Using Fig. 5-4, calculate the percentages for the phase changes for steel, carbon content 1.5 per cent, for temperatures (a) T_1, (b) 2500°F, (c) T_4, (d) T_E, and (e) T_6.

5-15. Using Fig. 5-4, calculate the percentages for the phase changes for steel, carbon content 0.3 per cent, for temperatures (a) T_1, (b) T_2, (c) T_e, (d) 2000°F, (e) T_5, (f) T_E, and (g) T_6.

5-16. Calculate the phrases at the eutectoid composition in Fig. 5-4 at tempera-tures (a) T_1, (b) T_3, (c) T_4, (d) T_E, and (e) T_6.

5-17. Discuss the physical properties of steel in relation to the carbon content. See Fig. 5-5(a) and (b).

5-18. (a) Compare the tensile strength of ferrite, cementite, and pearlite. (b) What effect does the percentage of each have on the hardness and tensile strength of steel?

5-19. What are the effects of alloying elements with carbon steel when they form (a) a solid solution, (b) an intermetallic compound, (c) or a combina-tion of the two?

5-20. What effect does the addition of manganese or cobalt have on the critical temperature when it is added to iron as an alloy? Relate your answer to Fig. 5-6(a).

5-21. What effect does the addition of molybdenum, vanadium, tungsten, or tantalum have on iron? Relate your answer to Fig. 5-6(b).

5-22. Referring to Fig. 5-7(a), what is the effect of increasing the percentage of the alloys shown in the graph upon the Brinell hardness in the α iron?

5-23. Compute the increase in the Brinell hardness number of a sample of steel when the following percentage of increase is combined with alpha iron: (a) tungsten, 8 to 20; (b) molybdenum, 6 to 16; (c) manganese, 1 to 8.

5-24. Will furnace cooling or air cooling increase the tensile strength of steel when chromium is added to the carbon content? Explain.

5-25. A 0.1 per cent carbon steel has a chromium alloy of 1 per cent. If it were possible to increase the chromium content to 4 per cent, (a) what would be the increase in tensile strength if the sample is air cooled? (b) furnace cooled?

5-26. Calculate the tensile strength increases for the three carbon steels graphed in Fig. 5-7(b) when the percentage of chromium increases from 1.5 to 4.5, when the samples are (a) furnace cooled, (b) air cooled.

5-27. Calculate the percentage of increase in tensile strength for each of the carbon content steels in Fig. 5-7(b) when the samples are furnace cooled versus air cooled for a chromium content of (a) 1.5, (b) 2.5, and (c) 4.0.

6 | The Iron–Iron Carbide Diagram: Cast Iron

6-1 CONSTITUENTS

In the iron–iron carbide diagram of Fig. 6-1 the cast iron portion lies between 2 and 6.67 per cent carbon. If a combination of iron and carbon is taken to have a composition at C, and is cooled from T_1 to T_e, the phases would be austenite and a eutectic of austenite and cementite. As the temperature drops from T_e to T_E, cementite precipitates from the austenite along the solubility line $e'E$. At T_E the austenite has a eutectoid structure that transforms to pearlite. The assumption here is that the only constituents present are iron and carbon.

If there are other elements present, as there most assuredly would be, the phases change and the diagram changes. If the constituents change, the physical properties of the material also change. If the cooling rate is increased, the physical properties are also affected. Thus, if phosphorus is present, a eutectic of iron phosphide and cementite, called *steadite*, is present.

It is possible to have the following phases present in cast iron at room temperature:

1. Free carbon, or graphite: soft, weak, no ductility, powder form.

2. Temper carbon: free carbon that has formed as almost circular particles.

3. Free ferrite: soft, ductile, T.S. about 40,000 psi, elongation of 40 per cent in 2 in.

4. Free cementite: hard, weak, brittle, highly wear resistant, T.S. of about 5000 psi.

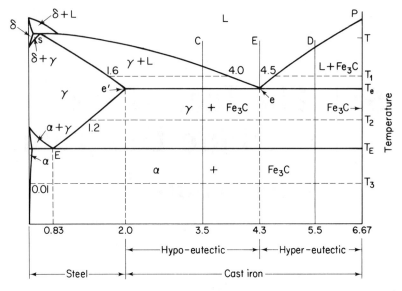

Figure 6-1

5. Pearlite, a eutectoid of cementite and ferrite: hardness between cementite and ferrite, strong with a T.S. of about 115,000 psi, elongation of 10 per cent in 2 in.

6. Steadite, a eutectic of iron phosphide and cementite: hard, brittle.

Alloying elements such as chromium and manganese are cementite stabilizers. Silicon, aluminum, and nickel are graphite formers.

6-2   TYPES OF CAST IRON

Gray Cast Iron.   It was stated that silicon, when alloyed with ferrite and carbon in amounts of about 2 per cent, makes the carbide of iron unstable. As a matter of fact, Fe_3C in cast iron is very unstable. At elevated temperatures, if slow cooling permits, the Fe_3C will decompose into ferrite and graphite. At room temperatures the decomposition takes place very slowly. This condition is referred to as *metastable*. Also, if the cooling rate is slow, the amount of free carbon in the structure increases. This free carbon collects as slivers or flakes of powdered graphite. Since graphite is very weak, the structure of gray cast iron is very weak. The characteristic gray appearance of a fractured piece of this material has given it the name of *gray cast iron* [Fig. 5-2(f)].

White Cast Iron.   As the cooling *rate* increases, the amount of free carbon decreases and the amount of combined carbon increases. Since cementite is very hard and since very rapid cooling will produce a high percentage of combined carbon in the form of cementite, the resulting structure is very hard. It is called *white cast iron* [Fig. 6-2(a)] because of its light appearance, which is due to the absence of graphite. When localized areas of a gray casting are cooled very rapidly from the melt, white cast iron is formed at the place that has been cooled. This type of white iron is called *chilled iron.*

Figure 6-2(b) shows a photomicrograph of a *hypoeutectic white cast iron,* and Fig. 6-2(c) shows the hypereutectic white cast iron structure. These are formed from the hypo- and hypereutectic gray cast irons as indicated in Fig. 6-1.

Malleable Cast Iron. If cast iron is cooled very rapidly, the graphite flakes needed for gray cast iron do not get a chance to form. Instead, white cast iron forms. This white cast iron is then reheated to about 1600°F for

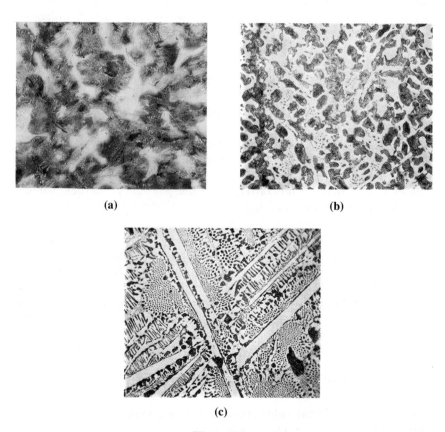

(a)

(b)

(c)

Figure 6-2

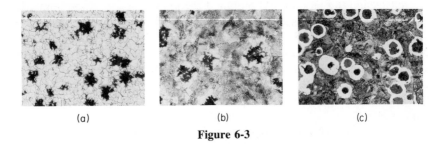

(a)   (b)   (c)

Figure 6-3

long periods of time in the presence of materials containing oxygen, such as iron oxide. At this elevated temperature the cementite formed in the white cast iron decomposes into ferrite and free carbon. Upon cooling, the combined carbon further decomposes according to the slope of the solubility line, and forms small compact particles of graphite instead of the flake-like form characteristic of gray cast iron. At the eutectoid temperature, slow cooling releases still more free carbon. All this free carbon is referred to as *temper carbon,* and the process is called *malleableizing. Ferritic malleable cast iron* [Fig. 6-3(a)] has a ferrite matrix into which are embedded the particles of temper carbon.

By controlling the heat-treating temperatures or by adding manganese to the structure, quantities of combined carbon in the form of cementite are retained, and the white cast iron may form a matrix of pearlite. A *pearlite malleable cast iron* [Fig. 6-3(b)] forms. A wide variety of physical characteristics may be achieved by controlling the matrix. Another method used to retain some cementite is the controlled cooling of the structure through the eutectoid temperature. As stated, slow cooling will cause the cementite to decompose. Fast cooling will retain some of the cementite. The amount retained, and consequently the physical properties, will depend on the rapidity of cooling.

Nodular Cast Iron.   This structure [Fig. 6-3(c)] is developed from the melt. The carbon forms into spheres when cerium, magnesium, sodium, or other elements are added to a melt of iron with a very low sulfur content, or which is low in any element that will inhibit carbon from forming. As is the case in malleable cast iron, the control of the heat-treating process can yield pearlitic and martensitic as well as ferritic matrices into which are embedded these spheres of carbon nodules. This type of iron is also known as *ductile cast iron.*

6-3 EFFECT OF ALLOYING ON CAST IRON

The elements of silicon, sulfur, phosphorus, and manganese generally are found alloyed with iron and carbon, since they have been retained from the iron ore.

The content of *silicon* in cast iron ranges up to about 3 per cent. Its effect is to make more carbon available to form graphite beyond that which will combine to form iron. As such, it is considered a graphite former. Higher percentages (5 per cent) of silicon will make the cast iron brittle but atmospheric corrosion resistant. Still higher percentages, over 10 per cent, makes the cast iron acid resistant.

The *sulfur* content is kept below 0.1 per cent. It makes cast iron brittle at high temperatures because it will combine with iron or manganese. Brittleness at high temperature is referred to as *hot shortness*. It also increases the shrinkage in iron upon solidification and thereby increases the tendency toward cracking.

Phosphorus also causes cast iron to be brittle if it is present in amounts over 2 per cent. It has the positive effect of increasing the fluidity of the melt and decreasing the shrinkage when the melt solidifies into thin sections. However, if thin and thick sections are cast as part of the same casting, a phosphorus content above 0.30 per cent causes porosity in the thick sections and also variations in shrinkage rates, which are undesirable. Since higher-temperature melting rates will also increase fluidity, phosphorous contents of less than 0.20 per cent will produce better results.

Phosphorus also forms the eutectic steadite. *Steadite* (Fig. 6-4) has a composition of iron, 91.2 per cent; phosphorus, 6.9 per cent; and carbon, 1.9 per cent. This eutectic has the effect of lowering the freezing range of cast iron. It also lowers the quantity of pearlite formed and therefore decreases the strength of the alloy.

Manganese increases the fluidity, neutralizes the sulfur present, and decreases the number of inclusions formed in a casting. It is used in amounts of less than 2 per cent. It will also combine with carbon and form a hard phase, so that together with cementite it increases the hardness and thus increases the difficulty of machining the castings.

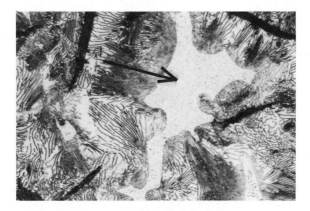

Figure 6-4

Other alloys that may be added to cast iron for varying effects are nickel, molybdenum, vanadium, chromium, copper, titanium, or aluminum.

Nickel, when dissolved in iron, promotes graphitization in cast iron, except that it does not decompose the cementite in pearlite. It does reduce the size of the graphite flakes and thereby creates a tougher, close-grained cast iron. This close-grained cast iron also has better machinability properties than does the cast iron with larger graphite flakes. The amounts added range from 0.10 to 2.5 per cent.

Molybdenum is added to pig iron in amounts of up to 1.25 per cent. It increases the resistance of the material to shock, and increases its hardness and strength. Unlike nickel, it has little effect upon graphitization or shrinkage. It also increases the effects favorable to the mechanical properties of castings that have thick and thin sections in the same casting.

Vanadium is a carbide former, thus opposing the formation of graphite. It is generally added in amounts of from 0.10 to 0.15 per cent.

Chromium is added to the melt in amounts of from 1.0 to 3.5 per cent. It opposes the formation of graphite, and as such it is a strong carbide former. The carbides deposit in the matrix, thus increasing the strength of the matrix. It is also added to cast iron to increase resistance to wear and heat. When used with copper and nickel, it increases the resistance of cast iron to corrosion.

Copper in amounts of up to 2 per cent in gray cast iron has the effects of slightly decreasing the combined carbon, preventing the formation of massive carbides, increasing the fluidity slightly, decreasing shrinkage slightly, and improving machinability. As indicated, when in the presence of nickel it adds corrosion resistance and heat resistance to the structure.

Titanium is a graphitizer that promotes graphitization and decreases the size of the graphite flakes. It also increases the wear resistance properties of cast iron.

Aluminum acts as a graphitizer. It is not used much at present as an alloy in cast iron.

6-4   PHASE CALCULATION FOR CAST IRON

The eutectic composition of the alloy E (Fig. 6-1) is 4.3 per cent carbon in combination with 95.7 per cent ferrite.

Example 1

Given a eutectic composition E (Fig. 6-1), calculate the various phases at (a) T, (b) T_e, (c) T_2, (d) T_E, and (e) T_3.

solution:

(a) At T the eutectic is all liquid.

(b) At T_e the transformation is eutectic. It transforms from a liquid directly to a solid ledeburite phase. The austenite and cementite are in the eutectic. The percentage of each is

$$\text{Eutectic austenite} = \frac{6.67-4.3}{6.67-2.0} \times 100 = 50.7\%$$

$$\text{Eutectic cementite} = \frac{4.3-2.0}{6.67-2.0} \times 100 = 49.3\%$$

(c) At T_2, because of the change of the solubility line $e\,'E$, there will be a change in the austenite composition, which will transform to the eutectoid at 1333°F. Thus at T_2 the *overall* austenite and cementite will be

$$\text{Austenite} = \frac{6.67-4.3}{6.67-1.2} \times 100 = 43.3\%$$

$$\text{Cementite} = \frac{4.3-1.2}{6.67-1.2} \times 100 = 56.7\%$$

(d) At T_E the austenite has a eutectoid composition and will transform into pearlite. The compositions of proeutectoid cementite and austenite are

$$\text{Eutectoid austenite} = \frac{6.67-4.3}{6.67-0.83} \times 100 = 40.6\%$$

$$\text{Proeutectoid cementite} = \frac{4.3-0.83}{6.67-0.83} \times 100 = 59.4\%$$

The austenite has a eutectoid composition, which is

$$\text{Eutectoid ferrite} = \frac{6.67-0.83}{6.67-0.025} \times 100 = 88\%$$

$$\text{Eutectoid cementite} = \frac{0.83-0.025}{6.67-0.025} \times 100 = 12\%$$

(e) Present at T_3 is ferrite and cementite. The percentage of each are

$$\text{Ferrite} = \frac{6.67-4.3}{6.67-0.01} \times 100 = 35.6\%$$

$$\text{Cementite} = \frac{4.3-0.01}{6.67-0.01} \times 100 = 64.4\%$$

Pearlite and proeutectoid cementite are also present.

$$\text{Pearlite} = \frac{6.67-4.3}{6.67-0.83} \times 100 = 40.6\%$$

$$\text{Proeutectoid cementite} = \frac{4.3-0.83}{6.67-0.83} \times 100 = 59.4\%$$

The composition of the pearlite is 88 per cent ferrite and 12 per cent cementite. The cementite present (64.4 per cent) is present in a free state and also combined with the ferrite (35.6 per cent) in the pearlite. The percentages in each are

$$
\begin{aligned}
\text{Eutectoid cementite} \quad &= 40.6 \times 12\% = 4.9\% \\
\text{Proeutectoid cementite} \quad &= \underline{59.4\%} \\
\text{Total} \quad & 64.3\% \; \textit{check} \\
\text{Eutectoid ferrite} \quad &= 40.6 \times 88\% = 35.7\% \; \textit{check}
\end{aligned}
$$

Example 2

Given the hypoeutectic cast iron that has a composition at C in Fig. 6-1, calculate the percentages of the phases at (a) T, (b) T_1, (c) T_e, (d) T_2, (e) T_E, and (f) T_3.

solution:

(a) At T the composition is entirely liquid and has 3.5 per cent carbon in solution with 96.5 per cent iron.

(b) At T_1 austenite has formed along the solubility line se. The percentages of each phase are

$$\text{Proeutectic austenite} = \frac{4.0-3.5}{4.0-1.6} \times 100 = 20.8\%$$

$$\text{Eutectic liquid} \quad = \frac{3.5-1.6}{4.0-1.6} \times 100 = 79.2\%$$

(c) At T_e, the eutectic temperature, the percentages of overall austenite and cementite are

$$\text{Austenite} \; = \frac{6.67-3.5}{6.67-2.0} \times 100 = 67.9\%$$

$$\text{Cementite} = \frac{3.5-2.0}{6.67-2.0} \times 100 = 32.1\%$$

The percentages of the eutectic–ledeburite, and austenite are

$$\text{Proeutectic austenite} = \frac{4.3-3.5}{4.3-2.0} \times 100 = 34.8\%$$

$$\text{Ledeburite} \quad = \frac{3.5-2.0}{4.3-2.0} \times 100 = 65.2\%$$

The percentages of austenite and cementite ledeburite are

$$\text{Eutectic austenite} \; = \frac{6.67-4.3}{6.67-2.0} \times 100 = 50.7\%$$

$$\text{Eutectic cementite} = \frac{4.3-2.0}{6.67-2.0} \times 100 = 49.3\%$$

(d) At T_2 the austenite is changing as a result of metastable cementite along solubility line $e'E$. The percentages of overall austenite and cementite are

$$\text{Austenite} = \frac{6.67-3.5}{6.67-1.2} \times 100 = 58\%$$

$$\text{Cementite} = \frac{3.5-1.2}{6.67-1.2} \times 100 = 42\%$$

The percentages of free austenite and ledeburite are

$$\text{Proeutectic austenite} = \frac{4.3-3.5}{4.3-1.2} \times 100 = 25.8\%$$

$$\text{Ledeburite} = \frac{3.5-1.2}{4.3-1.2} \times 100 = 74.2\%$$

The percentages of austenite and cementite in the ledeburite do not change. They are

$$\text{Austenite} = 50.7\%$$
$$\text{Cementite} = 49.3\%$$

(e) At T_E the austenite has a eutectoid composition that will transform into pearlite. The percentages of eutectoid austenite and cementite are

$$\text{Eutectoid austenite} = \frac{6.67-3.5}{6.67-0.83} \times 100 = 54.3\%$$

$$\text{Proeutectoid cementite} = \frac{3.5-0.83}{6.67-0.83} \times 100 = 45.7\%$$

The austenite has a eutectoid composition that does not change. It is

$$\text{Eutectoid ferrite} = 88\%$$
$$\text{Eutectoid cementite} = 12\%$$

(f) At T_3 the percentages of overall ferrite and cementite are

$$\text{Ferrite} = \frac{6.67-3.5}{6.67-0.01} \times 100 = 47.6\%$$

$$\text{Cementite} = \frac{3.5-0.01}{6.67-0.01} \times 100 = 52.4\%$$

The percentages of pearlite and cementite are

$$\text{Pearlite} = \frac{6.67-3.5}{6.67-0.83} \times 100 = 54.3\%$$

$$\text{Proeutectoid cementite} = \frac{3.5-0.83}{6.67-0.83} \times 100 = 45.7\%$$

The composition of the pearlite is 88 per cent ferrite and 12 per cent cementite. Thus

$$
\begin{array}{ll}
\text{Eutectoid cementite} & = 54.3 \times 12\% = 6.5\% \\
\text{Proeutectoid cementite} & = \underline{45.7\%} \\
& \text{Total} \quad 52.2\% \; check \\
\text{Eutectoid ferrite} & = 54.3 \times 88\% = 47.8\% \; check
\end{array}
$$

Example 3

Given the hypereutectic cast iron with the composition at D in Fig. 6-1, calculate (a) T, (b) T_1, (c) T_e, (d) T_2, (e) T_E, and (f) T_3.

solution:

(a)   At T the composition is entirely liquid and has a composition of 5.5 per cent carbon and 94.5 per cent iron.

(b)   At T_1 cementite forms along the solubility line pe in percentages of

$$
\text{Cementite} = \frac{5.5 - 4.5}{6.67 - 4.5} \times 100 = 46.1\%
$$

$$
\text{Liquid} \quad = \frac{6.67 - 5.5}{6.67 - 4.5} \times 100 = 53.9\%
$$

(c)   At T_e the overall percentages of austenite and cementite are

$$
\text{Eutectic austenite} \quad = \frac{6.67 - 5.5}{6.67 - 2.0} \times 100 = 25\%
$$

$$
\text{Proeutectic cementite} = \frac{5.5 - 2.0}{6.67 - 2.0} \times 100 = 75\%
$$

The percentages of the ledeburite and cementite are

$$
\text{Ledeburite} \quad\quad = \frac{6.67 - 5.5}{6.67 - 4.3} \times 100 = 49.4\%
$$

$$
\text{Proeutectic cementite} = \frac{5.5 - 4.3}{6.67 - 4.3} \times 100 = 50.6\%
$$

The percentages of austenite and cementite in ledeburite are

$$
\text{Eutectic austenite} \quad = \frac{6.67 - 4.3}{6.67 - 2.0} \times 100 = 50.7\%
$$

$$
\text{Eutectic cementite} = \frac{4.3 - 2.0}{6.67 - 2.0} \times 100 = 49.3\%
$$

(d)   At T_2 the overall austenite and cementite are

$$\text{Austenite } = \frac{6.67-5.5}{6.67-1.2} \times 100 = 21.4\%$$

$$\text{Cementite} = \frac{5.5-1.2}{6.67-1.2} \times 100 = 78.6\%$$

The percentages of cementite and ledeburite are

$$\text{Ledeburite} = \frac{6.67-5.5}{6.67-4.3} \times 100 = 49.4\%$$

$$\text{Proeutectoid cementite} = \frac{5.5-4.3}{6.67-4.3} \times 100 = 50.6\%$$

The percentages of austenite and cementite in the ledeburite do not change. They are

Austenite = 50.7%
Cementite = 49.3%

(e) At T_E the austenite is of the eutectoid composition and will transform into pearlite. The percentages of overall cementite and ferrite are

$$\text{Cementite} = \frac{5.5-0.025}{6.67-0.025} \times 100 = 82.4\%$$

$$\text{Ferrite } = \frac{6.67-5.5}{6.67-0.025} \times 100 = 17.6\%$$

The austenite has the eutectoid composition of

Eutectoid ferrite = 88%
Eutectoid cementite = 12%

(f) At T_3 the percentages of overall ferrite and cementite are

$$\text{Cementite} = \frac{5.5-0.01}{6.67-0.01} \times 100 = 82.4\%$$

$$\text{Ferrite } = \frac{6.67-5.5}{6.67-0.01} \times 100 = 17.6\%$$

The percentages of pearlite and cementite are

$$\text{Pearlite } = \frac{6.67-5.5}{6.67-0.83} \times 100 = 20\%$$

$$\text{Proeutectoid cementite} = \frac{5.5-0.83}{6.67-0.83} \times 100 = 80\%$$

The composition of the pearlite does not change; it is ferrite, 88 per cent, and cementite, 12 per cent. Thus

Eutectoid cementite = 20 × 12% = 2.4%
Proeutectoid cementite = 80.0%
Total 82.4% *check*
Eutectoid ferrite = 20 × 88% = 17.6% *check*

Problems

6-1. Explain the phase changes that take place when a 3.5 per cent carbon alloy cools from temperature T to T_3 in Fig. 6-1.

6-2. What effect does alloying have upon the physical properties of cast iron?

6-3. State the physical properties of (a) free cementite, (b) free ferrite, (c) pearlite, (d) graphite, (e) temper carbon, (f) ledeburite, and (g) steadite.

6-4. Name (a) two graphite formers and (b) two cementite stabilizers.

6-5. What is meant when Fe_3C is referred to as metastable?

6-6. Describe the formation of gray cast iron.

6-7. Describe the process of forming white cast iron. How does its structure differ from gray cast iron?

6-8. Describe the process known as "malleableizing," which produces malleable cast iron.

6-9. What is "temper" carbon?

6-10. Describe the processes used to produce (a) ferritic malleable cast iron and (b) pearlitic malleable cast iron.

6-11. Describe the heat-treating process used to produce nodular cast iron.

6-12. Describe the effects of varying amounts of silicon on the properties of cast iron.

6-13. What two adverse effects does sulfur have on cast iron?

6-14. Discuss the effects of phosphorus on fluidity, shrinkage, and porosity of iron when cast.

6-15. Discuss the effect of steadite on the physical properties of cast iron.

6-16. What effect does manganese have on cast iron?

6-17. Describe the effects of each of the following upon the formation of graphite, or carbide: nickel, chromium, copper, molybdenum, vanadium, titanium, and aluminum.

6-18. Given a hypoeutectic cast iron that has a 3.0 per cent carbon in combination with ferrite (Fig. 6-1), calculate the percentages of each phase at the following temperatures: (a) T, (b) T_1, (c) T_e, (d) T_2, (e) T_E, and (f) T_3.

6-19. Given the hypereutectic cast iron that has a 5.0 carbon content (Fig. 6-1), calculate the percentages of the phases at (a) T, (b) T_1, (c) T_e, (d) T_2, (e) T_E, and (f) T_3.

6-20. Given a eutectic composition E (Fig. 6-1), calculate the various phases at (a) T, (b) T_e, (c) T_2, (d) T_E, and (e) T_3.

7 | Classification of Steels

7-1 CLASSIFICATION OF CARBON STEELS

The Society of Automotive Engineers has established standards for specific analysis of steels. These SAE numbers are listed in Table 7-1. The series 10XX are the plain carbon steels, and the 11XX are those referred to as "free cutting." The latter have high sulfur content. The 12XX series has a high phosphorus as well as a high sulfur content.

In the 10XX series, the first digit indicates a plain carbon steel. The second digit indicates a modification in the alloys. Thus the second digit in the number 11XX denotes a change in the sulfur content. The last two digits indicate points of carbon. One point of carbon is defined as 0.01 per cent. Thus an SAE 1120 steel indicates a free-cutting steel with an average carbon content of 20 points, or 0.20 of 1 per cent carbon. The permissible carbon range in this SAE 1120 steel is 0.18 to 0.23 per cent carbon. In the 12XX series a popular basic open-hearth steel is one labeled 12L14. The L in this designation indicates about 0.25 per cent lead as an alloy.

The American Iron and Steel Institute (AISI) in cooperation with the Society of Automotive Engineers (SAE) revised the percentages of the alloys to be used in the making of steel, retained the numbering system, and added letter prefixes to indicate the method used in steelmaking. The letter prefixes are

A = alloy, basic open hearth
B = carbon, acid Bessemer
C = carbon, basic open hearth

153

D = carbon, acid open hearth

E = electric furnace

If the letter prefix is omitted, the steel is assumed to be open hearth. An AISI C1050 number indicates a plain carbon, basic open-hearth steel

Table 7-1   Classification of Steel

Classification	Number	Range of numbers
Carbon steel SAE–AISI	1XXX	
Plain carbon	10XX	1006–1095
Free machining (resulfurized)	11XX	1108–1151
Resulfurized, rephosphorized	12XX	1211–1214
Manganese (1.75% mm)	13XX	1320–1340
Nickel	2XXX	
3.5% Ni	23XX	2317–2345
5.0% Ni	25XX	2512–2517
Nickel–chromium	3XXX	
1.25% Ni, 0.65% Cr	31XX	3115–3150
1.75% Ni, 1.00% Cr	32XX	
3.50% Ni, 1.55% Cr	33XX	3310–3316
Corrosion-resist. stainless	30XX	
Austenitic stainless	303XX	(AISI 300 series)
Molybdenum	4XXX	
C–Mo (0.25% Mo)	40XX	4024–4068
Cr–Mo (Cr, 0.70%; Mo, 0.15%)	41XX	4130–4150
Ni–Cr–Mo (Ni, 1.8%; Cr, 0.65%)	43XX	4317–4340
Ni–Mo (1.75% Ni)	46XX	4608–4640
Ni–Cr (0.45%)–Mo (0.2%)	47XX	
Ni–Mo (3.5% Ni, 0.25% Mo)	48XX	4812–4820
Chromium	5XXX	
0.5% Cr	50XX	
1.0% Cr	51XX	5120–5152
1.5% Cr	52XXX	52095–52101
Corrosion-heat resistant	514XX	(AISI 400 series)
Chromium–vanadium	6XXX	
1% Cr, 0.12% V	61XX	6120–6152
Silicon–manganese		
0.85% Mn, 2% Si	92XX	9255–9262
Triple-alloy steels		
0.55% Ni, 0.50% Cr, 0.20% Mo	86XX	8615–8660
0.55% Ni, 0.50% Cr, 0.25% Mo	87XX	8720–8750
3.25% Ni, 1.20% Cr, 0.12% Mo	93XX	9310–9317
0.45% Ni, 0.40% Cr, 0.12% Mo	94XX	9437–9445
0.45% Ni, 0.15% Cr, 0.20% Mo	97XX	9747–9763
1.00% Ni, 0.80% Cr, 0.25% Mo	98XX	9840–9850
Low Alloy, High Tensile	950	
Leaded steel	XXLXX	
Boron (about 0.005% Mn)	XXBXX	

Boron is denoted by addition of B. Boron–vanadium is denoted by the addition of BV. Examples: 14BXX, 50BXX, 80BXX, 43BV14.
    TS denotes "tentative standard"; TS4150.

that has an average carbon content of 0.50 of 1 per cent (50 points of carbon).

Another prefix letter is the hardenability or H-value. Hardenability curves will be discussed in a subsequent section of this chapter. These numbers are written, followed by a capital H, such as 4340H.

The next series, SAE 13XX, constitutes a group of steels that contain approximately 1.5 to 2.0 per cent manganese. They are called the medium-manganese steels. A content of less than 1 per cent manganese is defined as "residual" and is considered normal when found in combination with other alloys in the plain carbon steels.

The addition of 1.5 to 2.0 per cent manganese increases the strength in the as-rolled state and increases the ductility in the heat-treated state if the carbon content is low. In some instances, some air hardening may take place upon cooling from elevated temperatures. Percentages of over approximately 0.8 per cent will affect hardenability, thus permitting the use of less severe quenching media. If the carbon content is high and the manganese ranges to about 8 per cent, the structure hardens very rapidly, because austenite will form at room temperature and the formation of carbides will result. If the manganese content is above 8 per cent, the steel develops high hardness when quenched. It can be machined only by grinding. It has high resistance to failure from fatigue with high plastic deformation properties. These steels also exhibit high resistance to abrasion.

The nickel steels, SAE 23XX and 25XX series, contain from 3 to 5 per cent nickel. It is also alloyed with chromium in the 3XXX series and with molybdenum in the 43XX, 46XX, 47XX, and 48XX series. This is one of the most important alloying elements. In percentages of approximately 5 per cent nickel with 1 per cent carbon, a martensitic* steel structure forms upon slow cooling. The addition of 5 per cent nickel to structural steels increases the tensile strength without reducing ductility.

The addition of from 8 to about 12 per cent nickel to steel has the effect of increasing the resistance of the steel to low-temperature impact.

When 12 to 15 per cent nickel is present with about 1 per cent carbon, an austenitic structure results. The effect of the latter is to reduce the internal stresses that result from severe quenching of carbon steel. The addition of the higher percentages of nickel also increases the hardness so that it requires grinding. When 8 to 12 per cent aluminum, 0 to 24 per cent cobalt, and 0 to 6 per cent copper is alloyed with percentages of 15 to 25 per cent nickel, the resulting materials are very hard and develop very high magnetic properties. They are called the *alnico* metals.

Steels containing 25 to 35 per cent nickel develop resistance to corrosion at elevated temperatures. Their use at elevated temperatures

*Martensite is the hardened structure that results when austenite is quenched.

(up to about 2000°F) is increased by the addition of chromium. Steels that have nickel in this range have a very low coefficient of thermal expansion. Because of this characteristic, these steels are used in the manufacture of measuring instruments, standards, etc. These metals are called *invar* metals.

Steels with nickel content of between 35 to 50 per cent have the same coefficients of thermal expansion as some glass materials and can, therefore, be banded together to prevent cracking of the glass as the temperature changes.

Higher nickel contents yield materials that may be used at low temperatures without destroying their mechanical properties. Contents of nickel in the 50 to 80 per cent range increase the permeability of iron from approximately 25 times for the former content to well over 150 times for the higher nickel content. These materials are called the *permalloy* materials.

A recently developed series of steels, not included in the AISI numbered listings because they are not machine steels, are those called *maraging steels*.* They are essentially carbon free and contain iron–nickel martensite. The essential difference between carbon steels and maraging steels is that the martensite that forms in carbon steels, when quenched, transforms to a softer more ductile material when tempered. The iron–nickel martensite starts out ductile and becomes hard and tough with aging. The advantages are evident. Machining or forming can take place in the ductile state, and hardening can be achieved by simple heat treatment, which consists of annealing to 1500°F for 1 hour, cooling in air, and reheating to 900°F for 3 hours.

Three series have been developed: 20 to 25 per cent nickel with 1.5 to 2.5 per cent aluminum and titanium was developed first, followed by 18 per cent nickel with cobalt, molybdenum, and titanium. The hardness results from the formation of the intermetallic compounds Ni_3Mo and Ni_3Ti. Another advantage over carbon steels is the very high resistance to stress corrosion, which is characteristic of this steel. The third class of maraging steels contains 12 per cent nickel alloyed with chromium and molybdenum, together with lesser amounts of aluminum and titanium. These steels develop notch strength higher than carbon steels.

The SAE 5XXX chromium series is generally alloyed with varying amounts of carbon to form carbides. Chromium is a ferrite strengthener in low-carbon steels. It increases the core toughness and the wear resistance of the case in carburized steel. The high carbon–high chromium (1.5 per cent Cr) steels have high wear resistance and hardness.

The SAE 3XXX, the nickel–chromium, series includes those steels

---

*The Making, Shaping and Treating of Steel, 8th ed. U.S. Steel, p. 1094.

in which chromium is alloyed with nickel to impart some of the physical characteristics of each element. Thus nickel–chromium steels are tough and ductile as a result of the nickel, and exhibit increased wear and hardenability qualities as well as corrosion resistance as a result of the chromium component. With high percentages of chromium, the various stainless steels are produced.

Contents of up to 3 per cent chromium have the effect of reducing crystal size in the structure. This percentage of chromium also has the effect of combining with both iron and carbon to form carbides. With percentages of up to 6 per cent chromium and 0.83 per cent carbon, a martensite structure is produced. Beyond 6 per cent and up to about 15 per cent chromium, the martensite is displaced by chromium carbide with a corresponding increase in hardness. Above 12 per cent chromium, all the intermetallic carbides are in the form of iron carbide and chromium carbide.

If a nickel–chromium (austenitic stainless) steel is heated to 1000°F, chromium carbides precipitate into the grain boundaries, which fosters grain boundary corrosion. These chromium carbides are not as corrosion resistant as the base austenitic structure. The procedure used to check the formation of chromium carbide is to use an element that has a greater affinity for the carbon than the chromium. Columbium or titanium, when added, combine with carbon, thus leaving the chromium free to prevent corrosion. Another method for stabilizing the structure is to heat the steel to about 2000°F for about $\frac{1}{2}$ hour; quenching follows. The heating causes the carbides to go back into solid solution. The quenching fixes the stabilized structure. This process is called *solution heat treatment.*

There are several special types of combinations of the steel alloys and chromium. One class of steels contains up to 2 per cent chromium and yields materials that have very high wear and impact resistance qualities. They produce steels used for ball bearings, tools, crushers, dies, crushing rollers, drawing steels, etc. Chromium contents of up to 30 per cent constitute the stainless steels discussed in Sec. 7-2.

The SAE 4XXX molybdenum series is generally used in combination with chromium, nickel, or both. Molybdenum is a very strong carbide former and has a pronounced effect upon hardenability and high-temperature hardness. As will be seen in Chapter 8, it has a very strong effect on hardenability by shifting the time–temperature transformation diagram to the right. This increases the quenching time and results in deep hardening of the material. The addition of about 1 per cent molybdenum to a low-carbon steel doubles its tensile strength. It also is used with tungsten to form complex carbides in tool steels. Since molybdenum is a decarburizer at high temperatures, it must be heat-treated in a controlled atmospheric furnace.

The 40XX series yields better mechanical properties than plain carbon steels. The 41XX series possesses good deep hardening qualities and is ductile. The 43XX and 47XX series add the property of deep hardenability to those properties inherent in the nickel–chromium 3XXX series. The 46XX and 48XX series add deep hardenability, toughness, and wear resistances to the properties inherent in a steel that contains nickel and chromium.

The SAE 6XXX chromium–vanadium series incorporates vanadium into the structures to enhance the properties of the other alloys added. It is a strong carbide former, decreases the tendency of grain growth, and improves fatigue strength. It also has a very high elastic limit and fatigue strength when heat-treated.

Silicon, in the SAE 92XX silicon–manganese series, is a deoxidizer, and only under certain conditions is it considered to be a carbide former. It also increases the elasticity of the material without loss of ductility, because it distributes itself throughout the ferrite. Steels with silicon contents of over 0.50 per cent are classed as silicon steels. In percentages of 1 to 2 per cent, silicon is used in a structural steel where high yield strength is needed. In percentages of 3 to 4 per cent it is used as an electrical steel because it reduces the solubility of carbon in ferrite, increases grain growth, reduces hysteresis losses, and increases the resistivity of iron. Care must be exercised when silicon in quantities greater than 4 per cent is used, because the shock resistance of the alloy declines rapidly.

The triple-alloy (low alloy chromium, nickel, and molybdenum) steels exhibit qualities that are inherent in each of the elements separately. They were developed during World War II, and retained and numbered AISI–SAE 86XX, 87XX, 92XX, 93XX, 94XX, 97XX, and 98XX.

The 950 series are high strength, low alloy steels developed to greatly increase the mechanical properties of steel. They may be welded without preheating and usually do not need to be stress relieved after welding. As a matter of fact, heat treating may change the mechanical properties of this type of steel. The high strength-to-weight ratio, their improved corrosion resistance, and their increased strength in general make them useful where weight must be conserved without affecting their strength. Typical applications are automotive parts, railroad cars, conveyors and shovels.

The boron (B) and boron–vanadium (BV) steels are used for heavy sections, or for materials that are to be subjected to particularly severe service. Boron was introduced as a substitute element for some of the more critical metals, such as manganese, molybdenum, and nickel. Small amounts of boron when coupled with the carbon content increase hardenability. The effect is very marked for the hypoeutectoid steels and practically nonexistent for eutectoid and hypereutectoid steels. Larger amounts (over 0.005 per cent B) cause both cold* and hot* shortness in materials. The B or BV addition to the AISI–SAE standard numbers indicates a modification of the physical properties because of the addition of boron.

7-2 STAINLESS STEEL

Resistance to corrosion in steels is primarily due to the amount of chromium present as an alloy with iron or with iron and nickel. Although other elements, such as aluminum and silicon, will increase the corrosion resistance of steel, the most effective element is chromium. The dividing line between chromium as an alloy in steel and chromium as a corrosion inhibitor is about 5 per cent. Steels with quantities of chromium over 5 per cent are called *stainless*.

The *iron–chromium* equilibrium diagram is shown in Fig. 7-1. It shows an austenite region to the extreme left. At about 12.5 per cent chromium the gamma (γ) phase disappears and is nonexistent above 12.5

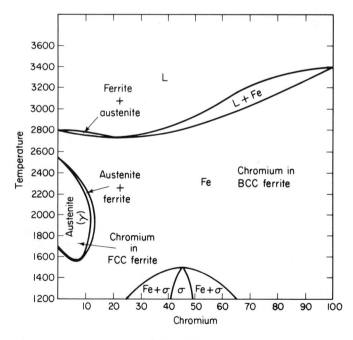

Figure 7-1

*Cold shortness: materials that cannot be worked at room temperatures. Hot shortness: materials that cannot be worked at elevated temperatures.

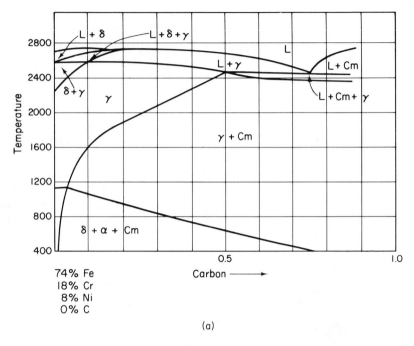

(a)

Figure 7-2

per cent chromium. In percentages greater than 12.5 per cent chromium, the solid solution is chromium dissolved in a matrix of body-centered, or alpha, iron. For high amounts of chromium in contact with iron and at temperatures of approximately 1000°F, a sigma (σ) phase may form. This phase forms as an intermetallic compound that is hard and brittle. The addition of carbon and nickel alters the diagram substantially.

This addition of alloys makes it necessary to consider the systems as ternary diagrams. Since a ternary constitution diagram is three dimensional, it may be more convenient to take a cross section through the ternary diagram in such a manner that two or more of the three constituents are held constant and one constituent is variable. The result of such a diagram is shown in Fig. 7-2(a).

Figure 7-2(b) shows a series of sections of the iron–chromium–nickel ternary constitution diagram for various percentages of chromium. The combinations are those of austenite (A) and ferrite (F). Note that, as the percentage of chromium increases, the gamma region gets smaller.

There are three general classifications of stainless steel: (1) martensitic, (2) ferritic, and (3) austenitic stainless steel.

Martensitic Stainless [Fig. 7-3(a)].   In this group, the 400 and 500 series, the chromium range is from about 11.5 to 18.0 per cent. The car-

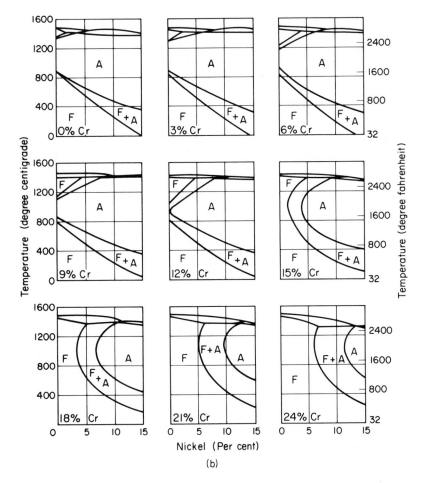

Figure 7-2 (continued)

bon content varies from 0.20 to 1.20 per cent. These steels, when quenched from the austenite, or austenite–cementite range, will form martensite. In some instances, the composition may contain nickel, molybdenum, or selenium. As indicated, these steels harden when heat-treated and exhibit very good corrosion resistance in the heat-treated state.

The martensitic series includes types 403, 410, 414, 416, 420, 431, 440, 501, and 502. They can be readily machined, are tough, and can be cold-worked. The 410, 416, and 420 are general-purpose martensitic steels. The 414, 420, and 440 series are used, among other items, in the manufacturing of cutlery, valves, bearings, surgical instruments, and springs. The 403 type is used in turbine blades, the 440 series when high

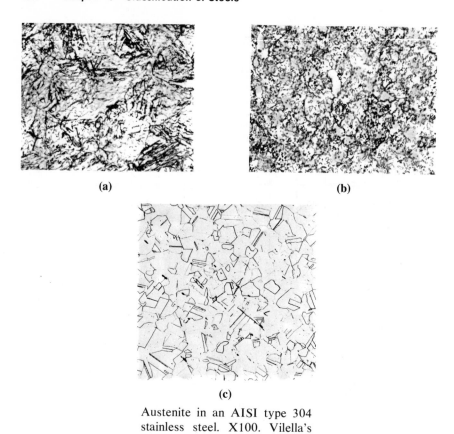

(a)   (b)

(c)

Austenite in an AISI type 304
stainless steel. X100. Vilella's
reagent.
Figure 7-3   Courtesy of U.S. Steel Corporation.

hardness is desired, and the 431 series when a material having high mechanical properties is desired. The 501 and 502 series both exhibit good strength characteristics at elevated temperatures.

Ferritic Stainless [Fig. 7-3(b)].   These are the 400 types, such as 405, 430, 442, 443, and 446, which will not harden when heated to elevated temperatures and quenched. The carbon content is generally below 0.20 per cent, except for 446, which is about 0.35 per cent C maximum. The chromium content ranges from 11.5 to approximately 28 per cent. As a result of the low carbon content, austenite will not form at elevated temperatures; therefore, ferrite, instead of martensite, forms upon quenching. They are ductile and may be hot- or cold-worked, but will become brittle at about 900°F.

The 430 and 443 series are easily worked and may be used in the manufacture of chemical equipment to be used at elevated temperatures. Where fabrication is a minimum, the 442 series may be used for high-temperature service. The 406 series has high electrical resistance but will air harden if care is not exercised during fabrication. The 405 series will not air harden when cooled, whereas the 446 series has very high corrosion resistance at elevated temperatures.

Austenitic Stainless [Fig. 7-3(c)]. These are the chrome–nickel 300 series and the chrome–nickel–manganese 200 series stainless steels. In these steels the nickel stabilizes the austenite that forms at elevated temperatures so that it is retained when the material cools to room temperature. Both series are not hardenable by heat-treating and are nonmagnetic. They may develop hardness by cold-working.

The most widely used stainless steel is the 18–8 chromium–nickel alloys in combination with varying amounts of carbon. Figure 7–2(a) is a constitution diagram of an 18–8 per cent stainless alloyed with 74 per cent ferrite and up to 1 per cent carbon. Normally, the constituents will be carbon, chromium, nickel, ferrite, and a carbide. If the carbon content is low, carbide will not form when the material is heated into a region in which the austenite is one of the constituents. In general, the austenitic stainless steels have better corrosion resistance than ferritic or martensitic stainless steels. If there is a carbide phase, it will transform upon heating. If heating is prolonged, carbides will precipitate into the grain boundaries and displace the chromium. Once displaced, there will be a lack of chromium at the grain boundaries, and corrosion will start there.

In general, the 300 series steels have a carbon content of from 0.08 to 0.25 per cent, a chromium content of from 16 to 26 per cent, and a nickel content of from 6 to 20 per cent. The majority of these steels have a maximum of about 2 per cent manganese as a fourth alloy and 1 per cent silicon for heat resistance. The series 301 and 302 are general-purpose austenitic stainless steels used for household utensils, structural trim, and general-purpose stainless trim for decorative purposes. Series 303 is a free-machining stainless. Series 305 is a low-work-hardening stainless. Series 304, 309, and 310 have high resistance to pitting during welding and high stability during welding. Series 308, 309, 316, and 317 have superior resistance to chemical corrosion and high-temperature corrosion. The 321 and 347 series are stabilized 18–8 stainless steels.

The 200 series steels have had some of the nickel of the 300 series replaced by manganese. Thus type 201, which has a composition of 17 per cent Cr, 4.5 per cent Ni, and 6.5 per cent Mn, and type 202, which has a composition of 18 per cent Cr, 5 per cent Ni, and 8 per cent Mn, are substitutes for 303, which has a nominal composition of 18 per cent Cr, 9 per cent Ni, and 2 per cent Mn.

Problems

7-1. List the significance of each of the digits in the following SAE numbers: (a) 1330, (b) 2350, (c) 2550, (d) 4170, (e) 4340.

7-2. Write the SAE number for the steels having the following compositions:
  (a) 4XXX series (Ni, Cr, Mo): 0.41 per cent C, 0.80 per cent Mn, 1.8 per cent Ni, 0.80 per cent Cr, 0.33 per cent Mo.
  (b) 5XXX series (1 per cent Cr): 0.51 per cent C, 0.80 per cent Mn, 0.20 per cent Ni, 0.80 per cent Cr.
  (c) 8XXX series (Ni–Cr–Mo): 0.50 per cent C, 1 per cent Mn, 0.50 per cent Cr, 0.20 per cent Mo, 0.60 per cent Ni.
  (d) 3XXX series (Ni–Cr): 0.20 per cent C, 0.70 per cent Mn, 1.25 per cent Ni, 0.65 per cent Cr.
  (e) 4XXX series (Cr–Mo): 0.37 per cent C, 0.80 per cent Mn, 0.90 per cent Cr, 0.20 per cent Mo.

7-3. (a) What is the significance of the prefix letters used with the SAE steel numbers? (b) List them and indicate what each letter represents. (c) What is assumed if the letter prefix is omitted?

7-4. What is the significance of the suffix letter H when used with an SAE number?

7-5. List the predominant alloy or alloys in the following SAE numbers: (a) 13XX, (b) 25XX, (c) 33XX, (d) 41XX, (e) 61XX.

7-6. (a) What are the effects of adding 1.5 to 2.0 per cent manganese to a plain carbon steel?

7-7. What effect does the addition of more than 8 per cent manganese have on plain carbon steel?

7-8. What effect does the addition of more than 8 per cent manganese have on the physical properties of plain carbon steel?

7-9. List the seven SAE number series that have nickel as one of the predominant alloys.

7-10. What effect does the addition of 5 per cent nickel steel have when added to a 1 per cent carbon steel?

7-11. What effect does the addition of 10 per cent nickel have on steel?

7-12. What is the effect on the physical properties of steel of alloying 15 per cent nickel with 1 per cent carbon?

7-13. List the component alloy percentages used in the manufacture of the alnico metals. What is the effect of nickel in this case?

7-14. What percentage of nickel is used in the manufacture of invar metal? What is the effect of nickel in this case?

7-15. When the nickel content is in the 35 to 50 per cent range, it may be bonded with glass. Why?

7-16. What are the permalloy metals?

7-17. (a) Describe the major characteristic of the maraging steels. (b) Describe the effect of heat-treating on carbon martensite and iron–nickel martensite in carbon-free steel.

7-18. Describe the three classes of maraging steel.

7-19. What is the effect of chromium when alloyed with plain carbon steels to form the SAE 5XXX series?

7-20. What is the effect of alloying chromium with nickel when forming the SAE 3XXX series?

7-21. Molybdenum is alloyed with chromium, nickel, or both to form the SAE 4XXX series. Discuss the effects of the various combinations of these elements.

7-22. What is the effect of adding vanadium to chromium when forming the SAE 6XXX series?

7-23. Discuss the effect of silicon in the SAE 92XX steels.

7-24. (a) What are the "triple-alloy steels"? (b) What determines the physical properties of these steels?

7-25. What effect does the addition of boron have on the SAE steels?

7-26. What percentage of chromium in steel is needed to classify the chromium as "a corrosion inhibitor" and the steel as "stainless"?

7-27. What is the maximum percentage of chromium that will yield austenite (γ) when iron and chromium are alloyed?

7-28. How does the sigma (σ) phase (Fig. 7-1) differ from the gamma (γ) phase and from the alpha (α) phase?

7-29. List (a) the three classifications of stainless steels discussed in this chapter and (b) the numbers assigned to each class.

7-30. Discuss the three types of stainless steels in light of the chromium–carbon content.

7-31. Discuss the three types of stainless steels listed in this chapter in terms of their physical properties.

7-32. What is the effect of prolonged heating of stainless steel upon grain boundary precipitation when the carbon content is low?

8 | Heat-Treating of Steel

There are several methods used to change or modify the properties of metals and their alloys. One method is to add alloys; another is to cold-work the metal. The method to be treated now deals with the heating and cooling of metals to modify or change their physical properties. The procedure of heating and cooling is called *heat-treating*. The processes that will be discussed are annealing, stress relieving, normalizing, hardening, austempering, tempering, spheroidizing, and case hardening.

In general, very slow cooling from elevated temperatures will follow the iron–iron carbide equilibrium diagram. Any rate of cooling interpreted as nonequilibrium may be analyzed by using the time–temperature transformation (TTT) curves to be studied in this chapter. It should be remembered that the iron–iron carbide equilibrium diagram (Fig. 8-1) is studied because of its comparative simplicity. The addition of any alloys could alter the diagram drastically.

The temperatures at which phase changes occur for the various percentages of carbon are labeled $A_1, A_3, A_{1.3}$, and A_{cm}. These are ideal temperatures and are approached only as the heating, or the cooling, time is increased. The slower the heating or cooling, the more nearly do the phase transformations approach these lines.

It was shown that a *eutectoid* steel (0.83 per cent C), when *heated* to the A_1, A_3, A_{cm} junction at 1333°F, will transform to austenite according to the relationship

$$Fe + Cm \rightleftharpoons austenite$$

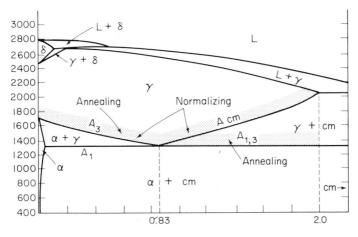

Figure 8-1

If this material is heated to a temperature *below* 1333°F, no phase changes occur. Since no phase changes have taken place on heating, none will take place on cooling. If austenite is heated *above* 1333°F and into the austenite (γ) region, no further phase changes will occur. Since it takes time for the change to take place, equilibrium heating is important.

If the material is hypoeutectoid, that is, if the carbon content is below 0.83 per cent, the pearlite phase will transform to austenite just above A_1. The proeutectoid ferrite will transform gradually to austenite as the temperature approaches the A_3 line. Above the A_3 line the entire structure will be austenite.

If a material that has a carbon content above 0.83 per cent and below 2.00 per cent (hypereutectoid) is heated to a temperature just above $A_{1,3}$, the pearlite will transform to austenite. As the temperature is raised, the proeutectoid cementite will dissolve into the austenite phase so that above A_{cm} the entire structural phase is austenite.

8-2 TIME–TEMPERATURE–TRANSFORMATION CURVES

As just discussed, when austenite transforms under equilibrium conditions upon very slow cooling to below the A_1 or $A_{1,3}$ temperatures, pearlite forms. As the cooling rate increases, the pearlite transformation temperature gets lower. The microstructure of the material is drastically altered as the cooling rate increases.

By heating and cooling a series of small samples, the history of the austenite transformation may be recorded, start to finish. Several idealized *time–temperature–transformation* (TTT) curves are shown in Fig.

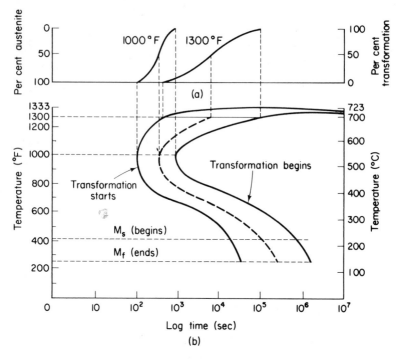

Figure 8-2

8-2(a). The abscissa is a record of time. The ordinate is a record of the percentage of transformation of austenite at a *particular temperature.*

In Fig. 8-2(b) the ordinate is now plotted as the temperature, and the start–finish of the austenite transformation become points on the curves as shown. An entire family of transformation curves [Fig. 8-2(a)] for a pearlite steel will yield the *isothermal transformation diagram,* also called TTT curves, shown in Fig. 8-2(c). These result from constant temperature curves.

Curves developed as a result of continuous cooling look somewhat like the TTT curves with a shift in some of the regions. The idealized differences are shown in Fig. 8-3. The dashed line is plotted as a result of continuous cooling. The curve has moved down and to the right, thus increasing the time required to transform the austenite. Figure 8-6 shows a *continuous-cooling transformation diagram* for SAE 4340 steel.

Cooling rates in the order of *increasing severity* are achieved by quenching from elevated temperatures as follows: furnace cooling, air cooling, oil quenching, liquid salts, water quenching, and brine. If these cooling curves are superimposed on the TTT diagram, the end-product structure and the time required to complete the transformation may be

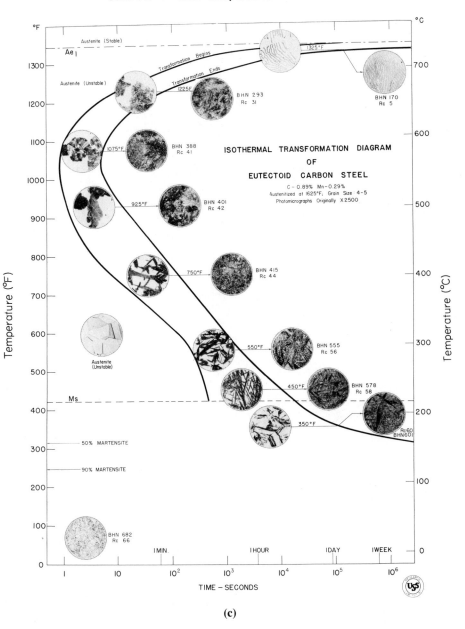

(c)

Figure 8-2 (continued)

found. Thus, in Fig. 8-4, if perpendiculars are dropped to the abscissa, the time to complete the transformation can be determined.

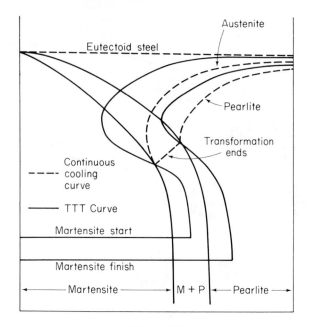

Figure 8-3

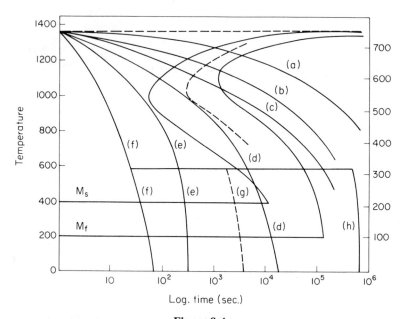

Figure 8-4

Example 1

In Fig. 8-4 name the room-temperature structure and calculate the time needed to complete the transformations for the curves *a* through *h*.

solution:

(a) Coarse pearlite
Finish $= 5 \times 10^3$
Start $= 1 \times 10^3$
Time $= 4 \times 10^3$ s

(b) Medium pearlite
Finish $= 2.0 \times 10^3$
Start $= 0.4 \times 10^3$
Time $= 1.6 \times 10^3$ s

(c) Fine pearlite
Finish $= 1.0 \times 10^3$
Start $= 0.2 \times 10^3$
Time $= 0.8 \times 10^3$ s

(d) 50 per cent pearlite–50 per cent martensite
Finish $= 500$
Start $= \underline{\ \ 90\ \ }$
Time $= 410$ s

(e) 100 per cent martensite
(f) 100 per cent martensite
(g) 100 per cent martensite (martempering)
(h) 100 per cent bainite (austempering)
Finish $= 6.0 \times 10^4$
Start $= 0.3 \times 10^4$
Time $= 5.7 \times 10^4$ s

At 600°F, curve *d* (Fig. 8-4), the structure is 50 per cent pearlite and 50 per cent austenite. Upon crossing the martensite M_s line, the austenite starts transforming to martensite. Once the cooling curve has crossed the M_f line, all the austenite has transformed to martensite.

Curves *e* and *f* (Fig. 8-4) will both transform to martensite at room temperature. They both miss the nose of the TTT curves.

Assume that curve *f* (Fig. 8-4) is quenched to 600°F above the M_s line and held to the point where curve *g* crosses both the M_s and the M_f curves. The result of quenching into a bath at an elevated temperature and holding the material at that temperature until it equalizes throughout the structure has many advantages. Two are that a more uniform structure and less distortion may result. This process is called *martempering*.

If cooling curve *f* is held longer and cooled to room temperature according to curve *h* (Fig. 8-4), *bainite* is formed. Since the transformation takes place at elevated temperatures (400 to 700°F), very little distortion or cracking takes place. Low-temperature bainite resembles martensite. It is called *acicular bainite*. High-temperature (700°F) bainite resembles pearlite. It is called *feathery bainite*. The heat-treating process is called *austempering*.

8-3 ANNEALING AND NORMALIZING

Full annealing is accomplished by heating a hypoeutectoid steel to a temperature above the A_3 line. In practice, the steel is heated to about 100°F above the A_3 line (Fig. 8-1). It is then cooled very slowly to room temperature. The formation of austenite destroys all structures that have existed before heating. As indicated, slow cooling yields the original phases of ferrite and pearlite. This is shown in Fig. 8-4, curve *a*. Machinability, ductility, and grain refinement are improved considerably.

Hypereutectoid steels consist of pearlite and cementite. The cementite forms a brittle network around the pearlite. Heating to just above the $A_{1,3}$ line and slow cooling will anneal these types of materials (Fig. 8-1).

Isothermal annealing (Fig. 8-4, curves *b* and *c*) refers to a faster cooling rate from the austenite range and yields a coarser pearlite than that achieved through full annealing. The material is not as soft, but neither is the required cooling time too long. Coarse pearlite may be achieved by cooling through the "nose" of the TTT curve.

Where coarse pearlite is too hard and the carbon content is high, a long-time *subcritical* annealing process is used to spheroidize the structure, or a combination of heating above and below the lower critical temperature is used. This process, called *spheroidize annealing,* is used for hypereutectoid steels.

Stress-relief annealing is a subcritical annealing process carried out by heating the material to a temperature below the lower critical temperature, followed by slow cooling. It is used to remove residual stresses in materials that result from cold-working or machining.

Process annealing is very similar to stress relieving in that the material is heated to below the lower critical temperature, allowing recrystallization to take place. This restores the ductility and softness of the material.

Normalizing is the process of heating material for the purpose of refining grain structure and reducing the size of carbides formed in as-rolled materials, in casting, or in forgings. The process involves heating the material well above the A_3 or A_{cm} critical temperature (100 to 150°F) and cooling in air (Fig. 8-1). Normalizing serves the purposes of stress relieving, strength and ductility improvement, and grain refinement in the low-carbon steels. In hypereutectoid steels, the A_{cm} temperature must be exceeded, followed by slow cooling in order to destroy the carbide networks that encase the pearlite in the structure. Rolled shapes, forgings, and casting that are normalized generally require tempering to relieve any stresses remaining in the structure.

8-4 HARDENING OF STEEL

Assume that a eutectoid steel is heated into the austenite range. The lattice structure of austenite is face-centered cubic (FCC). As such, it is able to hold carbon atoms that have diffused throughout the lattice. Under slow or moderately slow cooling, the atoms move to form body-centered cubic (BCC) lattices, and the carbon diffuses *out* of the austenite structure. The transformation to BCC takes place as a result of nucleation and growth. However, time is essential to allow this mechanism to take place.

Since time is all-important in the transformation from austenite to ferrite, a further increase in cooling rate does not permit this nucleation and growth to take place. It does not allow enough time for the carbon to diffuse out of solution. If the carbon is trapped in solution, the lattice structure cannot change to BCC.

The trapped carbon causes a shift in atoms to form a body-centered tetragonal structure. The shift of atoms and the trapped carbon create a stressed lattice structure. This stressed structure is responsible for the high hardness of the material. This structure is called *martensite*. The mechanism is the hardening process.

The formation of martensite is independent of the cooling rate and is a function of temperature reduction. The cooling rate at which martensite forms, and at which none of the transformations to softer phases take place, is called the *critical cooling rate*. Once the martensite starts to form, it continues to form as a function of reduction in temperature. It starts (M_s) to form slowly, increases as it proceeds, and decreases again at the finish (M_f). The graph [Fig. 8-5(a)] shows this process. The process starts at a given temperature for a given alloy and finishes at a fixed temperature. The starting and finishing temperatures are not affected by increasing the cooling rate. Figure 8-5(b) shows the starting and finishing temperatures of the martensite transformation as a function of the carbon content.

The hardness of the quenched material is a function of the carbon

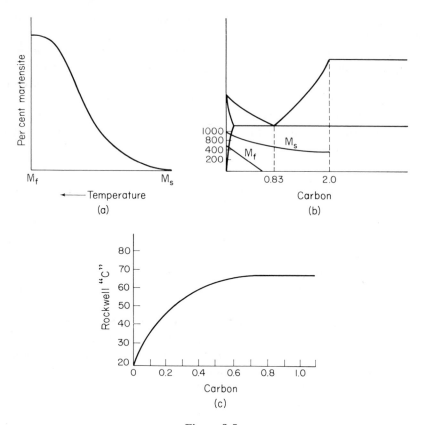

Figure 8-5

content. This is shown in the plot of carbon versus Rockwell hardness in Fig. 8-5(c). Appreciable hardness can be achieved when the carbon content is 0.30 per cent or *more*.

Figure 8-6(a) shows a continuous cooling transformation curve for the alloys carbon, manganese, nickel, chrome, and molybdenum, and the superimposition of several cooling curves on the diagram.

These curves show the various cooling rates (loss in temperature per hour) from a very slow quench of 40°F/h loss to 5400°F/h loss. Thus any cooling rate that decreases less than 40°F/h will result in a ferrite–pearlite structure as shown to the right of Fig. 8-6(a). The area enclosed by the 40 and the 150°F/h cooling curves will result in a room-temperature multiple structure of ferrite, pearlite, bainite, and martensite. In other words, from the top down, the area between the 40 and the 150°F/h cooling curves first crosses the *austenite-to-ferrite* transformation region (shaded). Some ferrite is formed. The next transformation region crossed by the cooling

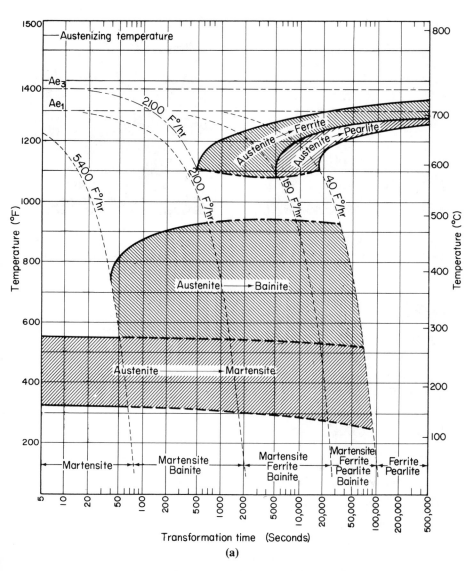

Figure 8-6 Courtesy of U.S. Steel Corporation.

curves is the shaded *austenite-to-pearlite* region. This accounts for the formation of pearlite. Bainite is formed from some of the austenite as the cooling curves cross the large *austenite-to-bainite* region. Finally, the remaining austenite transforms to martensite as the cooling curves cross the *austenite-to-martensite* region. The room-temperature structure of this material is ferrite–pearlite–bainite–martensite. The student should trace

(b)
Pearlite in an 0.8 percent carbon steel. Picral etch.

(c)
Bainite in an 0.8 percent carbon steel. White areas are unetched martensite. X1000. Picral etch.

(d)
Martensite in an 0.8 percent carbon steel. X1000. Picral etch.

Figure 8-6 (continued)

the martensite–ferrite–bainite, martensite–bainite, and the martensite cooling transformations in Fig. 8-6(a).

Figure 8-6(b) shows the photomicrograph of pearlite; Fig. 8-6(c), bainite; and Fig. 8-6(d); martensite.

It was shown in Fig. 8-4 that a cooling curve which missed the nose of the TTT curve and was quenched to room temperature resulted in a martensic structure. Therefore, the cooling rate and consequently the cooling medium could be selected on the basis of position of the nose with reference to the ordinate axis.

The effect of an *increase* in carbon content, the addition of alloys, or of an increase in the austenitic grain size will shift the nose of the curve to the right. This tends to increase the time needed for quenching, the hardness, and the depth of hardening.

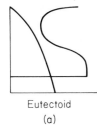
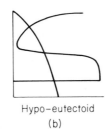
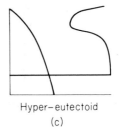

| Eutectoid | Hypo-eutectoid | Hyper-eutectoid |
| (a) | (b) | (c) |

Figure 8-7

As the carbon content decreases, the nose of the curve moves toward the ordinate axis and the time available for quenching decreases. The curves and carbon content in Fig. 8-7(a)–(c) are hypothetical.

The addition of alloys (except cobalt) will shift the nose of the curve to the right. Some alloys are more effective than others in shifting the curve. Thus vanadium and tungsten will shift the nose of the curve to the right if added to the structure and if they are present in the austenite phase when the material is quenched. The addition of small quantities of several alloys is more effective than the addition of larger amounts of one alloy.

Increase in grain size shifts the nose of the TTT curve to the right and increases the hardenability. However, large grain size is achieved by heating to high temperatures.

The relative grain sizes of two pieces of the same composition steel are shown in Fig. 8-8. The fundamental difference is the room temperature grain size. In Fig. 8-8(a), the grain is inherently *coarse*. It is refined as the temperature increases from T_1 to T_2 and then proceeds to grow uniformly to a size s at T_4. In Fig. 8-8(b), the grain is inherently *fine*. The refinement of the grain takes place in the same manner as before up to T_2. However, in the fine grain size, the growth is very slow up to T_3. Above T_3 it becomes

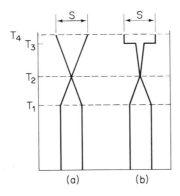

Figure 8-8

coarse rapidly up to T_4. Although an increase in grain size yields better hardness, it also produces a loss of toughness in the hardened steel. Martensite formed from fine austenite grains is very tough and more than offsets the slight loss of hardness.

8-5 HARDENABILITY

Hardenability is the capacity of a material to harden. It is a measure of the *depth of hardening* and not the maximum hardness obtained. It has been shown that a shift in the nose of the TTT curve to the right improves the steel's hardenability. Thus the addition of alloys or the coarsening of the austenitic grain structure will increase the hardenability of steel. Any steel with a low critical cooling rate will harden deeper than one that has a high cooling rate of quenching. The size of a piece of work has a direct effect upon the hardenability of the material.

There are several measures of hardenability. In one instance a bar is chosen that is approximately five times longer than the diameter. It is heated into the austenite range and quenched in a coolant such as water or oil. The process is repeated for test bars of many diameters. The bar is then broken and a hardness history is developed along the diameter of the bar. This may be done with a hardness tester or by etching to determine the depth of hardness. The depth of hardness determination is done by polishing the end of the piece and etching with 50 per cent aqueous hydrochloric acid at 180°F. A hardenability band (50 per cent martensite–50 per cent pearlite) is etched. The depth to which 50 per cent martensite is present is said to determine hardness. The core is taken to contain less than 50 per cent martensite. If the *center* of a bar contains 50 per cent martensite, the diameter of the bar is said to be the *critical diameter D*.

Figure 8-9(a) is a plot of a hardness band that produces 99.9 per cent martensite, and which shows the increased hardness when compared to the carbon content. Figure 8-9(b) is the same plot for a sample that produces 50 per cent martensite.

The quenching medium is then eliminated from consideration by standardizing the *effect* of the quenching medium on a steel sample regardless of the medium. This is designated by the ratio of the heat transfer factor F to the thermal conductivity K of the material. This ratio is called the H-factor.

$$H^* = \frac{F}{K}$$

F = heat transfer factor: Btu/in.^{2} s°F; cal/cm^2 s°C
K = thermal conductivity: Btu/in. s°F; cal/cm^2 s°C

*H-values are pure numbers and may be used in the metric system.

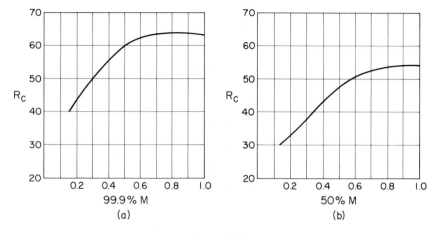

Figure 8-9

The *least* severe cooling rate H would be a decimal quantity like 0.2. The maximum severity would be represented by an H-value equal to infinity. The diameter of a test bar that has developed 50 per cent martensite at its center when quenched in an ideal coolant, $H = \infty$, is the *ideal critical diameter* and its designation is D_I. Some H-values are shown in Table 8-1.

Table 8-1

H-value	Quench conditions	Agitation
0.20	Poor oil quench	None
0.35	Good oil quench	Moderate
0.50	Very good oil quench	Good
0.70	Strong oil quench	Violent
1.00	Poor water quench	None
1.50	Very good water quench	Strong
2.00	Brine quench	None
5.00	Brine quench	Violent
∞	Ideal quench	

Source: U.S. Steel Carilloy Steel, p. 42. Carnegie-Illinois Steel Corp., Pittsburgh, Pa. (1948).

Table 8-2 shows the various cooling rates from the quenched end of a standard 1-in. hardenability test bar.

The plot in Fig. 8-10 is a graph for bars with a case-to-core ratio of 50 per cent. The bar diameter is shown at the left, the H-values at the right, and the distance from the water-cooled end (inches) as the abscissa. Additional curves are shown in Appendix 8-1.

One of the most useful tests to determine ideal critical diameters as

Table 8-2

Distance from quenched end		Cooling rate	
in.	mm	°F/s at 1300°F	°C/s at 705°C
$\frac{1}{16}$	1.6	489.0	272.0
$\frac{1}{8}$	3.2	307.0	171.0
$\frac{3}{16}$	4.8	195.0	188.0
$\frac{1}{4}$	6.4	124.0	69.0
$\frac{5}{16}$	7.9	77.2	42.9
$\frac{3}{8}$	9.5	56.3	31.3
$\frac{7}{16}$	11.1	41.9	23.3
$\frac{1}{2}$	12.7	32.3	17.9
$\frac{9}{16}$	14.3	25.0	13.9
$\frac{5}{8}$	15.9	21.4	11.9
$\frac{11}{16}$	17.5	19.5	10.8
$\frac{3}{4}$	19.1	16.3	9.1
$\frac{7}{8}$	22.2	12.4	6.9
$\frac{15}{16}$	23.8	11.5	6.4
1	25.4	10.0	5.6
$1\frac{1}{4}$	31.7	7.0	3.9
$1\frac{1}{2}$	38.1	5.1	2.8
$1\frac{3}{4}$	44.5	4.0	2.2
2	50.8	3.5	1.9

Source: U.S. Steel Carilloy Steel, p. 41. Carnegie-Illinois Steel
Corp., Pittsburgh, Pa. (1948).

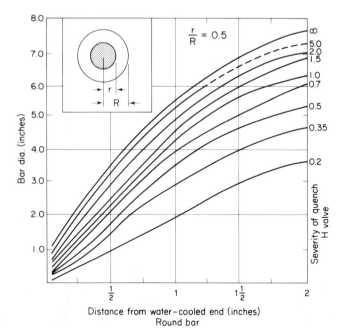

Distance from water-cooled end (inches)
Round bar

Figure 8-10

APPENDIX 8-1

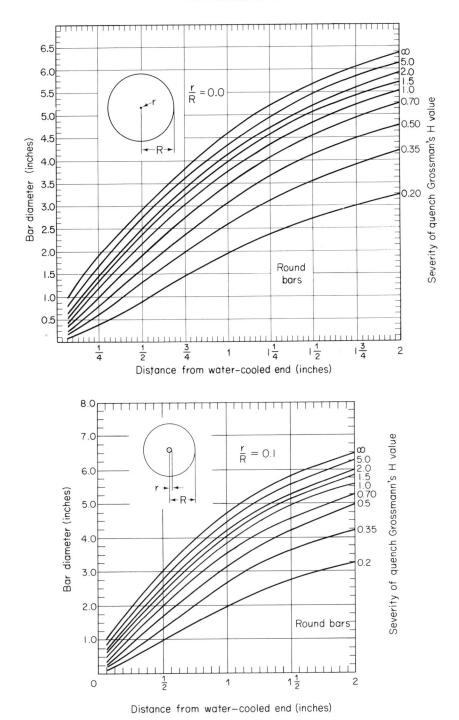

APPENDIX 8-1   *(continued)*

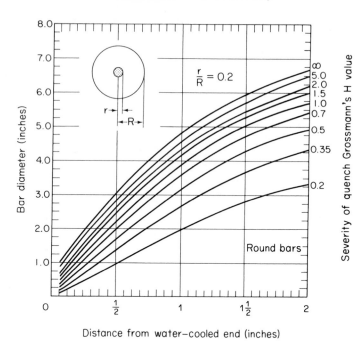

Distance from water-cooled end (inches)

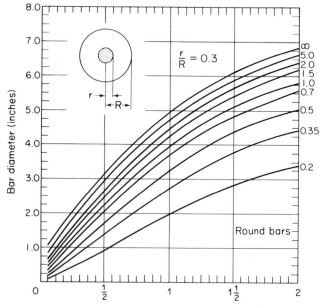

Distance from water-cooled end (inches)

APPENDIX 8-1 *(continued)*

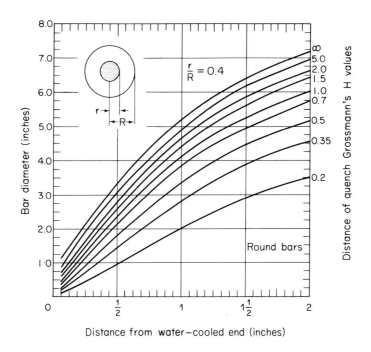

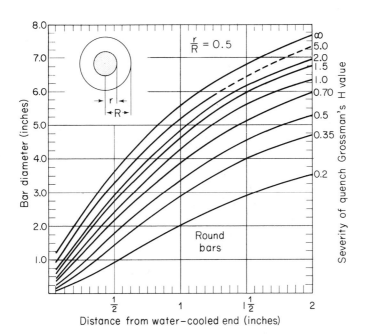

APPENDIX 8-1 *(continued)*

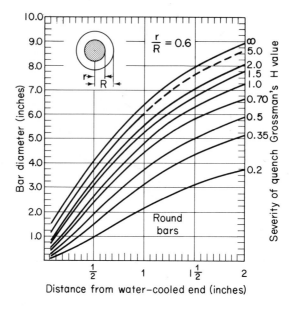

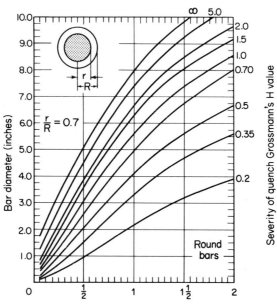

APPENDIX 8-1 (continued)

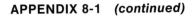

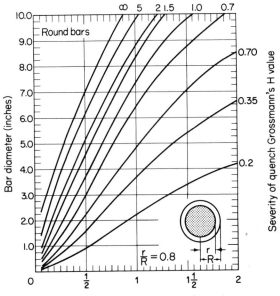

Distance from water-cooled end (inches)

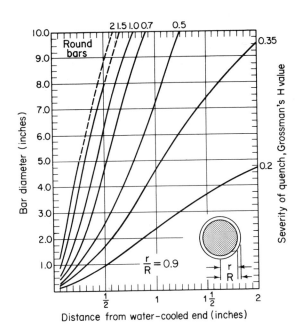

Distance from water-cooled end (inches)

APPENDIX 8-1 *(continued)*

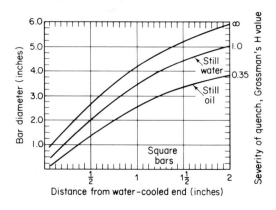

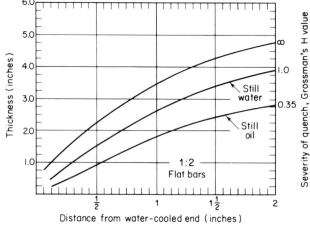

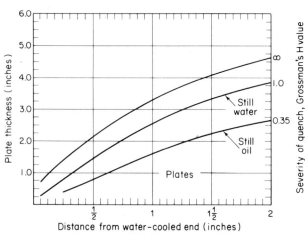

related to the quenched end is through the use of the plotted curves such as those in Fig. 8-11(c). Figure 8-11(a) shows the test specimen in the testing fixture. A valve is opened so that a free column of water is maintained to a height of 2 in. below the top of the fixture. The swivel baffle plate is pivoted to impede the water column. The specimen is heated to the desired temperature for the desired soaking period. It is removed from the furnace and quickly placed into the fixture as shown. The baffle is swiveled out of position so that the water impinges on the bottom of the specimen. The most severe quenching conditions take place at the end of the specimen near the water. Cooling rates become less severe along the specimen as the distance increases from the quenched end.

After a suitable time the specimen is removed and 0.015-in. flat surfaces ground as shown in Fig. 8-11(b). The specimen is placed in a fixture that is capable of moving in increments of $\frac{1}{16}$ in. Rockwell C readings are taken at $\frac{1}{16}$-in. increments along the length of the specimen. The readings are plotted on a graph such as those shown in Fig. 8-11. Additional end quench curves are shown in Appendix 8-2. Also shown are the respective cooling rates.

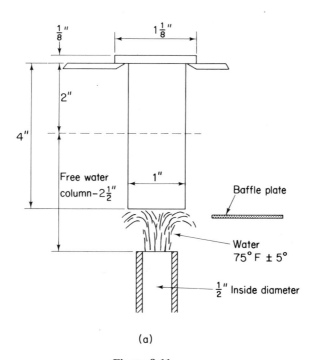

(a)

Figure 8-11

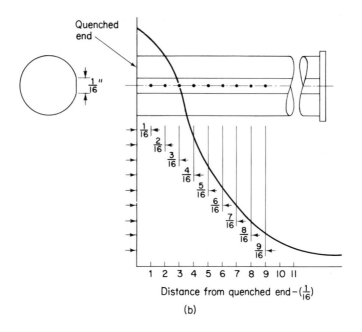

Distance from quenched end $-\left(\frac{1}{16}\right)$

(b)

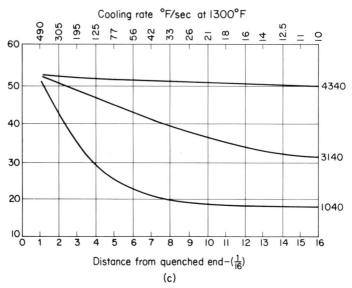

Distance from quenched end $-\left(\frac{1}{16}\right)$

(c)

Figure 8-11 (continued)

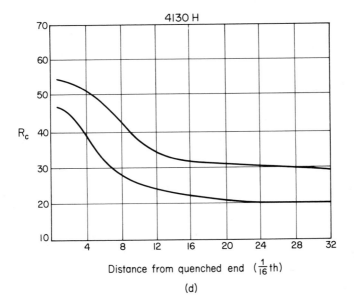

(d)

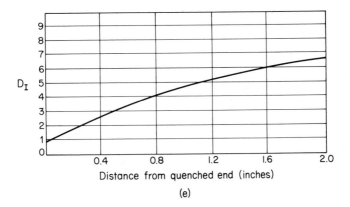

(e)

Figure 8-11 (continued)

APPENDIX 8-2

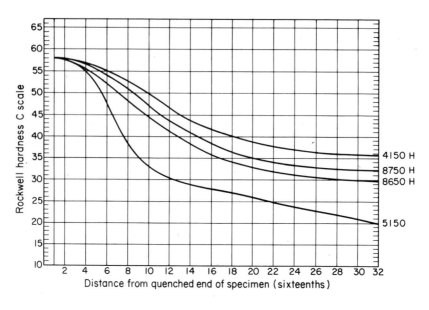

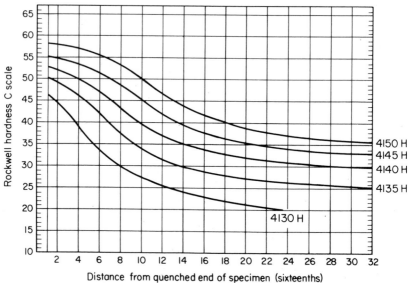

APPENDIX 8-2 *(continued)*

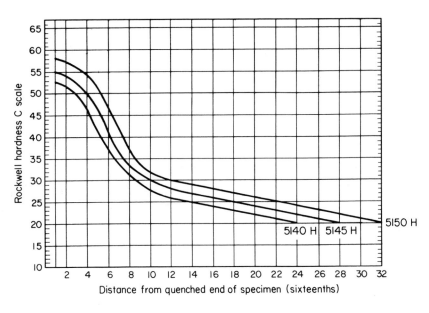

Distance from quenched end of specimen (sixteenths)

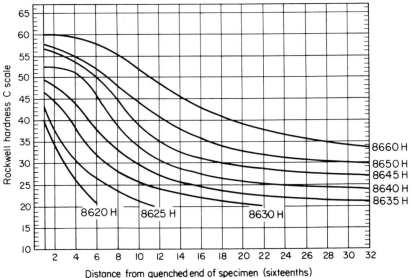

Distance from quenched end of specimen (sixteenths)

APPENDIX 8-2   *(continued)*

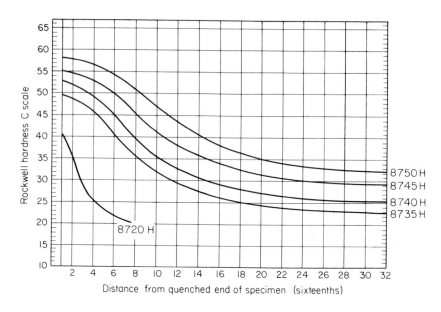

Distance from quenched end of specimen (sixteenths)

A Rockwell hardness of $45R_c$, $\frac{3}{16}$ in. from the quenched end, is written as a *Jominy number:*

$$J_{45} = 3$$

If there are maximum and minimum Rockwell hardness readings to be achieved, they are written

$$J_{35/45} = 3$$

Thus a minimum Rockwell hardness of $35R_c$ and a maximum of $45R_c$ is to occur $\frac{3}{16}$ in. from the quenched end. *Hardenability bands* give these maximum–minimum values. A hardenability band is shown in Fig. 8-11(d).

The ideal critical diameters (D_I) that will produce 50 per cent martensite structure at its center when quenched with a severity of $H = \infty$ may be determined from the end-quench data. One such curve is shown in Fig. 8-11(e).*

* From Asinow, Craig, and Grossman, "Correlation Between Jominy Test and Quenched Round Bars," *Trans. SAE,* **49** (1941), pp. 283–299. Modified by Hodge and Orchoski, "Hardenability Effects in Relation to the Percentage of Martensite," *Trans. AIME,* **167** (1946), pp. 502–512.

Example 2

Assume that a minimum hardness of $50R_c$ is desired at the mid-radius location of $1\frac{1}{2}$ in. (3-in. diameter) when quenched in an oil with good agitation. Determine the following: (a) the H-value, (b) the distance from the quenched end for these conditions, using Fig. 8-10, (c) the Jominy number, and (d) the steel that meets the above criteria, using Fig. 8-11(c).

solution:

(a) Mid-radius curves are shown plotted in Fig. 8-10, where the ratio of the radii $r/R = 0.5$.
The H-value from Table 8-1 is $H = 0.50$. The quench is oil. The severity is good.

(b) With reference to Fig. 8-10, the H-value 0.5 curve will yield $\frac{7}{8}$ in. ($\frac{14}{16}$ in.) when traced to the 3-in. diameter on the ordinate.

(c) Thus the same conditions will exist $\frac{14}{16}$ in. from the quenched end of a Jominy test bar; this is symbolized as

$$J_{50} = 14$$

(d) With reference to the Jominy curves, Fig. 8-11(c), the steel selected might be SAE 4340, which shows a $R_c = 50$, $\frac{14}{16}$ in. from the quenched end. A series of end-quench curves is shown in Appendix 8-2. It might be necessary to refer to a listing of other published curves.

To summarize:

(1) The H-value is determined from Table 8-1.

(2) Using this H-value and the diameter of the bar, the distance from the quenched end at which the desired conditions will prevail is determined from the appropriate r/R curves in Appendix 8-1.

(3) Having the distance from the quenched end, the desired Rockwell reading (R_c) makes it possible to write the Jominy number.

(4) Using the Jominy number and the desired Rockwell hardness number, the material may be selected from Appendix 8-2.

It has been pointed out that hardness readings may be taken across the surface of a polished end of a hardened sample. These readings when plotted may be related to the diameter of the sample. This is shown in Fig. 8-12(a), where the type of *alloy* used affects the hardness along the diameter of the cross section of the test bar.

The effect of mass on the cooling rate and consequently on the hardness of a quenched specimen is shown in Fig. 8-12(b). Since the larger mass cools more slowly than a thin mass, the center of a specimen will transform later than the outside. The mass effect upon hardenability is shown in Fig. 8-12(b).

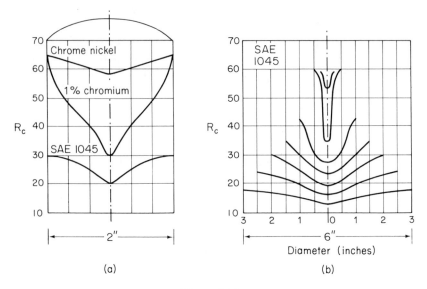

Figure 8-12

8-6   TEMPERING

Quenched martensite, although hard, has very little ductility or toughness. To achieve optimum mechanical properties, the martensite structure is heated to a temperature *below* the lower critical temperature (A_1) and cooled. The result is a structure that does not have the hardness, nor the strength, that it had initially. However, the ductility and toughness have improved. In addition, the part has been *stress relieved.* The end product is a better material for use in engineering design. The effect of tempering an SAE 1035 carbon steel after it has been quenched from 1450°F is shown in Fig. 8-13.

The right ordinate shows the structure of tempered martensite. Figure 8-14 shows as-quenched room temperature martensite. It etches almost brown when treated with a *nital* solution. It was pointed out that martensite is a supersatutated solid solution of carbon particles that have diffused at the austenite temperature throughout the FCC lattice. When quenched from the elevated temperature, the carbon is trapped within the BCC lattice, which it stresses into a tetragonal structure. Since the structure is metastable, the tempering temperatures cause the trapped carbon to precipitate as a transition iron carbide. This precipitation takes place by the process of nucelation and growth. The tetragonal lattice changes and becomes a BCC lattice. With increased temperature the precipitation

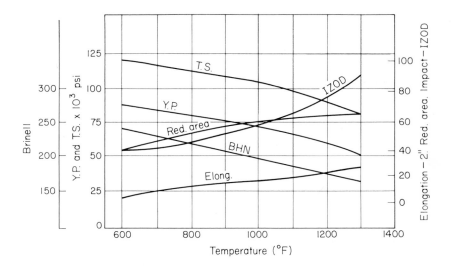

Figure 8-13

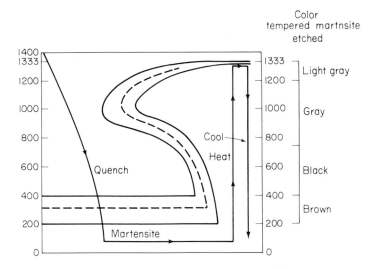

Figure 8-14

takes place in a matter of seconds at 400°F and minutes at 200°F; it could take years at room temperature.

If extreme hardness, a characteristic of martensite, is required, the as-quenched structure is reheated to a temperature below 400°F, where some transition carbide precipitates; but the tetragonal structure of the lattice appears. The structure is still very hard.

If toughness is desired, the as-quenched martensite is heated into the 400 to 750°F range. The temperature depends upon how much hardness is to be sacrificed in the interest of toughness. The specimen is then cooled slowly. The structure upon etching appears black as a result of the dispersion of fine transition carbide particles in the matrix of ferrite.

Within the tempering range of 750 to 1200°F, the transition carbide changes to cementite, and more transition carbides grow as fine spheroids in a matrix of ferrite.

Heating in the 1200 to 1300°F range causes the spheroids to become coarse and more globular.

The toughness of tempered steel is a function of the size of the carbides and their distribution. The tempering temperature and the time tempering are important factors that determine the size and distribution of the carbides.

When steels contain large amounts of carbide-forming alloys, and when high tempering temperatures and long tempering times are used, the structure *increases* in hardness. This may be caused when certain alloys cause retained austenite to be present in the as-quenched condition. Retained austenite results when an alloy is added which drops the M_f line *below* room temperature. If the part is quenched to room temperature, only part of the austenite transforms to martensite. The austenite remaining is called *retained austenite*. This may range from very little retained austenite in low-carbon steels to as much as 25 per cent in high-speed steels, and 45 per cent in steels containing 1.4 per cent carbon. A secondary tempering operation is needed to correct this condition, since the retained austenite will transform to martensite upon heating.

Hardness may also increase as a result of nucleation and growth of *alloy carbides*. Generally, the carbon and the alloys precipitate at the same time, facilitating growth. However, alloys precipitate and diffuse more slowly than carbon, thus causing a slowing down of the *alloy carbide* growth while the *iron carbide* continues to grow and temper hardness develops.

This mechanism is called *secondary hardness*. As just mentioned, a secondary tempering operation may also be needed.

Temper brittleness is the loss of notch-bar toughness in slow cooling from the tempering temperature. It results when the steel contains nickel, chromium, or manganese in appreciable quantities. It is thought to result from the precipitation of carbides, oxides, or nitrides from the ferrite into the grain boundaries on slow cooling. This condition may be corrected by cooling from the tempering temperature rapidly. The addition of molybdenum as an alloy may also correct this condition.

8-7 SURFACE HARDENING

Surface hardening refers to the process of creating a hard surface and a comparatively soft core. The processes involve, on the one hand, the impregnating of the outer structure of a sample with the elements that will facilitate hardening upon secondary heat-treating. This is necessary when the structure of the steel is such that it will not harden upon direct heat-treating. On the other hand, if the structure of the steel *does* contain the necessary elements for hardening, the heat-treating process is controlled so that the surface will harden but the core will remain soft. The methods available that will accomplish surface hardening are

1. Carburizing and case hardening.
2. Cyaniding.
3. Carbonitriding.
4. Nitriding.
5. Flame hardening.
6. Induction hardening.

Carburizing and Case Hardening. There are essentially two methods in use for carburizing steel surfaces. They are pack carburizing and fluid carburizing. In both instances, low-carbon steels with 0.15 to 0.25 per cent carbon are heated in contact with a carbonaceous material or a carbon gas.

In the first process, the material is sealed into a pack with a commercially prepared carbon material or some other material, such as coke, charcoal, or leather. The compound dissociates, releasing carbon, which diffuses below the surface of the steel to form iron carbides.

In the fluid process the steel is heated in a retort furnace. Carbon monoxide or hydrocarbons (ethane, methane, propane, or natural gas) are introduced into the furnace. When heated, the carbon will penetrate into the sample. Liquid cyanide is another fluid carburizing process.

In either process the materials are heated well above the critical temperature for a predetermined period of time. The carburizing temperature is generally about 1750°F, which is well up into the austenite range. The length of time employed depends upon the carburizing materials, the size of the part, and the depth of penetration desired. The soaking period generally takes from 4 to 10 hours at the carburizing temperature. If the part is heated, soaked, and removed from the furnace, the case is not too thick, and the change from high to low carbon is abrupt. If the part is furnace cooled, the penetration is deeper and more gradual. If the grain structure of the material is coarse, the penetration will be greater than if the grain is fine. The latter, however, yields a tougher overall structure. This is shown in Fig. 8-15.

The reaction desired is the release of carbon from the carburizing materials to form cementite, Fe_3C. The carbonaceous material releases CO, which reacts with Fe to form Fe_3C according to the reaction

$$2CO + 3Fe \rightarrow Fe_3C + CO_2$$

$$CO_2 + C \leftrightharpoons 2CO$$

The second equation shows the decomposition of the carbon dioxide CO_2. It should be noted that an atmosphere of almost all CO_2 will reverse the reaction at the surface, thus breaking down the cementite. This process is not acceptable. It is called *decarburization*.

Gas carburizing is a process in which the specimen is heated into the austenite range in the presence of a gas such as methane, ethane, propane, or natural gas. At the elevated temperatures the hydrocarbons decompose according to the equation

$$CH_4 + 3Fe \rightarrow Fe_3C + 2H_2$$

Cyaniding.   Liquid cyanide is used to produce a thin high-carbon case about 0.020 in. deep. Medium- or low-carbon parts are suspended in a 30

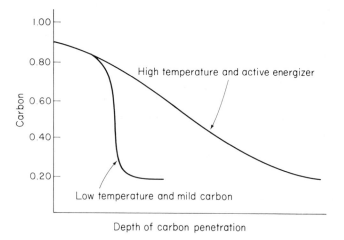

Depth of carbon penetration

Figure 8-15

per cent sodium cyanide bath for about $\frac{1}{2}$ to 1 hour at 1600°F. They are removed and quenched. This produces a very hard, fine-grained, shallow case. The presence of nitrogen also produces nitrides, which contribute to the hardness of the surface.

Sometimes ammonia gas is bubbled through the molten cyanide bath. This increases the release of nitrogen and consequently increases the quantity of nitrides formed. This latter process is called *chapmanizing*.

The above processes are being replaced by the *floating slag* calcium cyanide method. A floating slag of calcium cyanide, sodium chloride, calcium oxide, calcium cyanamide, and carbon is used to produce a deeper case, which contains higher amounts of cementite (carbon) and fewer nitrides (nitrogen).

Carbonitriding. This process uses a mixture of ammonia and a hydrocarbon gas. The thickness of the case produced is about 0.010 in. (minimum 0.002 to maximum 0.020 in.). When processed at high temperatures, the structure of the case is mainly composed of carbides. When processed at low temperatures, the structure of the case is mainly composed of nitrides.

Nitriding. This process uses an atmosphere of ammonia gas to provide the nitrogen at elevated temperatures, which form complex nitrides in the outer structure of the steel. In plain carbon steel at elevated temperatures, complex compounds in the form of needles are present. The thin case is very hard and brittle. Special steels containing up to 3 per cent total of aluminum, chromium, nickel, and molybdenum or combinations of these alloys have been produced. The nitriding process produces a very hard thin surface when a machined, hardened, and tempered surface is held at a temperature of from 900 to 1000°F in an ammonia gas atmosphere for about 50 hours.

Nitrided steels have an extremely hard (over $70R_c$) shallow case which is very stable at room temperature or at temperatures of up to 800°F. These steels are corrosion and wear resistant, and have a high endurance limit. Because the nitrided surface is carried out below the lower critical temperature, the steel may be hardened, tempered, machined, and polished before nitriding without causing cracking, distortion, or warpage.

Flame Hardening. This process was developed where localized heat treating is desired. The part or area that is to be hardened is heated above the upper critical temperature (about 1550°F) with an oxyacetylene flame. The flame is adjusted to produce a carburizing flame. The torch generally carries the coolant (water or air) under pressure a short distance behind the flame. Thus, as the part is heated by the flame, it is quenched. The sur-

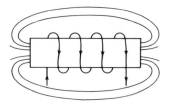

Figure 8-16

face of the steel should be medium or high carbon (0.35 to 0.60 per cent), and the maximum depth of hardness obtainable is approximately $\frac{1}{4}$ in. The metal does not pit, scale, or develop surface cracks.

Induction Hardening.   This heat-treating process (Fig. 8-16) is used for surface and localized hardening. Steel placed into a high-frequency field, 9000 to 300,000 Hz, acts as the secondary. Eddy currents are induced near the surface of the steel. The resistance to the buildup of these magnetic fields heats the workpiece. If the steel is subjected to these induced currents for a long enough period of time, it will reach a temperature at which it becomes nonmagnetic. The part is quenched. The heating operation takes only a few seconds.

In all cases just described, except for flame hardening and induction hardening, a secondary heating cycle was used. The first process introduced the carbon into the outer shell of the steel. The second heating process hardened the surface.

The photomicrograph, after carburizing, is shown in Fig. 8-17(a). The outer structure could have a carbon content of 1.10 per cent, depending upon the extent of the process. The core will have the original carbon content of about 0.10 to 0.20 per cent. Since the carbon content decreases progressively as one proceeds from the outside of the specimen to the core, heating and quenching will yield all degrees of hardness when a radius of the cross section is examined. The outer extremities of the case, where the carbon content is high, may be Rockwell $60R_c$, and the core may be so soft that no R_c reading can be taken. Figure 8-17(b) shows this duplex structure.

The inherently fine-grained steels do not carburize or harden as deeply as the coarse-grained steels. Fine-grained steels possess superior toughness and usually require one subsequent heat treatment to case harden them. They may, however, develop soft spots. Coarse-grained steels carburize and harden deeper than fine-grained steel and do not develop soft spots. They are not as tough, take much longer to carburize, and require additional heat-treating to refine the coarse-grained core.

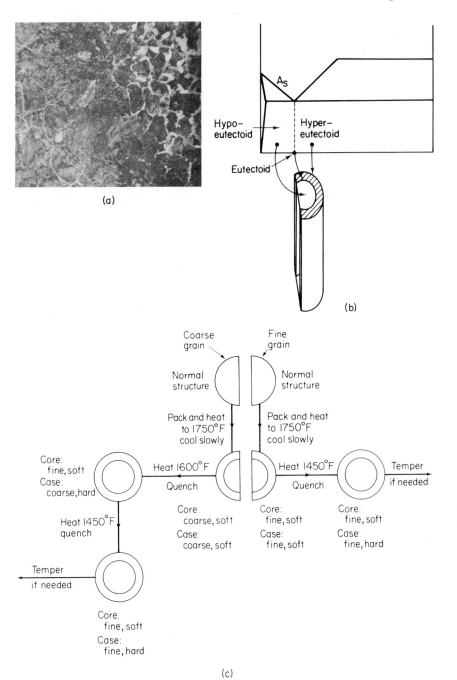

(a)

(b)

(c)

Figure 8-17

In Fig. 8-8(b) it was seen that, for inherently fine-grained steels, the grain remains fine at higher temperatures after refinement than for coarse-grained steels [Fig. 8-8(a)].

In Fig. 8-17(c) both coarse- and fine-grained low-carbon steels are packed, carburized, and cooled slowly, thus creating the high-carbon case. At this point the core and case hardness and grain sizes are much the same as they were originally. If the steel is inherently fine, the core need only be heated and quenched to achieve a fine-grained soft core and a fine-grained hard case.

If the steel is inherently coarse, an additional process is required. The carburized steel is heated to 1600°F and quenched. The hypoeutec-toid core will be refined, since 1600°F is just above the A_3 line [see Fig. 8-17(b)]. The hypereutectoid case will harden but will be coarse when quenched from 1600°F. Reheating the steel to 1450°F will not affect the core, but the grain of the case will be refined so that upon quenching the resulting structure will show the existence of a fine-grained core, and a hard, fine-grained case.

It should also be remembered that the carbon content affects the position of the nose of the TTT curves. Thus Fig. 8-7 (idealized) shows the cooling curves that result when a specimen is quenched in water. The high-carbon case [Fig. 8-7(c)] shifts the nose of the TTT curve to the right, resulting in a martensite case after quenching. The low-carbon core [Fig. 8-7(b)] shifts the nose of the TTT curve far to the left so that quench-ing results in a pearlite core. If the effect of subsurface cooling *rate* is con-sidered, the nose of the TTT curve for the core would be "cut" even fur-ther to the right than shown.

Problems

8-1. In terms of usage, explain the general applications of an equilibrium dia-gram as compared with the TTT curves.

8-2. Draw the iron–iron carbide equilibrium diagram. Label the lines A_1, A_2, A_3, and A_{cm}. What does each represent?

8-3. Define the following terms: (a) ferrite, (b) cementite, (c) pearlite, (d) hy-poeutectoid steel, (e) hypereutectoid steel, (f) austenite.

8-4. Assume a 0.50 per cent carbon steel. This steel is heated to below the lower critical temperature (A_1) and allowed to cool slowly to room tem-perature. What is the structure?

8-5. Same as Problem 8-4, except that the carbon content is 1.10 per cent.

8-6. Assume that the steel in Problem 8-4 is heated above the upper critical temperature A_3. (a) What is the structure? (b) What is the room-tempera-ture structure, if the steel is cooled very slowly?

8-7. Assume that the steel in Problem 8-5 is heated above the A_{cm} line. (a) What is the structure? (b) What is the room-temperature structure if the steel is cooled very slowly to room temperature?

8-8. Describe the sequence of phase changes that takes place as (a) a hypoeutectoid steel is heated into the austenite range, (b) a hypereutectoid steel is heated into the austenite range, and (c) a eutectoid steel is heated into the austenite range.

8-9. Describe the method used to develop TTT curves from cooling curves. Support your description, using a diagram.

8-10. How does a *continuous cooling curve* differ from a *time–temperature-transformation* curve?

8-11. List quenching media in order of their increasing severity.

8-12. In Fig. 8-18 name the room temperature structures and the percentages of the phases present if the steel is eutectoid for cooling rates *a, b, c, d,* and *e.*

8-13. In Fig. 8-18 calculate the time needed to complete the transformations for the curves *a* through *e.*

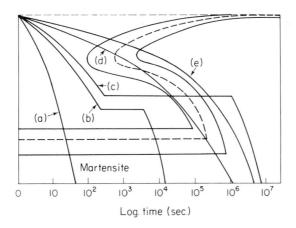

Figure 8-18

8-14. What is martempering? Describe the procedure for developing a martempered structure.

8-15. How do acicular and feathery bainite differ?

8-16. Describe the austempering process.

8-17. Describe the full annealing of (a) hypoeutectoid steel, (b) hypereutectoid steel, and (c) eutectoid steel. What does it accomplish in each of these steels?

8-18. Describe the process known as *isothermal annealing.*

8-19. Describe the process known as *spheroidizing annealing*. When is it used?

8-20. Describe *stress-relief annealing*.

8-21. Describe *process annealing*.

8-22. Describe the *normalizing* process. What does it accomplish?

8-23. Describe the hardening process of a eutectoid steel. Concentrate on the effect of heating on the lattice structure and the carbon atoms in the material.

8-24. Describe the term "critical cooling rate."

8-25. What effect does an increase in cooling rate have on the M_s and M_f curves?

8-26. What effect does the carbon content have on the hardness of steel when quenched?

8-27. Trace the transformation that takes place in Fig. 8-6 for a material which cools at the rate of 1600°F/h.

8-28. Repeat Problem 8-27 for a cooling rate of 2500°F/h.

8-29. Repeat Problem 8-27 for a cooling rate of 5500°F/h.

8-30. Given the continuing cooling curve in Fig. 8-19, identify the room temperature structure for each of the superimposed cooling curves.

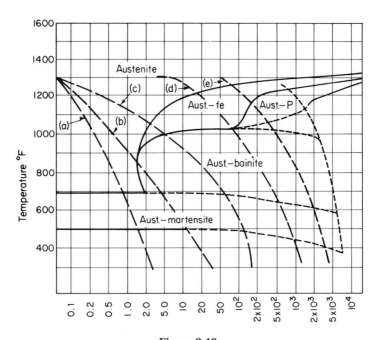

Figure 8-19

8-31. Describe grain growth for hypo- and hypereutectoid steels.

8-32. What effect does an increase in carbon content have upon the nose of the TTT curves?

8-33. (a) What is the effect of an increase in the percentage of alloys on the position of the nose of the TTT curve? (b) What does the addition of cobalt do to the nose of the TTT curve?

8-34. What is the effect on the position of the nose of the curve if the grain size is increased?

8-35. What is the effect of a right and left shift in the nose of a TTT curve on hardenability? Relate your answers to the addition of alloys, grain coarsening, low critical cooling rate, and the size of the workpiece.

8-36. Define the *critical bar diameter D*.

8-37. Describe the method discussed when a hardenability history of a steel is developed. Discuss Fig. 8-9(a) and (b).

8-38. How is an *H*-value determined? What is it?

8-39. What is meant by the term "ideal critical diameter" D_I?

8-40. Describe the end-quench Jominy test and its curve.

8-41. Given the following symbols, interpret their meaning: (a) $J_{25} = 8$; (b) $J_{50} = 2$; (c) $J_{10} = 21$; (d) $J_5 = 24$; (e) $J_{25/30} = 5$.

8-42. Given the following data, write the Jominy symbol: (a) a specimen Rockwell of 60, $\frac{3}{16}$ in. from the quenched end; (b) $1\frac{5}{8}$ in. from the quenched end, a specimen Rockwell of 20; (c) $1\frac{1}{8}$ in. from the quenched end, a specimen Rockwell of 28; (d) $2\frac{1}{4}$ in. from the quenched end, a specimen Rockwell of 5; (e) the Rockwell reading is 15, $1\frac{1}{4}$ in. from the quenched end.

8-43. A sample steel bar with a diameter of 2 in. is to yield a Rockwell reading of $40R_c$ at a distance 50 per cent from the center when quenched in a "strong oil quench–violent agitation." Determine (a) the *H*-value, (b) the distance from the quenched end for these conditions, (c) the Jominy number, and (d) the steel that meets the above conditions, selected from Fig. 8-11(c) cr Appendix 8-2.

8-44. The minimum hardness desired at 30 per cent of the distance from the center of a 4-in. round bar is to be $35R_c$ when quenched in a poor water quench that is not agitated. Determine (a) the *H*-value, (b) the distance from the quenched end for these conditions, (c) the Jominy number, and (d) the steel needed to meet these conditions, selected from Fig. 8-11(c) or Appendix 8-2.

8-45. From Fig. 8-12(a), determine the hardness of the bar $\frac{1}{16}$ in. below the surface if the steel has 1 per cent chrome.

8-46.   Same as Problem 8-45, but the bar is $\frac{3}{8}$ in. below the surface and the steel is SAE 1045.

8-47.   Using Fig. 8-12(b), determine the hardness at the surface and center of a 1-in.-diameter bar.

8-48.   Same as Problem 8-47, except that the bar has a diameter of 3 in.

8-49.   In your own words, analyze and describe each of the curves in Fig. 8-13

8-50.   Describe the effect of hardening and tempering on the lattice structure of steel.

8-51.   Explain the effect of increased tempering temperatures on the microstructure of a piece of hardened steel.

8-52.   Some alloys drop the M_f line below room temperature. What is the result when steel is hardened? How is the condition corrected?

8-53.   Explain the two mechanisms in the heat-treating of alloy steels that result in *secondary hardness*. How is this condition corrected?

8-54.   What is *temper brittleness*? How may it be corrected?

8-55.   Define surface hardening.

8-56.   Describe the process of pack carburizing and fluid carburizing of low-carbon steels.

8-57.   During the carburizing process, describe some of the conditions that result from (a) the soaking time, (b) the cooling rate, and (c) a CO_2 atmosphere.

8-58.   Describe the three *cyanide* processes discussed in this chapter.

8-59.   Describe the carbonitriding process of case hardening.

8-60.   Describe the nitriding process, and describe the physical properties that result in the steel.

8-61.   Describe the flame hardening process.

8-62.   Describe the induction hardening process of case hardening steel.

8-63.   Describe the microstructure of a piece of carburized and case-hardened low-carbon steel.

8-64.   Describe the heat-treating process used to harden and refine both the case and core in (a) an inherently fine-grained, low-carbon steel, and (b) a coarse-grained steel.

8-65.   Relate the hardness of the case and the softness of the core to the equilibrium diagram and to a set of TTT curves.

9 | Tool Steels

9-1 ALLOYS AND THEIR EFFECT ON TOOL STEELS

Plain carbon steels, if used for cutting tools, lack certain characteristics necessary for high-speed production, such as red hardness and hot strength toughness. The effect of alloying elements in steel is of great advantage and yields tool steels that overcome many of the shortcomings of the plain carbon steels. The effects of the various alloys when added to tool steels are as follows:

Chromium increases the hardenability, wear and abrasion resistance, and toughness of steel. The addition of chromium has the effect of raising the critical temperature and causes the steel to distort if water quenched to the same degree as plain carbon steel. When added to steel, its effectiveness as a carbide former places it between tungsten and manganese, and therefore the steel's resistance to tempering is mild. As already indicated, it increases corrosion and oxidation resistance in the steel to which it is added.

Manganese increases the depth of hardness. It lowers the critical temperature of steel so that approximately 1.5 per cent manganese in combination with approximately 1 per cent carbon will make the steel oil harden. Its carbide-forming ability is greater than that of ferrite but less than that of chromium. It also counteracts brittleness caused by sulfur. It has little effect on the tempering characteristics of the steel to which it is added.

Molybdenum, when used with chromium, silicon, and manganese, increases the strength, toughness, and hardness of steel. It greatly increases the hardenability of alloy steels. This effect is greater than that of chromium. It raises the hot strength and red hardness of steel when alloyed with chromium and vanadium. It also raises the grain-coarsening temperature of the austenite phase. Its carbide-forming tendency is strong and therefore opposes softening during tempering by promoting secondary hardness in steel.

Silicon in percentages of up to 0.25 has little effect as an alloy. In percentages up to 2, it intensifies the effects of molybdenum, manganese, and chromium. It increases the hardenability of steel, but this effect is not as great as that of manganese. Its carbide-forming effects are less than ferrite. It increases the resistance of steel to oxidation.

Nickel tends to retain austenite in high-carbon steels and thus has some effect on hardenability. This is especially true if nickel is alloyed in steels that have a high chromium–iron content. It is a weak carbide former. It has practically no effect on the tempering operation.

Phosphorus increases the hardenability of steel about the same as manganese. It increases machinability of low-carbon steels and increases their corrosion resistance. It has no carbide-forming tendency and has no effect on the tempering operation.

Cobalt is one of the few elements that decreases hardenability when it is alloyed with other elements. However, when alloyed with vanadium, chromium, or tungsten, it greatly increases the red hardness of the material. Its carbide-forming tendency is about the same as ferrite.

Tungsten, when used in small amounts in low- or medium-carbon steels, increases hardenability slightly. When used in high-carbon steel in amounts of about 4 per cent, it imparts hardness and wear resistance to steel. When used in amounts of about 18 per cent, it imparts red hardness and hot strength to steel. It also forms abrasion-resistant particles in tool steel. It is a strong carbide former and opposes softening during tempering by secondary hardening.

Titanium as dissolved increases the austenitic hardenability. It reduces martensitic hardness in medium-chromium steel.

Vanadium increases hardenability very greatly and is a strong carbide former. It also elevates the grain-coarsening temperature, thus yielding a fine-grained steel. It resists tempering by aiding secondary hardening.

Aluminum is a deoxidizer. When used in alloy steels it restricts grain growth. In nitriding steels, when alloyed with chromium, it aids in the formation of complex nitrides.

9-2 CLASSIFICATION

In general, tool steel may be classified as plain high-carbon, low-alloy, intermediate-alloy, and high-speed steels. The purposes of the alloys

(Sec. 9-1) are to increase wear resistance by forming hard carbides, to reduce the steel's tendency to soften because of tempering, and, probably of greatest importance, to increase the hardenability of the steel.

Plain high-carbon tool steels have already been discussed. In review, they are the high-carbon steels composed, in general, as follows:

Carbon	0.50–1.50 per cent
Manganese	0.12–0.40 per cent
Silicon	0.10–0.30 per cent
Sulfur and phosphorus	up to 0.50 per cent

Carbon is the important element. Hardness increases as the carbon content is increased. Up to about 0.85 per cent carbon, the effect on hardness is marked. Above 0.85 per cent carbon, the hardness increases very slowly. The effect is to increase the cutting properties and resistance to wear.

Manganese, in conjunction with carbon, forms carbides (Mn_3C) and together with iron carbide (Fe_3C) increases the wear resistance, tensile strength, and hardness of the material. It also has a strong effect on the quenching characteristics of steel. In quantities of about 0.30 per cent manganese and with eutectoid carbon, it becomes possible to quench the material in water without cracking. Higher quantities of manganese will cause cracking when the specimens are quenched in water.

If the silicon, sulfur, and phosphorus are not allowed to increase beyond those quantities shown above, they have little effect on the properties of plain high-carbon tool steels. Quenching these steels will cause distortion when internal stresses are generated and when the cross sections of the part change.

Low-alloy tool steels will harden deeper and faster than plain high-carbon tool steels. Since the quenching medium need not be as severe as that required for the plain high-carbon tool steels, the low-alloy steels will harden with less distortion and to a greater depth.

Medium-alloy tool steels will form hard, wear-resistant carbides. They are generally used for cutting tools for taking finishing cuts.

High-speed tool steels are those that possess high amounts of carbides, thus making the steel highly wear resistant. Their ability to achieve secondary hardness when heated makes them especially useful in high-speed cutting. Since they have extremely high hardenability characteristics, they may be quenched at slow cooling rates. Thus oil, or air, may be used as the quenching medium. This reduces the possibility of cracking, distortion, or warping. If the *carbon* content is low, the steels have good impact resistance. If the carbon is high, they have high abrasion resistance.

Tool steels have also been classified according to some special purpose. Several tool steels are listed in Table 9-1. The symbols used to identify the various classifications are shown below:

High-speed tool steel
    M, molybdenum type
    T, tungsten type

Hot-work tool steel
    H: H1–H19, chromium type
         H20–H39, tungsten type
         H40–H59, molybdenum type

Cold-work tool steel
    D, high-carbon, high-chromium type
    A, medium-alloy, air-hardening type
    O, oil-hardening type

Shock-resisting tool steels
    S, shock-resisting type

Mold steel
    P, mold type

Water-hardening tool steels
    W, water-hardening type

Special-purpose Tool Steel (L, low-alloy type)
    F, carbon–tungsten type

Table 9-1   AISI Identification and Type Classification of Tool Steels

High-speed Tool Steel (M, molybdenum type)

Type	C	Mn	Si	W	Mo	Cr	V	Co	Ni	Al
M1	0.80	0.20	0.20	1.55	8.50	4.00	1.00			
M4	1.30	0.20	0.20	5.50	4.50	4.25	4.00			
M6	0.75	0.20	0.20	4.00	5.00	4.00	1.25	12.00		
M10	0.90	0.20	0.20		8.00	4.00	2.00			
M30	0.80	0.20	0.20	1.50	8.25	4.00	1.10	5.00		
M41	1.10	0.20	0.20	6.75	3.75	4.25	2.00	5.00		
M43	1.25	0.20	0.20	1.75	8.75	3.75	2.00	8.25		

High-speed Tool Steel (T, tungsten type)

Type	C	Mn	Si	W	Mo	Cr	V	Co	Ni	Al
T2	0.80	0.20	0.20	18.00	0.60	4.00	2.00			
T6	0.80	0.20	0.20	20.00	0.80	4.50	1.60	12.00		
T8	0.75	0.20	0.20	14.00	0.75	4.00	2.00	5.00		
T9	1.25	0.20	0.20	18.00	0.75	4.00	4.00			
T15	1.50	0.20	0.20	13.00	0.50	4.50	5.00	5.00		

Hot-work Tool Steel (H, chromium type)

Type	C	Mn	Si	W	Mo	Cr	V	Co	Ni	Al
H10	0.40	0.60	1.00		2.50	3.25	0.40			
H14	0.40	0.30	1.00	5.00	0.25	5.00	0.25	0.50		
H19	0.40	0.30	0.30	4.25	0.45	4.25	2.20	4.25		

Table 9-1 *(continued)*

Hot-work Tool Steel (H, tungsten type)

	C	Mn	Si	W	Mo	Cr	V			
H20	0.35	0.20	0.40	9.00	9.50	2.00	0.50			
H23	0.30	0.20	0.50	12.00		12.00	1.00			
H26	0.60	0.20	0.20	18.00		4.00	1.00			

Hot-work Tool Steel (H, molybdenum type)

	C	Mn	Si	W	Mo	Cr	V			
H41	0.65	0.20	0.20	1.60	8.75	4.00	1.00			
H43	0.55	0.20	0.20		8.00	4.00	2.00			

Cold-work Tool Steel (D, carbon–chrome type)

	C	Mn	Si	W	Mo	Cr	V			
D1	1.00	0.30	0.25		0.80	12.00	0.50			
D2	1.50	0.30	0.25		0.75	12.00	0.60			
D3	2.25	0.30	0.25		0.80	12.00	0.60			
D7	2.35	0.40	0.40		1.00	12.50	4.00			

Cold-work Tool Steel (A, air-hardening type)

Type	C	Mn	Si	W	Mo	Cr	V	Co	Ni	Al
A2	1.00	0.60	0.25		1.00	5.00	0.30			
A3	1.25	0.50	0.25		1.00	5.00	1.00			
A4	1.00	2.00	0.25		1.00	1.00				
A7	2.25	0.55	0.55	1.20	1.00	5.25	4.75			
A8	0.55	0.25	1.00	1.25	1.50	5.00	0.25			
A9	0.50	0.40	1.00		1.50	5.00	1.00		1.50	

Cold-work Tool Steel (O, oil-hardening type)

	C	Mn	Si	W	Mo	Cr	V			
O1	0.90	1.25	0.25	0.50		0.50	0.25			
O2	0.90	1.60	0.25		0.30	0.25	0.20			
O6	1.45	0.65	1.00		0.25	0.20				
O7	1.20	0.25	0.25	1.55	0.25	0.50	0.25			

Shock-resisting Tool Steel (S, shock-resisting type)

	C	Mn	Si	W	Mo	Cr	V			
S1	0.50	0.25	0.25	2.50		1.50	0.20			
S2	0.55	0.50	1.00		0.50		0.20			
S4	0.55	0.85	2.00			0.20	0.20			
S6	0.45	1.40	2.25		0.40	1.25	0.30			

Weld Steels (P)

	C	Mn	Si	W	Mo	Cr	V	Co	Ni	Al
P2	0.07				0.20	2.00			0.50	
P5	0.10					2.25				
P21	0.20								4.00	1.20

Special-purpose Tool Steel (L, low-alloy types)

	C	Mn	Si	W	Mo	Cr	V	Co	Ni	Al
L1	1.00					1.25				
L3	1.00				0.25	0.75			1.50	
L7	1.00	0.35			0.40	1.40				

Table 9-1 (continued)

Water-hardening Tool Steel (W)

W1	0.60-1.4		
W2	0.60-1.4		0.25
W4	0.60-1.4	0.25	
W5	1.10	0.50	

Special-purpose Tool Steel (F, carbon–tungsten type)

F1	1.00	1.25	
F3	1.25	3.50	0.75

Source: AISI Product Manual, *Tool Steels,* April 1963.

A classification proposed in the third edition of *Tool Steels* published by the American Society for Metals has certain advantages over the method shown in Table 9-1. This classification and the AISI classification are shown in Table 9-2.

Table 9-2   Proposed Tool Steel Classification

Class	AISI	Tool Steel
110	W1	Carbon
120	W2	Carbon–vanadium
130	W4, 5	Carbon–chromium
140		Carbon–chromium–vanadium
210	L1, 2, 7	Chromium up to 3% and carbon greater than 0.65%
220	L2	Chromium up to 3% and carbon less than 0.65%
230	L6	Nickel and carbon greater than 0.65%
240		Nickel and carbon less than 0.65%
310	S	Silicon
320	S1	Tungsten chisel
330		Nontempering chisel steel
340	F2, 3	Tungsten finishing steel
350	F1	High-carbon low-alloy
360		
370	P	Mold steels for carburized cavities
380	P20, 21	Mold steels for machined cavities
390	0 and A10	Graphic tool
410	0	Oil-hardening cold-work die
420	A	Air-hardening cold-work die
430	D	High-carbon high-chromium cold-work die
440	A7, 0	Special wear-resistant cold-work die
510	S7	3 to 4% chromium die steel (hot work)
520	H, A8–9	Chromium–molybdenum hot-work die
530	H-14	Chromium–tungsten hot-work die
540	H-20 series	Tungsten hot-work die
550	H15–H4 D series	Molybdenum hot-work die

Table 9-2 *(continued)*

Class	AISI	Tool Steel
610	T1, 9, 7	Tungsten
620	T4, 5, 6, 8, 15	Tungsten–cobalt
630	M1, 10, 7	Molybdenum
640	M30	Molybdenum–cobalt
650	M2, 3, 4	Tungsten–molybdenum
660	M6, 15, 35, 36	Tungsten–molybdenum–cobalt
710		Wortle die and self-hardening steel

Source: Roberts, Hamaker, and Johnson, *Tool Steel,* 3rd ed. Metal Park, Ohio: American Society for Metals.

In Table 9-3 are shown several representative practices used in the heat treatment of tool steels. It is important that the student understand that many of the temperatures and temperature ranges are appropriate only under the most ideal conditions. The heat treatment of tool steels follows a very precise procedure. Table 9-3 should be applied with great care.

9-3 PROPERTIES OF TOOL STEELS

Another method used to classify tool steels is by the effect that alloys have upon the physical properties. Thus physical properties such as deformation, toughness, resistance to wear, red hardness, decarburization, depth of hardness, and machinability are discussed here.

Deformation. Many steels go through substantial volume changes during the heating and quenching cycle when being heat-treated. Severe quenching or abrupt changes in shape will cause cracking or distortion and cracking. In other instances, deformation could be so severe that too much metal must be left on the surface to be machined so that subsequent machining operations (i.e., subsequent to heat-treating) will ensure "cleaning up" the finished surface. Other tools are designed so that they can be hardened within dimensional tolerances required. In the latter case, deformation must be held to a minimum.

Carbon steels (W) are essentially water hardening and as such are subject to maximum distortion. If *manganese* is added to a carbon steel, it will retard the austenite transformation. This steel will be capable of hardening in oil. *Oil-hardening steels* (O) deform less than water-quenched steels (W), and with but few exceptions air-hardening steels (A) deform the least. In Table 9-4 the various types of steel, graded from best to poorest, are shown.

Table 9-3   Heat-treating Conditions

AISI series	ASM* class	Preheat temp. °F	Hardening H — Aust. temp. °F	Hardening H — Holding time min	Hardening H — Quench medium†	Hardening H — Hardness Rc	Temper Temp. °F	Annealing Temp. °F	Annealing Hardness BHN	Normalize Temp. °F	Type of tool steel
M—	630–660	1425	2150–2275	2–5	O, A, S	63–66	1000–1100	1600–1650	210–295	None	High-speed molybdenum
T—	610–620	1550	2200–2375	2–5	O, A, S	63–67	1000–1100	1600–1650	220–295	None	High-speed tungsten
H10	520	1500	1850–1900	15–40	A	56–60		1550–1650	192–229	None	Hot-work chromium
H16	530	1500	2050–2150	2–5	O, A	55–58	1050–1250	1600–1650	212–241	None	
H19	530	1550	2000–2200	2–5	A, O	52–55	1050–1250	1600–1650	207–241	None	
H20	540	1500	2000–2200	2–5	A, O	53–55		1600–1650	207–235	None	Hot-work tungsten
H23	530	1500	2200–2300	2–5	O	33–35	1100–1250	1600–1650	212–255	None	
H26	540	1600	2150–2300	2–5	A, O, S	63–64	1050–1200	1600–1650	217–241	None	
H41	550	1350	2000–2175	2–5	A, O, S	64–66	1050–1200	1500–1600	207–235	None	Hot-work molybdenum
H43	550	1350	2000–2175	2–5	A, O, S	54–58	1050–1200	1500–1600	207–235	None	
D—	430	1500	1750–2000	15–45	A	61–65	400–1000	1600–1650	200–260	None	Cold-work carbon–chromium
D3	430	1500	1700–1800	15–45	O	64	400–1000	1600–1650	215–250	None	
A—	420	1200	1450–1850	15–90	A	60–65	300–1000	1550–1600	200–260	None	Air hardening
A10	390	1200	1450–1500	30–60	A	62–64	300–400	1410–1460	235–270	1450	
O	410	1200	1400–1650	5–30	O	63–65	300–500	1375–1500	180–215	1600	Cold-work oil hardening
S—	310–320	1300	1600–1750	5–45	O	55–60	300–1200	1400–1550	187–230	None	Shock resisting
S2,4	310	1200	1550–1700	5–20	B, W	60–63	300–800	1400–1450	192–230	None	
S7	513	1250	1700–1750	15–45	A, O	60–61	400–1000	1500–1550	187–223	None	
P	370–380		1450–1600	15	O, W	62–65	300–500	1350–1650	100–135		Mold steel
P20	380		1450–1600	15	A, O	60–62	300–500	1400–1450	150–180		
L	210–230		1450–1700	10–30	O	60–65	300–600	1400–1500	160–215	1650	Special purpose, low alloy
L1,2,3	210		1450–1500	10–30	O, W	63–65	300–600	1400–1450	160–200	1650	
F	340–350	1200	1450–1600	15	W, B	65–67	300–500	1400–1500	180–240	1650	Special purpose, tungsten
F3	340	1200	1450–1600	15	W, B, O	65	300–500	1450–1500	210–250	1650	
W	110–120 / 130–140		1400–1450	10–30	B, W	50–65	300–600	1350–1450	160–200	None	Water hardening

Source: Taylor Lyman, ed., *Metals Handbook,* Vol. 2, 8th ed. Metal Park, Ohio: American Society for Metals. For specific values the handbook *Tool Steel* (American Society for Metals) should be consulted.

*Proposed.

†W, water; O, oil; B, brine; S, fused salts; A, air.

Table 9-4   Nondeforming Properties

Type Steel	A, D	O, D3, H11	F, H, T, M	S	L, W, P
	Best	V. Good	Good	Fair	Poor

Toughness. This is a necessary physical property in tool steels and refers to the resistance to breaking. The correct combination of silicon with other elements, such as molybdenum or manganese, will yield the correct balance between toughness and hardness, and provide a good *shock-resisting tool steel* (S). Table 9-5 shows the toughness of the tool steels as graded.

Table 9-5   Toughness

Type	S	W	H10-, H20-, P	O, A, L, H40-	D, T, M, F
	Best		Good	Fair	Poor

Resistance to Wear. This property is in general related to hardness and refers to the resistance to abrasion. Plain high-carbon steels may be hardened and used as cutting tools. However, their resistance to wear is low when compared to tool steels, which contain alloys such as tungsten, chromium, molybdenum, and vanadium. Table 9-6 shows the classes of tool steel and their positions from best to fair.

Table 9-6   Resistance to Wear

Type	D	A, T, M, F	O, H20-, H40-, L, P	W, S, H10-
	Best	V. Good	Good	Fair

Red Hardness. Cutting tools that must operate at high temperatures, generally above 900°F, are said to possess red hardness. Plain high-carbon steels will harden when heated and quenched to Rockwell hardness comparable to alloy tool steels. However, they will also soften and become worthless as cutting tools when heated to temperatures of 600 to 800°F. When alloys are added, the tools will retain their cutting edges even when the temperature of the tool exceeds 900°F. Alloys that will yield red hardness are tungsten, molybdenum, and chromium. Table 9-7 shows a graded list of tool steel types from best to poor.

Table 9-7   Red Hardness

Type	T	H20-, H40-, M	D, H10-	S, A	W, O, L, F, P
	Best	V. Good	Good	Fair	Poor

Decarburization.   Since the carbon content of steel is probably the most important element, any change in the amount of this element will change the physical properties of a heat-treated steel sample. In some instances the surface of steel will lose carbon when heated above 1333°F. This surface decarburization will affect the hardness of the surface drastically when the steel is quenched from elevated temperatures. If, subsequent to heat-treating, it is necessary further to machine the surface of the steel, it is important to know to what depth this decarburization has taken place. If the steel is not to be machined after heat-treating, it is equally important to know whether any decarburization has taken place. The types of steel, and the degree to which they decarburize, are shown in Table 9-8.

Table 9-8   Resistance to Decarburization

Type	W	O, P, L, F	A, M, H, T	S, A, D, T	M, H40-, S
	Best	V. Good	Good	Fair	Poor

Depth of Hardening.*   Hardenability refers to the depth to which hardening takes place. Generally, if the austenite-to-pearlite transformation is retarded, the depth of hardness increases. Manganese, molybdenum, tungsten, and chromium have strong influence on the depth of hardening. Vanadium, silicon, nickel, and copper also influence to a lesser degree the depth to which steel will harden. Cobalt decreases the depth to which hardness will occur. It is to be noted that fine-grained steels are generally shallow hardening because the austenite transformation starts at the grain boundaries. Since fine-grained steels have more grain boundaries than coarse-grained steels, the austenite-to-pearlite transformation is accelerated and the depth of hardness is shallow. Coarse-grained steels will harden to greater depths because the large grains will remain austenitic for longer periods of time, and they will be affected by mild quenching. The hardenability of various types of steel is shown in Table 9-9.

Table 9-9   Depth of Hardness

Type	A, D, H, T, M	S, O, L	W, F, P
	Deep	Medium	Shallow

*Also see Sec. 8-5.

Machinability.   This refers to the ease with which an alloy may be machined relative to the *water-hardening alloy steels* (W). Of all the alloy tool steels, the water-hardening steels have the best machinability characteristics. Thus in Table 9-10(a) the water-hardening *tool* steels are assigned the arbitrary value of 100. In Table 9-10(b) the machinability ratings for several *carbon alloy* steels are listed. The B1112 carbon alloy steels have about 75 per cent better machinability characteristics than the W tool steels.

Table 9-10(a)   Machinability of Alloy Tool Steels (Represents average values)

Type								
	|——— P ———|							
	L	A	F	H(W)	H(W)			
W	O	S	H(Cr)		H(Mo)	T	M	D
100	90	80	70	60		50		40

Table 9-10(b)   Machinability of Carbon Steel

AISI	Rating	AISI	Rating
B1112	100	A5140	60
C1010	55	A5150	55
C1015	50	A6145	50
C1118	80	A8620	60
C1020	65	A8650	50
C1045	60	A9140	60
C1070	45	Malleable	120
B1113	130	Cast iron: soft	80
A3120	60	medium	65
A3130	55	hard	50
A3140	55	Wrought iron	50
A3145	50	Stainless (12% Cr)	70
A4130	65	18–8 (Aust.)	25
A4145	55	Cast steel (0.35 C)	70
A4150	50	Manganese, oil-hard.	30
A4340	45	Tool steel (low W,	
A5120	65	Cr, C)	30
		Tool steel (high C,	
		Cr)	25
		High-speed steel	30

The greater the quantity of carbon and of alloying elements, the more difficult it is to machine the steel. The combination of carbon with tungsten, vanadium, molybdenum, or chromium forms carbides that make machining more difficult.

The physical properties just discussed are summarized in Table 9-11.

Table 9-11   Physical Properties

		Quench	Non-def.	Tough-ness	Resist. to wear	Red hard.	Resist. to decarb.	Depth hard.	Mach. rate ave.
M	High speed Mo	A O S	F G F	P	V.G.	V.G.	G–P	D	52
T	High speed W	A O S	F G F	P	V.G.	B	G–F	D	48
H$_{10}$	Hot work Cr	A O	G F	G	F	G	G	D	75
H$_{20}$	Hot work W	A O	G F	G	G	V.G.	G	D	55
H$_{40}$	Hot work Mo	A O S	F G F	F	G	V.G.	P	D	55
D	Cold work C–Cr	A O	B V.G.	P	B	G	F	D	45
A	Air hard.	A	B	F	V.G.	F	G–F	D	85
O	Cold work oil hard.	O	V.G.	F	G	P	V.G.	M	90
S	Shock resist.	O W A	F P G	B	F	F	F–P	M	85
L	Spec. purp. low alloy	O W	F P	F	G	P	V.G.	M	80
F	Spec. purp. W	W B O	F G G	P	V.G.	P	V.G.	S	75
W	Water hard.	W B	P P	G	F	P	B	S	100
P	Mold steel	O W A	P	G	G	P	V.G.	S	90–75

B, brine	B, best	D, deep
W, water	V.G., very good	M, medium
O, oil	G, good	S, shallow
S, molten salt	F, fair	
	P, poor	

9-4 TOOL STEEL TYPES

High-Speed Tool Steels (M and T Types).   These are the classes of steel that deep harden, retain that hardness at elevated temperatures, and have a high resistance to wear and abrasion. The M-type tool steels are high in molybdenum content in combination with chromium or vanadium. The T-type tool steels are high in tungsten and are alloyed with chromium, vanadium, and in some cases with high percentages of cobalt. These steels are used for lathe centers, blanking dies, hot forming dies, lathe cutting tools, drills, taps, etc. They are used in almost all cutting tools.

The M-type high-speed tool steels were produced by replacing tungsten in the T-type steel with molybdenum. They are comparable in performance to the tungsten-type tool steels. The typical molybdenum high-speed tool steel contains about 4 to 8.5 per cent molybdenum, 4 per cent chromium, 1.5 to 6 per cent tungsten, and about 2 per cent vanadium. Since this material is very susceptible to oxidation, these tool steels must be heated in a fused salt bath or coated with borax. They must be preheated to approximately 1400°F and then heated quickly to 2200°F. They are quenched in air, oil, or fused salts, depending on the type of steel used. They also have a strong tendency to decarburize. These steels may be tempered at 1050°F.

The T-type high-speed tool steels' hardness is controlled by the carbon content. Those with high carbon content have high wear resistance and are very hard. Those with lower carbon content are tougher but not as hard as the former group. The most widely used tool steel and the best combination of tungsten, chromium, and vanadium seems to be the 18–4–1 tool. As the amount of tungsten increases, the toughness decreases. The 4 per cent chromium controls the hardenability. As the chromium content decreases, the steel loses all the qualities that make it an acceptable high-speed steel. The vanadium increases the cutting qualities of high-speed steel, especially where high resistance to abrasion is desired.

These steels are hardened by preheating slowly to 1500°F, then quickly to 2350°F, and then quenching in air, oil, or fused salts, depending on the type of material being used. It may be tempered at 1050°F.

Many T-type high-speed tool steels are sometimes alloyed with substantial quantities of cobalt. Because cobalt reduces hardenability, the carbon content is generally increased. They are susceptible to cracking on quenching and decarburizing on heating. They must be quenched from temperatures of about 2400°F. However, their red hardness is substantially increased and they possess high toughness.

Because this class of tool material has a substantial amount of wear-resistant carbides in a very high heat resistant matrix, it is generally used in all types of machine-cutting tools such as tool bits, milling cutters, taps, reamers, drills, broaches, and hobs. In some instances it is used where high-temperature structural steel is needed.

Hot-Work Die Steels (H Types).   These steels are classified into three categories: H-chromium, H-tungsten, and H-molybdenum types, or combination of these elements.

The H-chromium-type die steel (H10 series) must be resistant to cracking that may result from high-temperature use during operation or from thermal shock. When used in hot-work dies they must not deform while at elevated temperatures. This class of tool steel is subject to decarburization; therefore, preheating to 1500°F is recommended. If deep hardening is desired, these steels are then heated to 1850°F and air quenched or oil quenched. However, if the latter is used, the parts must be tempered immediately to avoid cracking. The tempering range is about 1000°F. These steels are used in aluminum, magnesium, or zinc extrusion dies and die castings, hot shear blades, plastic molds, header dies, punches and dies for piercing shells, hot press dies, etc.

The H-chromium–molybdenum hot-working die steels (H11–H15 series) are the same as the H-chromium-type steels with decreased quantities of molybdenum. These are the most widely used of the H-type steels. They are hardened at relatively low temperatures, are less subject to cracking with change of temperature, and resist heat checking. They are, in general, very tough materials with high shock-resistant qualities. These steels are used in extrusion, die casting, and forging dies. They are especially useful in dies that are subjected to shock loading when used at normal or elevated temperatures. They are preheated to 1500°F, then heated to about 1800°F, and quenched in air. The annealing temperature is about 1600°F.

The H-11 series hot-work tool steels are easily formed and worked, and have the advantages of possessing exceptional weldability, some resistance to corrosion, a lower coefficient of thermal expansion than most tool steels, and coupled with high strength, a resistance to softening during exposure. It has wide applications in structures that are subject to high stress at relatively high temperature, such as supersonic aircraft.

The H-chromium–tungsten-type hot-work die tool steels (H16) possess high heat hardness and wear resistance, which increase with increased tungsten content up to 7 per cent. They have good machinability qualities. They are not particularly shock resistant, especially when the chromium content is high. They are also very susceptible to decarburization. Hardenability increases with the increase in the carbon content. In all cases these steels are considered deep-hardening tool steels. They must be preheated to about 1500°F, then heated rapidly to the elevated temperatures of between 2000 and 2200°F, and quenched in air or oil. Immediate tempering should take place in the range of 1100 to 1250°F. These steels may be used at working temperatures of 1100°F and therefore are suitable for use in hot dies for forging or extruding brass, die-casting dies for aluminum, permanent molds for molding brass, hot blanking or forming punches and dies, and piercing points or mandrels used in shell forging.

The H-tungsten-type hot-work die steels (H-20 series) use high quantities of tungsten, which ranges from 9 to 18 per cent in combination with 2 to 4 per cent chromium and about 0.50 per cent vanadium. They have the highest red hardness of any of the hot-working steels and also possess good shock-resisting qualities. They have poor thermal shock resistance and therefore cannot be cooled rapidly during use without risking the development of cracks. They should be preheated to 1500°F, then heated to 2000 to 2300°F, and quenched in air or oil. Tempering should occur at a temperature range of 1050 to 1500°F, depending on the qualities desired. They are used essentially for forging, extruding, and die-casting dies for nonferrous materials, and in hot blanking and forming dies.

These steels are used in hot forming or blanking, hot or cold header, thread-rolling, hot trimming, or shearing dies. They are also used in extrusion dies for copper-base alloys.

The H-molybdenum-type hot-work die steels (H-40 series) replace the tungsten component with molybdenum. They have good red hardness, fair wear-resistant qualities, and fair toughness. They are very susceptible to decarburization and therefore must be preheated to about 1400°F, heated rapidly to 2000 to 2200°F, and quenched in air, oil, or fused salts, depending upon the physical properties desired. Tempering takes place at about 1050°F. They are used for essentially the same operations as the H-tungsten materials.

Cold-Work Tool Steels (D, A, O Types).   These steels are classified as follows: high-carbon, high-chromium D steels; the medium alloy in combination with medium- to high-carbon air-hardening A steels; and the oil-hardening O steels. In all cases, these classes of steel require that they be preheated to 1100–1500°F and then raised to the austenite temperature of from 1400 to 1900°F. Most D and O steels require quenching in oil. The A steels are quenched in air. They must be tempered immediately in a temperature range of 300 to 1000°F, depending on the composition of the class. The D and the S steels are deep hardening. The O steels have medium hardenability. They all have excellent resistance to deformation, fair-to-good red hardness, and high wear-resistant properties.

The D-type steels have a high percentage of retained austenite after quenching. To ensure the austenite-to-martensite transformation, the steels are tempered at about 300°F, cooled in air, and then cooled to about −100°F to allow the martensite transformation to take place. They are then tempered to achieve desired properties such as hardness, toughness, etc. Since austenite is unstable, it will eventually transform to a softer structure, causing distortion during the process.

The D and A groups, because of their excellent deformation and wear resistance, are used for gages, blanking, drawing and piercing dies, shears, forming and banding rolls, wear plates, lathe centers, mandrels, and broaches. The O group is used for making reamers, taps, holes, and threading dies, blanking and forming dies, gages, plastic molds, cold trimming and coining dies, and knurling and burnishing tools.

Shock-Resisting Tool Steels (S Type).   These steels have from 45 to 50 per cent carbon content. The alloys, silicon and nickel, are ferrite strengtheners. Chromium increases wear resistance and depth of hardness. They should be preheated to 1300°F. The hardening range is 1550 to 1750°F, depending upon the alloys; they are generally quenched in oil. The tempering range of 400 to 800°F will yield the desired hardness, toughness, wear resistance, etc.

This class of steel has very good shock-resistant qualities with excellent toughness. Their wear resistance and red hardness is somewhat low, and they possess medium hardenability.

They are used in form tools, chisels, punches, cutting blades, springs, crimping, trimming, and swaging dies. When tungsten is present in the S steels, they develop red hardness and can be used in cold- or hot-working tools, such as concrete and rock drills, bolt cutters, pipe cutters, and hot or cold header dies.

Mold Tool Steels (P Type).   These steels are classified into two general types: (1) those which are machined by hubbing (pressing a form into the material), carburizing, and then hardening the surface, and (2) those which are machined or hubbed and then machined. The first class has a low carbon content; the latter has higher carbon content. Both groups possess high toughness. The low-carbon group has low red hardness, whereas the higher-carbon group has poor-to-medium red hardness. Both possess poor-to-medium depth of hardening properties.

The low-carbon group is carburized and then quenched from a hardening temperature range of 1450 to 1750°F, and tempered from a 300 to 500°F range. The higher-carbon group is quenched from a hardening temperature range of 1500 to 2000°F and tempered from 300 to 800°F.

Special-Purpose Tool Steels (L and F Types).   The L-type steels are low-alloy steels with a chromium content that makes them a good low-cost substitute for the cold-work steels. The F-type steels are high in carbon and tungsten. They have high wear resistance, good toughness, and poor-to-medium red hardness with medium hardenability. They are hardened from temperatures of 1450 to 1700°F by quenching in oil. They are tempered from a range of 300 to 900°F, depending on the physical properties desired.

The L-type steels are used in gages, broaches, drills, taps, threading dies, ball and roller bearings, feed fingers, clutch plates, knurls, files, and all instances where high wear resistance and toughness are required. The F-type steels are used as finish machining tools. They have good wear resistance and will maintain a sharp cutting edge. They may be used in dies, other tools used for machining light cuts or cutting soft materials such as aluminum, wire drawing dies, form tools, knives, etc.

Water-Hardening Tool Steels (W Type). These tool steels are the plain carbon tool steels that have carbon content in the range of 0.60 to 1.40 per cent. These plain carbon tool steels contain small amounts of manganese and silicon. In addition, one group has vanadium added, another group has chromium added, and another group has all the alloys mentioned in combination with about 1.00 per cent carbon.

They are used in cutting tools. They have average wear resistance, toughness, and shallow hardenability. Their red hardness is poor, so that although they may be used as cutting tools, this application of these materials applies only when the heat generated by cutting is low. These tool steels are used in woodworking tools of all kinds, punches, chisels, hammers, files, reamers, etc.

As indicated, these materials should be quenched from a temperature range of 1450 to 1550°F. The tempering range is 300 to 650°F.

9-5 SPECIAL MATERIALS

Three materials used as cutting tools are cast alloys, carbides, and ceramics.

Cast tools are alloy tools that possess high hardness, excellent red hardness, and excellent wear and corrosion resistance. When used to cut steel, they have a low coefficient of friction as compared to other cutting tool materials. The alloys used in these tools are 1 to 3 per cent carbon, 10 to 20 per cent tungsten, and 25 to 30 per cent chromium. The remainder is cobalt.

These materials have a Rockwell range of up to $60R_c$ depending upon the tungsten and carbon content. Heat-treating will not appreciably affect the strength or hardness of these materials. Cast and wrought materials show almost no plastic deformation at temperatures of 1500°F. They are cast to the desired size and shape, and may be used at higher machining speeds than the high-speed steels. Because they are cast, they are not as tough as high-speed steels. Their performance and life are generally better than high-speed steel because of their ability to retain their hardness at high cutting temperatures. Their ability to resist wear is due to the fact that they do not undergo phase changes and therefore their structures do not transform.

These materials are used in all types of cutting tools to machine steel, cast iron, stainless steel, cast steel, all nonferrous metals, and plastics. Their machine capabilities lie somewhere between high-speed steel and carbide tools. It is recommended that they be used at twice the surface speed as high-speed tools with the same or greater depth of cut and feed. Other uses are in hard facing dies, gages and cams, crushing and Two of the better-known cast tools are manufactured under the trade grinding machinery, wear strips, antifriction bearings, hard facing, etc. names of Stellite and Rexalloy.

Table 9-12 lists the general physical properties for the cobalt-based cast tool materials.

Table 9-12   Properties of Cobalt-Based Cast Tools

Grade	Tungsten–carbon content %	Hardness R_c	Tensile st. psi × 10³	Impact resist. ft-lb	Cast- ability	Machin- ability
Hard	18 W, 2.5 C	62	50	2–3	Poor	Finish by grinding only
Medium	11 W, 2 C	53	78	3–4	Fair to good	Simple machine with carbide tool
Soft	4 W, 1 C	41	133	8–10	Good	Relatively easy to machine or grind

Source: Taylor Lyman, ed., *Metals Handbook*, Vol. 1, Properties and Selection, 8th ed., p. 669. Metal Park, Ohio: American Society for Metals.

The *cemented carbides* are made by blending the carbides of tungsten, titanium, or tantalum with a binder that has a lower melting point, such as cobalt, iron, or nickel. In addition, the carbides mentioned and, in rare instances, the carbides of columbium, vanadium, chromium, zirconium, and molybdenum may be added to tungsten to yield special effects. The powders are cold-pressed into the desired shape at pressures of up to 30 tons/in². They are then sintered at temperatures of 2500 to 2900°F for about an hour. In some instances the pressing and heating are done at the same time. At these pressures and temperatures, the cobalt forms a eutectic with the tungsten carbides. Once cooled, further heat treatment does not affect the structure. These carbide materials are used to machine cast-iron and nonferrous and nonmetallic materials.

The Rockwell is generally over $70R_c$ with compressive strengths greater than 7×10^5 psi. Because of this the carbide material is very brittle, lacks toughness, and has low tensile properties. This material has very high red hardness.

When tantalum carbide is added to tungsten carbide, the coefficient of friction between the tool and the material being machined is greatly reduced. When titanium carbide is added to tungsten carbide, the ability of the tool to cut steel is greatly improved. These grades are used primarily to machine ferritic steels. Besides being used in cutting tools such as tool bits, milling cutters, reamers, and dies in the machining of ferrous metals and tool steels, they are also used in the machining of nonferrous metals such as aluminum, brass, and copper. Small form tools are shaped in the mold, whereas in other larger tools the carbide materials are used as inserts, slugs, or disposable tips.

Table 9-13 shows nine groups of carbide materials. Groups 1 through 3 are the *straight tungsten carbides*. Groups 4 through 8 contain *titanium carbide* and *tantalum carbide* and are those used in the cutting of steel. Groups 9 and 10 contain only *tantalum* and are used in heat-resisting and special items.

Cermet materials are those materials that result from the combination of a metal and a nonmetal. Thus Al_2O_3 and nickel produces a heat-resistant material. The combination of Teflon with iron produces a self-lubricating material that may be used as bearings. These materials are called the cermets.

Another class of cermets, referred to as *ceramics,* are those which sinter an oxide material, such as aluminum, titanium, or chromium oxide, with a binder material. After compacting and sintering, the material is rolled or shaped as desired. Because the oxides are extremely hard, such materials have great potential as cutting tools. The relative hardnesses of various materials are shown in Table 9-14.

Table 9-13   Carbide Materials

Carbide group	Composition (% remaining WC)		Rockwell hardness R_A
	Cobalt	TaC + TiC	
Straight tungsten carbide			
1	2.5– 6.5	0–3	93–91
2	6.5–15.0	0–2	92–88
3	15.0–30.0	0–5	88–85

Table 9-13 (continued)

Carbide group	Composition (% remaining WC)		Rockwell hardness R_A
	Cobalt	TaC + TiC	
TiC predominantly added			
4	3–7	20–42	93.5–92.0
5	7–10	10–22	92.5–90.0
6	10–12	8–15	92.0–89.0
TaC predominantly added			
7	4.5–8		
8	8.0–10		
TaC added exclusively			
9	5.5–16	18–30	91.5–84.0
10	12.0–16		

Source: Taylor Lyman, ed., Metals Handbook, 8th ed., p. 660. Metal Park, Ohio: American Society for Metals.

Table 9-14   Hardness Numbers

Material	Knoop hardness no.
Tungsten carbide and cobalt	1050–1500
Topaz	1250
Aluminum oxide	1625–1680
Silicon carbide	2130–2140
Boron carbide	2250–2260
Sapphire	1600–2200
Diamond	6000–6500

Source: Taylor Lyman, ed., Metals Handbook, 8th ed., p. 660. Metal Park, Ohio: American Society for Metals.

Problems

9-1.   (a) List the various elements used as alloys in steel. (b) List at least three effects each has on the physical properties of steel.

9-2.   List the alloys that are strong carbide formers in steel.

9-3.   List the alloys that increase red hardness in steel.

9-4.   List the alloys that increase hardenability when used as alloys in steel.

9-5.   List the alloys that affect the tempering characteristics of steel.

9-6. One method used to classify *tool* steels is plain carbon, low alloy, medium alloy, and high speed. Discuss each category.

9-7. List the symbols used by the American Iron and Steel Institute (AISI) to classify tool steels. State the categories (7) and subcategories (13) for each symbol.

9-8. Study Table 9-3. Discuss the heat-treating characteristics of the following steels: (a) T-type, (b) one H-type, (c) an O-type, (d) an A-type, (e) W-type.

9-9. Discuss the nondeforming property of steel.

9-10. Discuss (a) toughness and (b) resistance to wear of the various tool steels.

9-11. Discuss red hardness and decarburization in tool steels.

9-12. Depth of hardening in steel is extremely important. Discuss this property of steel thoroughly.

9-13. Compare the machinability of carbon alloy steels and tool steels.

9-14. Discuss the physical properties of the M- and T-type tool steels.

9-15. Discuss the various H-type tool steels.

9-16. The D, A, O-types of steels are referred to as cold-work tool steels. Discuss them fully.

9-17. What makes the S-type steels shock resistant? How are they used?

9-18. Discuss the two classes of P-type steels.

9-19. Discuss the special-purpose L- and F-types of steels.

9-20. The most important class of steel is the water-hardening (W-type) tool steels. Discuss them.

9-21. Discuss the physical properties of the "cast tools."

9-22. Compare the physical properties of the cobalt cast materials as the tungsten and carbon content increases.

9-23. Discuss the physical properties of carbide materials. What effect does the addition of tantalum carbide have on the cutting operation? the addition of titantium carbide?

9-24. Discuss the cermet materials fully. Compare them with carbides and cast materials.

10 | Nonferrous Light Metals

10-1 SOLUTION HEAT TREATMENT AND AGE HARDENING

Aluminum, as well as some alloys, is subject to a two-stage phenomenon known as solution heat treatment and aging. Thus, one type of aluminum, when tested a very short time after being rapidly cooled from an elevated temperature, will have a tensile strength of approximately 40,000 psi. If tested at various intervals over a period of time, the tensile strength will have increased to 50,000 psi. Non-heat-treatable alloys, or those heat-treated and annealed, may have tensile strengths of approximately 15,000 psi.

Materials may be hardened by working. Thus, if copper wire is bent several times, it will harden at the place where it is bent. If the process is continued, the wire will fracture. Another method of hardening is by alloying a metal in solid solution, so that the solute distorts the lattice or distributes itself as a fine precipitate to cause hardening. Lattice distortion and diffused precipitates provide a keying effect, which offers a resistance to ductility and which we term hardness.

Figure 10-1 shows the solubility line *ps,* which indicates a decrease in solubility of the β phase in the α phase as the temperature decreases. If a metal *X,* which has a composition of 90 per cent A, 10 per cent B, is heated above *ps* to a temperature T_2, which is lower than T_1 and higher than T_3, the β phase will dissolve and uniformly disperse into the homogeneous solid α solution. Upon *slow cooling,* the phase will reform and be-

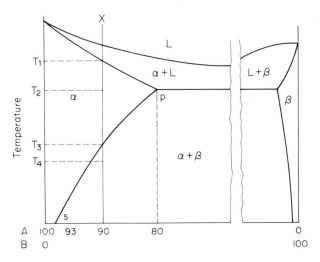

Figure 10-1

low T_4 the metal X will once again consist of two distinct phases, α and β.

If metal X is heated to T_2 and *quenched,* the dispersed submicroscopic phase is trapped in the α solution. The α solution is said to be supersaturated, because it contains more β particles at room temperature than it can hold in its lattice structure. The process is called *solution heat-treating.*

This supersaturated solid solution is unstable and, if left alone, the excess β, above that which the α phase will hold at room temperature, will precipitate out of the α phase. When the process occurs at room temperature, it is called *natural aging.* If the material that has been solution heat-treated requires a slight heating to speed up the precipitation, the process is called *artificial aging.* It should be noted that freezing the solution heat-treated material will retard the aging process.

In many instances aging will occur without precipitation. The particles may actually diffuse within the lattices and distort them. This type of aging is called *abnormal aging.*

After solution heat treatment the material is ductile, since no precipitation has occurred. Therefore it may be worked. After a time the solute material precipitates and hardening develops. As the composition reaches its saturated normal state, the material reaches its maximum hardness. The precipitate, however, continues to grow. The fine precipitates disappear. They have grown larger, and as a result the tensile strength (T.S.) of the material decreases. This is called *overaging.* The stages are shown in Fig. 10-2.

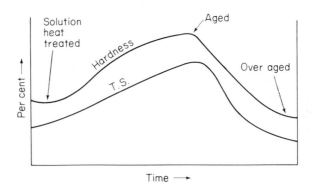

Figure 10-2

10-2 RECRYSTALLIZATION

The mechanisms of slip and twinning were discussed in Sec. 3-3 and should be reviewed by the student before continuing.

Metals become harder and stronger when stressed. The greater the deformation of a lattice, the greater the resisting force to deformation. This resistance to deformation defines the degree of hardness.

There are several reasons put forth for the greater hardness of a metal that is *strain-hardened.*

1. During the process of slip, as the planes move over one another some of them may become dislodged from their parent lattices and deposit in the slip planes, where they act to impede the movement of one plane over another. This is known as the *amorphous theory.*

2. The *fragmentation theory* states that large blocks of atoms and particles break away and lodge in the slip planes.

3. The *lattice-distortion theory* says that lattices under stress kink and distort. The slip planes thus create a keying effect.

4. Another theory deals with lattice imperfections in materials which interfere with each other, and which multiply when certain materials are cold-worked. This is the *dislocation theory.*

When ductile metals are cold-worked, they strain-harden. If cold-working continues long enough, internal stresses build up and fracture results. If the cold-working stops, the atoms may diffuse again into a more stable pattern. This takes a long time. The process, called *annealing,* can be speeded up by controlled heat-treating procedures.

Figure 10-3(a) is an idealized set of curves of a material that shows the effects of heating a severely cold-worked sample. At temperature T, slip planes are shown in severely flattened crystals. At T_1 a few small crystals appear. Some of the stresses are relieved, and the tensile and yield strengths of the material have increased slightly. This is called *recovery*. At T_2 more recrystallization occurs. Hardness and strength fall off rapidly, and ductility increases rapidly. As the temperature increases, these physical properties continue to be affected. At T_3 the stressed crystals no longer exist. They have been replaced by smaller, unstressed crystals. The space on the graph between the two vertical lines T_1 and T_2 is called the *recrystallization* zone. From T_3 to T_5 the crystals grow. Ductility does not increase as rapidly as before. The hardness and strength level off somewhat. The grain growth is rather rapid. This takes place at the expense of the other grains.

It should be noted that recrystallization preserves the room temperature crystal structure. The recrystallization temperature is below the critical range. The recrystallization range is not necessarily synonymous with the phase change temperature. For example, steel recrystallizes below 1330°F, its critical temperature. Cold-working of metals is defined as deformation that takes place below the recrystallization temperature. Hot-working, therefore, takes place above the recrystallization temperature. The recrystallization of various metals is given in Table 10-1.

Table 10-1　Recrystallization Temperatures

Metal	Temperature, °F	Metal	Temperature, °F
Aluminum	300	Molybdenum	1655
Beryllium	1300	Nickel	1115
Cadmium	75	Platinum	850
Copper	390	Silver	390
Gold	390	Tin	below 65
Iron	850	Tantalum	1830
Lead	below 65	Titanium	1220
Magnesium	300	Wolfram	2190
		Zinc	75

Figure 10-3(b) shows a photomicrograph of a severely worked sample of brass. Figure 10-3(c) shows the same sample heated to 750°F. Recrystallization has already taken place. Figure 10-3(d) shows the same sample heated to 1000°F. The grains have started to grow. Figure 10-3(e) shows very large grains when this sample has been heated to 1500°F. All magnifications are 100×.

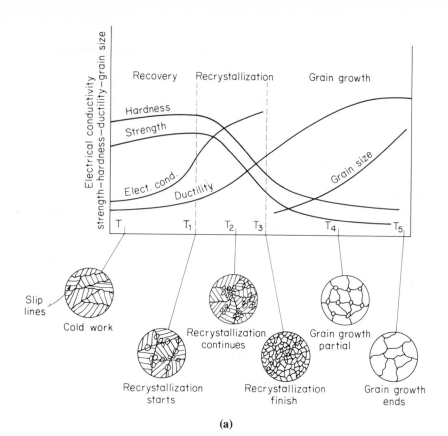

Recovery   Recrystallization   Grain growth

Hardness

Strength

Elect. cond.   Ductility   Grain size

T T_1 T_2 T_3 T_4 T_5

Electrical conductivity
strength–hardness–ductility–grain size

Slip lines

Cold work

Recrystallization starts

Recrystallization continues

Recrystallization finish

Grain growth partial

Grain growth ends

(a)

(b)

Figure 10-3

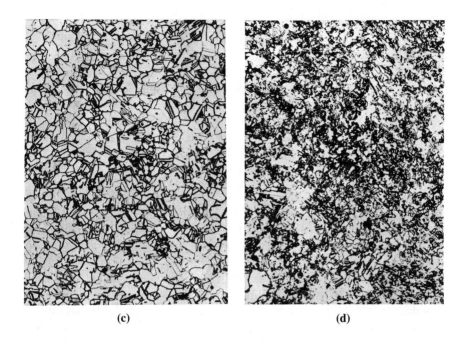

(c)   (d)

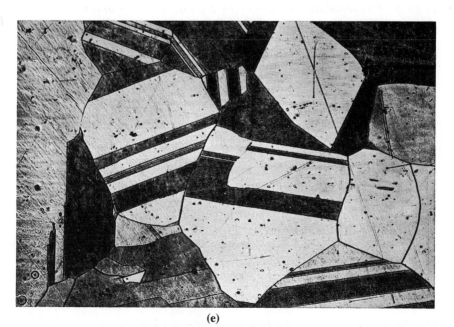

(e)

Figure 10-3 (continued)

10-3   ALUMINUM

Pure aluminum has a specific gravity of 2.70, a coefficient of expansion of 1.4×10^{-5}, a melting point of about $1218°F$, and a modulus of elasticity of 10^7 psi, which is about one-third that of steel. This means that for similar cross sections aluminum will deflect three times more than steel. It has resistance to corrosion. It is nonmagnetic and has excellent electrical and thermal conductivity, about 60 per cent that of copper.

High-purity aluminum, about 99.9 per cent pure, is very ductile, which is characteristic of most face-centered-cubic lattice structures. In the annealed condition it has a tensile strength of about 7000 psi with an elongation of about 50 per cent. Cold-working increases its strength to about 15,000 psi and its elongation by about 6 per cent. Its creep strength is good below a temperature of 400°F. However, with temperatures higher than this, its strength decreases rather rapidly. Table 10-2 shows some specific values for wrought aluminum.

Impurities in very minute quantities affect the mechanical properties of aluminum markedly. Thus impurities such as copper, silicon, and iron in amounts of less than 1 per cent may increase its tensile strength by as much as 40 per cent and may increase its yield strength by as much as a factor of 3. Aluminum containing small quantities of these impurities is said to be *commercially pure.* Its *load-to-weight* ratio makes it comparable to steel. As indicated, its *load-to-cross-sectional area ratio* is much less than steel.

10-4   DESIGNATION SYSTEMS

The American Standards Association adopted a numbering system for wrought aluminum and its alloys in 1957, which had previously been established by the Aluminum Association. The numbering system designates series as shown in Table 10-3.

The 1XXX series, which indicates a commercially pure aluminum, has a minimum of 99.00 per cent aluminum. The last two digits of a number in this series indicate the minimum percentage of aluminum in hundredths. The second digit from the left indicates the special control of the given number of impurities present. Integers 1 through 9 indicate the number of impurities that are controlled.

Example 1

Explain the significance of the number 1035 for wrought aluminum.

Table 10-2   Wrought Aluminum

Alloy	Per cent composition							Cond.	Strength, 1000 psi					
	Cu	Mg	Si	Mn	Zn	Cr	Other		BHN	U.S.	Y.S.	S.S.	Fatigue endur.	Elong., % 1/16-in. dia.
E.C.*							99.45 Al	0		12	4	8		
1060							99.60 Al	0	19	10	4	7	3	43
1100							99.00 Al	0	23	13	5	9	5	35
2011	5.5						0.5 Pb	T3	95	55	43	32	18	
2014	4.4	0.5	0.8	0.8				0	45	27	14	18	13	21
2017	4.0	0.5		0.7				0	45	26	10	18	13	
2024	4.4	1.5		0.6				0	47	27	11	18	13	
2218	4.0	1.5					2.0 Ni	T7	70	48	37	30		20
3003	0.12			1.2				0	28	16	6	11	7	30
3004		1.0		1.2				0	45	26	10	16	14	20
4032	0.9	1.1	12.2				0.9 Ni	T6	120	55	46	38	16	
5005		0.8						0	28	18	6	11		25
5050		1.4						0	36	21	8	15	12	24
5056		5.1		0.12		0.12		0	65	42	22	26	20	
5154		3.5					0.25 Cr	0	58	35	17	22		
5454		2.8		0.8			0.12 Cr	0	62	36	17	23	17	27
5456		5.1		0.8			0.12 Cr	0		45	23			22
6003		1.2	0.7				0.2 Cr	0						
6061	0.27	1.0	0.6					0		17	7	11	8	25
6063		0.7	0.4					0	25	13	7	10		
6066	0.9	1.1	1.3	0.9					43	22	12	14		

Table 10-2   *continued)*

| | Per cent composition | | | | | | | | Strength, 1000 psi | | | | | |
Alloy	Cu	Mg	Si	Mn	Zn	Cr	Other	Cond.	BHN	U.S.	Y.S.	S.S.	Fatigue endur.	Elong.,% $\frac{1}{16}$-in. dia.
6151		0.6	0.9				0.15 Cr							
6262	0.27	1.0	0.6				0.09 Cr, 0.55 Pb, 0.55 Bi	T9	120	58	55	35	13	
7001	2.1	3.0			7.4	0.3		T6	160	98	91		22	
7075	1.6	2.5			5.6	0.3		T6		76	67	46		11
7079	0.6	3.3		0.2	4.3	0.2		T6	145	78	68	45	23	
7178	2.0	2.7			6.8	0.3		T6		88	78			10

Source: *Aluminum Standards and Data*, 2nd ed., pp. 14 and 27. Aluminum Association of America (Dec. 1969).

*Electrical conducting.

Mod. of elasticity = 10×10^6 psi (tension) = 4×10^6 psi (compression).

Table 10-3

Material	Designation
Aluminum, 99.00% minimum	1XXX
Major alloys	
Copper	2XXX
Manganese	3XXX
Silicon	4XXX
Magnesium	5XXX
Magnesium and silicon	6XXX
Zinc	7XXX
Other	8XXX
Unused series	9XXX

Source: *Aluminum Standards and Data*, 2nd ed., p. 6.
Aluminum Association of America (Dec. 1969).

solution:

(a) 1XXX indicates a minimum of 99.00 per cent pure aluminum.
(b) The 35 in the number indicates that there is 99.35 per cent aluminum present in this material.
(c) The 0 in 1035 indicates no special control of the remaining $1.00 - 0.35 = 0.65$ per cent of the impurities.

The 2XXX to 8XXX series indicate the alloys as shown in Table 10-3. The first digit to the left indicates the alloy type, and the second digit the modifications. The original alloy is designated when the second digit from the left is zero. As indicated, if the second digit is 1 through 9, it designates the number of impurities that are controlled, which are registered with the Aluminum Association. The last two digits serve no purpose other than possibly to identify the different alloys in the group as registered. The 9XXX series is used while new alloys are being developed. Once the new alloy loses its experimental status, it loses its 9XXX designation. While it is being commercially evaluated, one of the above four-digit designations is assigned and prefixed with an X. The prefix is dropped when the new material joins other standard aluminum alloys.

Example 2

Explain each digit in the wrought aluminum number 7075.

solution:

(a) 7XXX: aluminum with zinc as a major alloy.
(b) The second digit, zero, indicates no special control of impurities.
(c) The 75 in this case indicates major alloys of zinc and magnesium as registered.

Aluminum alloys fall into two general categories: those that are *heat-treatable* and those that are *non-heat-treatable*. In general, the 2000, 4000, 6000, and 7000 series may be strengthened by heat treatment. The 3000 and 5000 series cannot. These series depend upon the alloys of manganese and magnesium for their strength. They may be further strengthened by strain-hardening procedures or by cold-working.

A system of letters and numbers to indicate *temper* has also been approved by the American Standards Association. In this system the basic temper has been designated with a letter, whereas the modification of this temper is indicated with a number. The list in Table 10-4 and the following text show the letters and numbers, and their meaning.

Table 10-4

Symbol	Meaning
O	Annealed
F	As-fabricated
H	Strain-hardened
W	Unstable followed by solution heat treatment
T	Heat-treated

–EC, *electrical conductor.* Indicates aluminum used in the electrical field. It contains a minimum of 99.45 per cent aluminum, with traces of boron and copper, and a very close control of impurities.

–O, *annealed, recrystallized.* Indicates the softest temper.

–F, *as-fabricated.* Indicates the as-cast condition with no guarantee that the properties of the material will be other than those that result from the normal fabricating processes. In casting it refers to the as-cast condition. In wrought materials, there is no intent to guarantee the properties of the material.

–H, *strain-hardened.* Refers to the aluminum products that do not have their mechanical properties affected by heat treatment alone. The mechanical properties may be changed by cold-working and subsequent heat treatment to produce partial softening. This H is always followed by two or more digits. The first digit designates the heat-treating process, and the second, or subsequent, digits, the degree of cold-working.

–H1, *strain-hardened material.* A *second-digit* range of 1 through 8 indicates the degree of hardness in *eights.* The digit 1 indicates "dead" softness. The digit 8 indicates "full"* hardness following a full anneal.

*"Full" hardness means a temper having an ultimate tensile strength equivalent to 75 per cent cold reduction after full annealing.

Sometimes the digit 9 is used to indicate "extrahard" temper. Thus a *half-hard* material would have the number –H14 ($\frac{4}{8} = \frac{1}{2}$); a quarter-hard material would have the number –H12 ($\frac{2}{8} = \frac{1}{4}$); etc.

Sometimes a third digit is used to indicate a special set of properties. Assume a set of properties that fulfill the minimum but not the maximum requirements to be designated as –H12, which makes it quarter-hard. If the material does not fit an –H13 or an –H11 number, then a third digit is added and the number becomes –H121. The H designates a strain-hardened material, the 1 following the H indicates that this is the only hardening process, the 2 that the material is quarter-hard, and the last 1 that this material has not been hardened to achieve the exact properties of a quarter-hardened aluminum.

–H2, *strain-hardened and partially annealed.* Indicates a method for achieving desirable physical and mechanical properties by first work-hardening the material and then reducing the hardness by partial annealing. The H indicates "strain hardening," and the 2 indicates partial annealing. A second numeral indicates the degree of hardness in the end product. Thus the 6 in –H26 indicates $\frac{6}{8}$ or $\frac{3}{4}$-hardened end product.

–H3, *strain-hardened and stabilized.* Strain-hardened alloys containing magnesium, when heat-treated at low temperatures, will achieve structural stability. Once stabilized, these materials will have better ductility than before they were processed. The 3 in –H34 indicates this type of processing. The 4 indicates a $\frac{4}{8}$, or half-hardened end product.

Example 3

(a) What is the relative hardness of an –H16 temper suffix? (b) What is the percentage of cold reduction that can be expected?

solution:

(a) The hardness designation due to cold-working is

$$\frac{6}{8} = \frac{3}{4} \text{ hard}$$

(b) The cold reduction is

$$\frac{3}{4} \times 75\% = 56\% \text{ approx. (see footnote to –H1)}$$

–W, *conditions following solution heat-treating.* Indicates the condition of a material subject to natural age hardening after a fixed period of time. Since the age-hardening process is continuous, the time period following solution heat-treating is specified. Thus 6061–W–8 hrs has specific properties as a result of solution heat-treating and precipitation hardening after 8 hours.

–T, *thermally treated products.* Indicates a stable tempered state with or without cold-working. Age-hardening may take place after the indicated operations. The digit that follows the T temper indicates a specific operation or operations. Any variation of properties from that indicated by the first digit requires a second digit that is assigned and registered with the Aluminum Association.

–T1, *hot-worked, cooled, and naturally aged to a stable condition.* Applied to those products that have been cast or hot extruded and whose strength properties are increased by room temperature aging.

–T2, *annealed (cast products only).* The purpose here is to increase ductility and volumetric stability.

–T3, *solution heat-treated and cold-worked.* Cold-working, when applied to wrought aluminum, is for the purpose of increasing strength after solution heat treatment. The process is allowed to proceed by age-hardening.

–T4, *solution heat treatment and natural aging to a stable condition.* The same as T3, except that no cold-working is required.

–T5, *artificially aged only.* Applies when there is rapid cooling from elevated temperature but no solution heat treatment has taken place.

–T6, *solution heat treatment and then artificial aging.* Applies when there is no cold-working after solution heat-treating.

–T7, *solution heat treatment and stabilizing.* Applies to materials where the temperature and time required for maximum hardness are exceeded and where there is control of growth and residual stress.

–T8, *solution heat treatment, cold-working, and artificial aging.* Applies when intermediate cold-working is needed to improve the strength.

–T9, *solution heat treatment, artificial aging, and cold-working.* Cold-working after solution heat-treating and artificial aging to improve strength.

–T10, *partial solution heat treatment, artificially aged, and cold-worked.* Used when the cold-working improves strength.

Example 4

Assume a material number of X1035–H24. State the meaning of each digit.

solution:

(a) The X indicates a material for which all experimental work has been completed and the material is being commercially tested and evaluated.

Table 10-5 Cast Aluminum

Alloy	Proposed	Cu	Mg	Si	Mn	Zn	Cr	Other	Cond.	BHN	U.S.	Y.S.	S.S.	Fatigue end.	Elong., $\frac{1}{16}$-in. dia.	Process
13	413			12.0					F	40	39	21	25	8	19	DC
43	443			5.0							19	8	14	8	8.0	SC, PM, DC
108	208	4.0		3.0					F	55	21	14	17	11	2.5	SC
113	213	7.0		2.0			1.7		F	70	24	15	20	9	1.5	SC, PM
122	222	10.0	0.2						T2	80	27	20	21	9.5	1.0	SC, PM
142	242	4.0	1.5					2.0 Ni	T2	70	27	18	21	6.5	1.0	SC, PM
195	295	4.5		0.8					T4	60	32	16	26	7.0	8.5	SC, PM
212	412	8.0		1.2					F	65	23	14	20	9.0	2.0	SC
214	514		3.8						F	50	25	12	20	7.0	9.0	SC, PM
220	520		10.0						T4	75	48	26	34	8.0	16.0	SC
319	319	3.5		6.3					F	70	27	18	22	10.0	2.0	SC, PM
355	355	1.3	0.5	5.0					T6	80	35	28	28	9.0	3.0	SC, PM
356	356		0.3	7.0					T6	70	33	24	26	8.5	3.5	SC, PM
360	460		0.5	9.5							44	27	28	20.0	3.0	DC
380	380	3.5		9.0							43	26	28	20.0	2.0	DC
384	384	3.8		12.0							46	27	29	21.0	4.0	DC
750	850	1.0						1.0 Ni 6.5 Sn								SC, PM

Source: F. W. Wilson, ed., *ASTME Tool Engineer's Handbook,* 2nd ed. New York: McGraw-Hill Book Company (1959). SC, sand casting; PM, permanent mold; DC, die casting.

(b)   The 1 indicates a commercially pure aluminum.
(c)   The 0 indicates no special control in the aluminum.
(d)   The 35 indicates a 99.35 per cent pure aluminum alloy.
(e)   The –H indicates a cold-worked aluminum.
(f)   The 2 indicates a strain-hardened and partially annealed aluminum.
(g)   The 4 indicates a $\frac{4}{8}$ or half-hardened material which has a tensile strength due to $\frac{1}{2} \times 75$ per cent = 37.5 per cent cold-working.

10-5   CAST ALUMINUM

So far, the characteristics of wrought aluminum have been discussed. The following are comments that relate to the *casting* of aluminum. The alloys of aluminum used in the casting processes in general divide into two types: (1) those that owe their mechanical properties to the addition of alloys, and (2) those that are heat-treated to improve their engineering properties. They are used in sand casting, permanent mold casting, and die casting. Table 10-5 shows the composition and mechanical properties of several aluminum alloys used for casting. Table 10-6 shows some values for carrying out the processes of solution heat treatment and aging of aluminum alloys.

Castings, as a rule, do not have the same mechanical properties as the counterpart wrought alloy products. Thicker sections are usually designed in castings to compensate for this factor. However, one advantage of casting over machining from the wrought material is that it is far less expensive to cast intricate sections than it is to machine them.

Fluidity is generally not a problem. Where high fluidity is required, aluminum–silicon alloys are used. Such alloys are 43 or 355 aluminum.

The determining factors in choosing between wrought and cast processes reduce to one of cost. The cost of a finished casting is related to (1) the geometric shape of the casting, (2) the casting process chosen, (3) the alloy chosen, (4) the inherent problems encountered when attempting to cast that material, (5) the number of parts to be cast, and (6) the dimensional tolerances, surface finish, and the amount of machining needed to achieve these tolerances.

The following numbering system for sand casting, permanent molding, and die casting materials has been proposed by the Aluminum Association.* The number designation for aluminum castings and ingots, recently developed, is based on four digits, 1XX.X. For aluminum castings and ingots the 1 indicates commercially pure aluminum, as seen in Table 10-7.

*The Aluminum Association, *Standards and Data,* 1970–1971.

Table 10-6 Solution Heat Treat

Aluminum Wrought Materials

Alloy	Annealing			Solutionizing		Precipitation heat-treat		
	Temp °F	Time h	Temper	Temp °F	Temper	Temp °F	Time h	Temper
1100	650		–0					
2011	775	2–3	–0	950	–T4	320	12–16	–T6
2014	775	2–3	–0	940	–T4	340	8–12	–T6
2017	775	2–3	–0	940	–T4	–	–	–
2024	775	2–3	–0	920	–T4	375	11–13	–T81
2218	775	2–3	–0	950	–T4	460	5–8	–T72
3003	775		–0					
4032	775	2–3	–0	950	–T4	340	8–12	–T6
5005	650		–0					
5050	650		–0					
5154	650		–0					
6061	775	2–3	–0	970	–T4	⎰320 ⎱350	16–20 6–10	–T6
6063	775	2–3	–0			⎰450 365 ⎱350	⎰1–2⎱ ⎰4–6⎱ 6–8	–T5 –T6
6151	775	2–3	–0	960	–T4	340	8–12	–T6
7075	775	2–3	–0	870	–W	250	24–28	–T6

Sand-Casting Materials

Alloy	Pro-posed	Temp °F	Time h	Temp °F	Time h	Temper
122	222	950	12	310	10–12	–T61
142	242	960	6	650	1–3	–T77
195	295	960	12	310	3–5	–T6
319	319	940	12	310	2–5	–T6
355	355	980	12	310	3–5	–T6
356	356	1000	12	310	2–5	–T6

Permanent Mold-Casting Materials

Alloy	Pro-posed	Temp °F	Time h	Temp °F	Time h	Temper
122	222	950	8	340	7–9	–T65
132	332	960	8	340	14–18	–T65
142	242	960	6	400	3–5	–T61
195	295	950	8	310	5–7	–T6
319	319	940	8	310	2–5	–T6
355	355	980	8	310	3–5	–T6
356	356	1000	8	310	3–5	–T6

Source: F. W. Wilson, ed., *ASTME Tool Engineer's Handbook,* 2nd ed., pp. 14–51. New York: McGraw-Hill Book Company (1959).

Table 10-7

Material	Designation
Aluminum — 99.00% min	1XX.X
Major alloys	
Copper	2XX.X
Silicon, with Cu and/or Mg	3XX.X
Silicon	4XX.X
Magnesium	5XX.X
Zinc	7XX.X
Tin	8XX.X
Unused series	6XX.X
Other major alloys	9XX.X

Source: *Aluminum Standards and Data,* p. 11. Aluminum Association of America.

The second two digits indicate the percentage of 1 per cent beyond 99.00 per cent pure aluminum. Castings should be indicated by the number 0 after the decimal point, and ingots by the number 1 after the decimal point. A letter X used as a *prefix* for experimental alloys and other letters would be used arbitrarily, assigned, and registered.

For alloys (2XX.X, 3XX.X, etc.) the first digit designates the major alloy; the next two digits are assigned when the material is registered. The 0 after the decimal point indicates a casting; a 1 indicates an ingot with the same composition as had been assigned to the casting (with some specified exceptions), and a 2 indicates a variation from the 0 category.

Example 5

What does the aluminum number 332.0 indicate?

solution:

The zero indicates a casting.
The 3XX.X indicates the major alloys as silicon with copper and/or magnesium.
The 32 has no significance.

10-6   ALLOYS OF ALUMINUM

Copper, magnesium, zinc, silicon, and manganese are elements usually alloyed with aluminum. These alloys increase the response of aluminum to precipitation hardening. The composition and physical properties of some of these alloys are shown in Tables 10-2 and 10-5.

Figure 10-4(a) shows the aluminum-rich end of the aluminum–copper equilibrium diagram. The maximum solubility of copper in aluminum is 5.65 per cent at the eutectic temperature of 1018°F. This solubility decreases to about 0.25 per cent. This decrease in solubility indicates that between these two percentages this alloy will respond to precipitation hardening. If the alloy selected has a copper content of between 0.25 and 5.65 per cent, and if the alloy is heated into the α phase and cooled rapidly, the θ (CuAl$_2$) phase will be trapped. Artificial or natural aging will cause the θ phase to precipitate and harden the aluminum.

Intergranular corrosion may result. This may be impeded by *anodizing,* which is an oxidizing process that creates a very dense thin oxide coating when the aluminum is treated with sulfuric, chromic, or oxalic acid. A process called "alclad" coats the surface of the material with high-purity aluminum that is very corrosion resistant.

A very common wrought aluminum–copper alloy is that which contains about 4.0 per cent Cu, 0.5 per cent Mg, 0.6 per cent Mn, other impurities, and the balance aluminum. This combination is referred to as a duraluminum 2017. Alloy 2024 also contains about 4.5 per cent copper, 1.5 per cent magnesium, and 0.6 per cent manganese. This latter alloy of aluminum was developed by the Aluminum Company of America. It is their version of duraluminum. It is one of the aluminum–copper alloys that develops the highest strength through the mechanism of age hardening. Both 2014 and 2218 have higher tensile and yield strength than 2017. Alloy 2218 contains nickel and has high-temperature characteristics not possessed by the other types mentioned.

There has been a series of aluminum–copper alloys developed that are suitable for casting. They are essentially in the 100 and 300 number series (see Table 10-5).

Figure 10-4(b) shows the aluminum-rich end of the aluminum–magnesium equilibrium diagram. This alloy also responds to solution heat-treating and aging. The diagram shows the maximum solubility of the β phase in α to be 14.9 per cent magnesium at 844°F. As the temperature drops, the solubility reduces to about 1 per cent at room temperature.

However, since almost all alloys of aluminum do not contain more than 5 per cent magnesium, they do not respond to solution heat-treating. This alloy is lightweight, and offers maximum resistance to corrosion by the atmosphere and salt water. It also has good machinability characteristics. The addition of higher percentages of magnesium to aluminum does make the alloy solution heat-treatable. The advantage gained by alloying magnesium with aluminum is not sufficient to warrant the manufacture of such alloys for commercial use.

Figure 10-4(c) shows the aluminum-rich end of the aluminum–zinc equilibrium diagram. Zinc, by itself, has little effect on the age-hardening

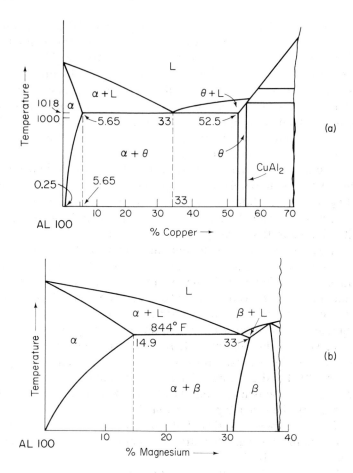

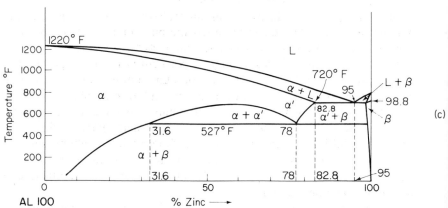

Figure 10-4

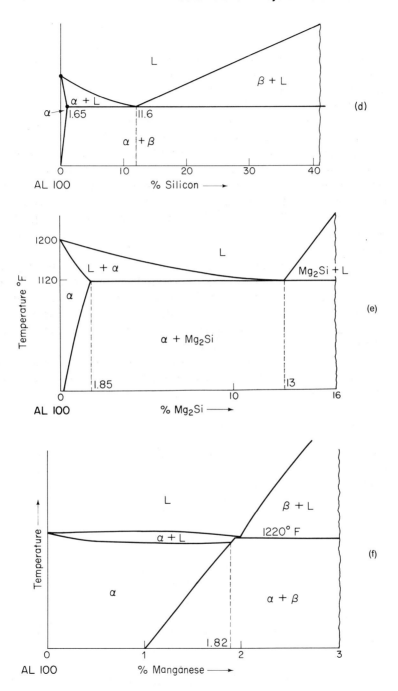

Figure 10-4 (continued)

processes, even though the solubility of the β phase drops off sharply as the temperature of the alloy drops when cooled to room temperature. When zinc is alloyed with aluminum together with magnesium, the possibilities for solution heat-treating and precipitation hardness increase through the formation of AlMgZn.

When zinc is added to aluminum, it reduces the corrosion resistance of the aluminum. Under certain conditions this alloy is subject to intergranular corrosion or stress corrosion.

Figure 10-4(d) is the aluminum-rich end of the aluminum–silicon equilibrium diagram. Normal amounts of silicon, up to about 0.5 per cent, are found in raw aluminum combined with iron. In percentages of from 5 to 25 per cent, its fluidity makes it acceptable for pouring into molds or for making castings. The wear resistance of the resulting castings is greatly improved. The hardness is also improved, but the machinability of the material is reduced.

When magnesium is added to the silicon as a second alloy with aluminum, magnesium silicide (Mg_2Si) is formed. The diagram in Fig. 10-4(e) shows the formation of the eutectic of aluminum and Mg_2Si. This Mg_2Si precipitates during aging. The 6061, 6063 wrought alloys and the 355, 356 casting alloys have the appropriate percentages of silicon and magnesium to precipitate Mg_2Si.

These alloys have excellent corrosion resistances and good strength characteristics. They have good pressure-tight qualities and good fluidity for casting purposes. The silicon, however, has the effect of making the wrought material more difficult to machine, form, or work.

The aluminum-rich end of the aluminum–manganese equilibrium diagram is shown in Fig. 10-4(f). Aluminum containing manganese as a major component will not age-harden. It does have good resistance to corrosion. The wrought alloy 3003 is the material generally used when an aluminum manganese alloy is desired.

10-7   MAGNESIUM AND ITS ALLOYS

Magnesium is produced by electrolysis from seawater, dolomite, or brines. It has a minimum purity of approximately 99.8 per cent. The impurities generally found with magnesium are aluminum, zinc, silicon, iron, and manganese. Aluminum, when added in excess of 10 per cent as an alloy, causes embrittlement. Zinc, in amounts of approximately 3 per cent, increases the resistance of magnesium to saltwater corrosion. Higher amounts of zinc, when used as an alloying element, increase the brittleness of the magnesium. When both aluminum and zinc are present, manganese will aid the resistance to corrosion of magnesium. Iron drastically reduces magnesium's resistance to corrosion because of the galvanic action between the two. The electromotive series in Table 12-1 on page 298 shows that magnesium has a potential of +2.34 and iron a +0.44. Silicon in small amounts forms a Mg_2Si compound, which increases its hardness.

When magnesium is alloyed with one of the above elements, it is susceptible to age-hardening (350°F) after it has been given a long solution heat treatment at about 700°F. As is generally the case, the tensile strength and ductility are increased after this solutionizing, and the hardness and yield strength are increased after aging with a retention of the tensile strength and the ductility.

Magnesium has a hexagonal close-packed lattice structure, and as such has a stiffness that makes it difficult to shape at room temperatures. Any cold-forming operation performed on magnesium causes it to work-harden. At elevated temperatures, between 400 and 600°F, the structure transforms to a body-centered cubic lattice structure. The material in this configuration possesses many more slip planes, which makes it more ductile. Slow working (press forging) is always preferred to fast working (drop-hammer forging).

Several classes of magnesium have excellent properties for machinability. Since its conductivity of heat is very good, heat is dissipated rapidly during the machining process, and long tool life is possible. Magnesium powder generated by cutting may catch fire easily. Tools must be kept sharp to avoid building up of cutting pressures, heat, and sparking. It is also possible to weld magnesium alloys by inert gas and electric resistance welding methods. Brazing and soldering is also possible when one is forming magnesium.

Magnesium has a modulus of elasticity of about 6.5×10^6 psi as compared to steel's 29×10^6 psi and aluminum's 10×10^6 psi. It has a specific weight* of about 60 per cent of the weight of aluminum, 25 per cent of the weight of steel, and about 20 per cent of that of copper. Its corrosion resistance is good in dry atmospheres, but extremely low in damp or salt atmospheres. When used under the latter conditions, the surface of the material should be treated with a dichromate dip.

Figure 10-5 shows the magnesium-rich end of the magnesium–aluminum equilibrium diagram. The solubility of aluminum in magnesium at about 820°F is about 11.6 per cent, which reduces to about 1.5 per cent at about 200°F. The eutectic forms at 32.2 per cent aluminum, and a delta (δ) phase forms at approximately 42 per cent aluminum.

*Weight per unit volume.

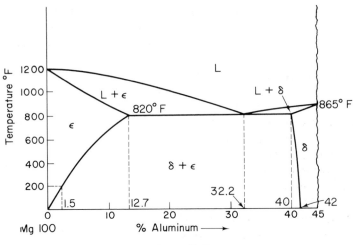

Figure 10-5

10-8   MAGNESIUM NUMBERING SYSTEM

The American Society for Testing Materials (ASTM) has standardized the designation of the alloys of magnesium, using a four-part system. The designation works as follows:

1. The first part of the number consists of two *code letters,* which indicate the two main alloys. The first letter indicates the alloy with the higher percentage; the second the alloy with the next higher percentage. The letter code is shown in Table 10-8.

Table 10-8

A – Aluminum	L – Beryllium
B – Bismuth	M – Manganese
C – Copper	N – Nickel
D – Cadmium	P – Lead
E – Rare earth	Q – Silver
F – Iron	S – Silicon
H – Thorium	T – Tin
K – Zirconium	Z – Zinc

2. The second part consists of two *numbers* corresponding to the rounded-off percentages to the nearest whole number of the two principal alloys. If the last digit of the percentage is a 5, then the nearest *even* whole number is used.

3. The third part consists of a letter indicating how many alloys have been developed that have the same percentages of the first two alloys. Thus an A would mean that the percentages developed under step 2 are the first alloy of magnesium which has the indicated percentages. All letters of the alphabet are used except I and O.

4. The fourth part consists of a letter and a number that indicates the temper heat treatment. This system is the same as that used for aluminum, Sec. 10-4. They are repeated here for convenience.

> O – Annealed
> F – As-fabricated
> H – Strain-hardened
> W – Unstable followed by solution heat treatment
> T – Heat-treated

Example 6

Given a magnesium number AZ61A–F, indicate the meaning of each of the letters and numbers for this alloy.

solution:

(a) A indicates that the principal alloy is aluminum.

(b) Z indicates that the second most prevalent element used as an alloy is zinc.

(c) The 6 indicates that a percentage of approximately 6 per cent aluminum is used. A table of compositions indicates this percentage to be 6.5 per cent aluminum.

(d) The 1 indicates that a percentage of between 0.6 and 1.4 per cent zinc is used. The amount shown in composition tables is actually 1 per cent.

(e) The A indicates that this is the first alloy developed having these percentages of alloying materials.

(f) The F indicates an "as-fabricated" condition.

Example 7

Write the identification number for the magnesium that has a composition of 5.5 per cent zinc and 0.45 per cent zirconium, and which has been partially solution heat-treated and artificially aged. Assume this to be the second magnesium alloy developed of this type.

solution:

(a) Since zinc has the highest alloy percentage, it will be the first letter of the desired identification number. The letter will be Z. The letter for zirconium will be K. The designation is

ZK

(b)   The indicated percentage of zinc is 5.5 per cent. When rounded off to the nearest even whole number, 5.5 becomes 6 per cent. Since the zirconium percentage is less than 0.5 per cent (0.45 per cent), it rounds off to 0. The designation to this point is

ZK60

(c)   Since this is the second alloy that has these percentages, the letter B will be added, and the designation will become

ZK60B

(d)   This alloy is to be partially solution heat-treated and artificially age-hardened. The designation for this treatment is T5 (see Sec. 10-3). The designation is

ZK60B–T5

Several alloys of magnesium and their physical properties are shown in Table 10-9.

10-9   TITANIUM AND ITS ALLOYS

Titanium is considered one of the light metals. It has a specific gravity of 4.5, which makes it the heaviest of the light metals. It has a weight density of 0.16 lb/in.^{3}, about two-thirds that of steel (0.28 lb/in.^{3}). It is also one of the strongest of the light metals, especially at elevated temperatures (1000°F), which compares with austenitic stainless steel. It also has excellent corrosion resistance.

The hexagonal close-packed lattice structure of titanium at room temperature makes it brittle. At 1625°F it transforms to a body-centered cubic lattice structure. Alloying elements affect the structure. Thus

1.   Aluminum stabilizes the alpha structure but *raises* the alpha–beta transformation temperature.

2.   Iron, chromium, molybdenum, and vanadium, when used as alloys, stabilize the alpha structure by *lowering* the alpha–beta transformation temperature.

3.   Tin, as an alloy with titanium, has very little effect.

Since the microstructure has a direct effect on the properties of a

Table 10-9   Magnesium Alloys

Alloy	Composition						Strength × 10³ psi				
	A	*Mn*	*Zn*	*Th*	*Zr*	*Rare earth*	*Tensile yield*	*Comp. yield*	*Shear*	*Bear yield*	*BHN*
AM 100A–T6	10	0.1					22	22	21	52	73
AX 63A–T6	6	0.15	3				19	19	19	34	45
EK 30A–T6					0.3	3.3	16	16	21	40	50
EZ 33A–T5			2.7		0.6	3.3	15	15	21	47	65
ZE 51A–T5			4.6		0.7		24	24	23	37	57
HZ 32A–T5			2.1	3.3	0.7		14	14	20		60
AZ 91A	9.0	0.13					22	22			44
M1A–F		1.2					26	12	18	28	
HM31A–F		1.2		3.0	0.45		33	27	22	50	
ZK 60A–T5			5.5				44	36	26	59	88
ZE 10A–H24			1.3			0.17	28	24			

Source: Taylor Lyman, ed., *Metals Handbook,* 8th ed., Table 3, p. 1069. Metal Park, Ohio: American Society for Metals.

metal, the recognition of the existence of an alpha, alpha–beta, or beta phase is important. Single-phase alloys can be welded to produce ductile welds. The two-phase alloys that can be welded are not as ductile. Two-phase alloys may be heat-treated by aging or heating and quenching. Alpha–beta alloys are stronger than alpha alloys, apparently because close-packed hexagonal alpha is not as strong as body-centered cubic beta.

Table 10-10 shows the composition and physical properties of titanium and its alloys. They are divided into four categories: (1) commercially pure titanium, (2) the alpha alloys, (3) the alpha–beta alloys, and (4) the beta alloys. Also shown are several semicommercial titanium products that have not been fully classified.

The 99.2Ti commercially pure titanium finds use where ductility is needed so that it may be fabricated, but where strength is not a serious consideration. Other commercially pure titaniums are used where intermediate strength and fabricability are required. Aircraft tailpipes, fire-walls, bulkheads, etc., are a few applications. Accessories (valves, pipes, etc.) for chemical-processing machinery are another application.

The alpha titanium alloys (only one is shown in Table 10-10) are also used extensively in aircraft components, where an operating temperature of up to 900°F is required. Examples are casting, tailpipes, and compressor blades. Chemical-processing equipment that must operate up to 900°F is also made from this alloy of titanium.

The alpha–beta titanium alloys are used in aircraft structural members and skins, which operate at temperatures of up to 600°F, airframe forging, fasteners, gas turbines, and compressor blades, disks, rings, etc.

The beta alloy is weldable in the heat-treated or annealed state. It is solution heat-treatable and is subject to age-hardening by a titanium chromate compound. Alpha precipitates, yielding ultimate strengths of up to 200,000 psi with 6 per cent elongation.

Problems

10-1. Describe the mechanism of solution heat treatment.

10-2. Describe the following: (a) natural aging, (b) artificial aging, (c) abnormal aging, (d) overaging.

10-3. Assume a metal with 85 per cent of A in Fig. 10-1; trace the solid solution structure when this metal is heated to T_2 and (a) cooled slowly, (b) quenched.

10-4. Assume the room-temperature structure of the metal in Problem 10-3(b). Trace the solid solution structural changes that occur (a) if the material is allowed to remain at room temperature until its structure stabilizes; (b) if the material is reheated to T_4 in Fig. 10-1.

Table 10-10   Commercial Titanium Alloys

Classification	Composition					Physical Properties			
	Ti	Al	Cr	Mo	Other	T.S. $\times 10^3$	Y.S. $\times 10^3$	Elong. %	Process
Commercially pure									
4902	99.2					59	40	28	Anneal
4900A	99.0					79	63	27	Anneal
4901B	99.0					95	88	25	Anneal
Alpha alloy									
4926	Bal.	5.0			2.5 Sn	125	120	18	Anneal
Alpha–beta alloys									
4923	Bal.		2.0	2.0	2 Fe	137	125	18	Anneal
						179	171	13	Heat-treated
4908	Bal.				8 Mn	138	125	15	Anneal
4925	Bal.	4.0			4 Mn	148	133	16	Anneal
						162	140	9	Heat-treated
		4.0		3.0	1 V	140	95	15	Heat-treated
						195	167	6	Heat-treated
4929	Bal.	5.0	1.4	1.2	1.5 Fe	154	145	16	Anneal
						195	184	9	Heat-treated
4911	Bal.	6.0			4 V	135	120	11	Anneal
						170	150	7	Heat-treated
4969	Bal.	7.0		4.0		160	150	15	Anneal
						190	175	12	Heat-treated
Beta alloy									
	Bal.	3.0	11		13 V	135	130	16	Heat-treated
						180	170	6	Heat-treated

Table 10-10 (continued)   Semicommercial Titanium Alloys

	Al	V	Zr	Other				
Alpha alloys	6	1.0	4	1.0 Mo	143	138	17	Anneal
	8		1.0	1 Ta	147	135	16	Heat-treated
	8			2 Cb	126	120	17	Anneal
	8		8	1 (Cb + Ta)	135	125	16	Anneal
	4		12		140	133	12	Anneal
Alpha–beta	3	2.5			100	85	15	Anneal
	5			2.7 Cr	160	154	15	Anneal
				1.3 Fe	190	175	6	Heat-treated
	2.5	16			110	55	16	Heat-treated
					180	165	6	Heat-treated

Source: Taylor Lyman, ed., Metals Handbook, 8th ed., Properties and Selection, p. 1179, Tables 4 and 5. Metal Park, Ohio: American Society for Metals.

10-5. In your own words, explain one of the theories that attempts to relate hardness of a material to lattice distortion.

10-6. In your own words, expand on the amorphous theory of strain-hardening.

10-7. Repeat Problem 10-6 for the fragmentation theory.

10-8. Repeat Problem 10-6 for the lattice-distortion theory.

10-9. Repeat Problem 10-6 for the dislocation theory.

10-10. Figure 10-3(a) is a very important graph. Explain each of the curves and relate each curve to the *properties* of a metal at each of the temperatures shown.

10-11. In Fig. 10-3(a), describe the effect of slow heating upon the crystal structure of a severely worked material.

10-12. Discuss (a) the recrystallization temperature range, (b) cold-working, and (c) hot-working. Relate your answers (b) and (c) to your answer in part (a).

10-13. Refer to Table 10-1. Compare the recrystallization temperatures of lead and tin with that of aluminum. What is the significance of these temperatures on their room-temperature structures? Is it possible to cold-work lead? Explain.

10-14. List the physical properties of pure aluminum.

10-15. What is the lattice structure of pure aluminum?

10-16. What effect does the addition of 1 per cent copper as an impurity have on the tensile strength of pure aluminum?

10-17. In your own words, explain your understanding of the terms "load-to-weight" and "load-to-cross-sectional area" ratios.

10-18. Write each of the number designations for the various series of aluminum.

10-19. Indicate the meaning of the wrought aluminum numbers (a) 2017, (b) 6063, (c) 1060, (d) 7178, and (e) 5056.

10-20. (a) In general, which aluminum series may be heat-treated? (b) Which cannot be heat-treated? (c) How are the latter strengthened? Explain.

10-21. Identify the temper designations for aluminum that follow: (a) –O, (b) –F, (c) –H, (d) –W, (e) –T.

10-22. (a) What does the temper designation –F, when affixed to an aluminum number, guarantee about that material? (b) What does –O mean?

10-23. (a) What is the significance of the –H suffix? (b) Explain the meaning of the digits 1, 2, and 3 when they follow the suffix –H.

10-24. (a) What is the relative hardness of an –H14 temper suffix? (b) What is the percentage of reduction that may be expected?

10-25. (a) What is the relative hardness of an –H36 temper suffix?

10-26. Repeat Problem 10-24 for an –H12 temper suffix.

10-27. What does the temper suffix –W2 hrs mean?

10-28. Explain each of the –T temper designations when followed by the numbers 1 through 10.

10-29. Explain the meaning of the wrought aluminum number 1060–H18.

10-30. Repeat Problem 10-29 for the number 6061–T6.

10-31. Repeat Problem 10-29 for the number X2024–F.

10-32. Repeat Problem 10-29 for the number X5454–H15.

10-33. Explain the meaning of the wrought aluminum number 1060–T7.

10-34. In general, how may the alloys of aluminum be classified?

10-35. Discuss the relative merits of casting versus machining of aluminum parts.

10-36. What are the two types of aluminum that are used in the casting process?

10-37. Upon what factors does the cost of an aluminum casting depend?

10-38. Refer to Table 10-6. Discuss the heat-treat temperature and time required for the (a) 2014–O, (b) 2014–T4, and (c) 2014–T6 designations.

10-39. Repeat Problem 10-38 for an aluminum sand casting made from a 355–T6 material.

10-40. Repeat Problem 10-38 for a permanent mold casting from aluminum material 355–T6. Does it differ from the sand-casting material?

10-41. Assume an aluminum number 132.O–T2. Explain each digit.

10-42. List the five elements referred to in this text that increase the precipitation hardening properties of aluminum.

10-43. Refer to Fig. 10-4(a). (a) What are the eutectic temperature and composition of the aluminum and copper alloy? (b) What is the maximum solubility of copper in aluminum? (c) What is the low limit of this solubility? (d) Assume an alloy of 4 per cent copper and 96 per cent aluminum; explain the processes of solution heat treatment. (e) Explain artificial aging. (f) Explain natural aging.

10-44. List the elements and the physical properties of duraluminum. From Table 10-6, how may it be heat-treated?

10-45. (a) What is the maximum solubility of magnesium in aluminum? (b) What is the minimum solubility? (c) If the addition of magnesium to aluminum makes the alloy subject to solution heat-treating, why are most of these alloys not produced?

10-46. What effect does zinc have when alloyed with aluminum? Is this alloy subject to solution heat treatment?

10-47. (a) What are the results of using silicon as an alloy with aluminum in percentages of 5 to 25 per cent silicon? (b) What are the results when magnesium is added as a third alloy?

10-48. Is it possible to solution heat-treat any of the combinations of the aluminum–manganese alloy?

10-49. List five impurities found in magnesium. Indicate one major effect of each impurity.

10-50. Is magnesium susceptible to solution heat treatment and age hardening? Explain.

10-51. What effect does the lattice structure have on the properties of magnesium?

10-52. State and discuss some of the properties of magnesium.

10-53. (a) What is the maximum solubility of aluminum in magnesium? (b) At what temperature does this occur?

10-54. Explain the meaning of each of the letters or numbers in the magnesium number HZ32A–T5.

10-55. Repeat Problem 10-54 for the magnesium number M1A–F.

10-56. Repeat Problem 10-54 for the magnesium number AM100A–T6.

10-57. Given a magnesium alloy with the following conditions, write the number symbol: 2.7 per cent zinc, 0.6 per cent zirconium, 3.3 per cent rare earth. The alloy has been solution heated and artificially aged. This is the second of this series of alloys developed.

10-58. Repeat Problem 10-57 for the following conditions: 1.3 per cent zinc, 0.17 per cent rare earth; strain-hardened and partially annealed to $\frac{3}{4}$ hardness. This is the first alloy of the series developed.

10-59. State some of the properties of titanium.

10-60. Discuss the effect of alloying on the physical properties of titanium.

10-61. Discuss the relationships that exist between the properties of titanium and its phase structures — alpha, alpha–beta, and beta.

11 | Nonferrous Heavy Metals

11-1 CLASSIFICATION OF COPPER AND ITS ALLOYS

Copper and its alloys may be classified in several ways. One classification is by three categories: (1) tough pitch, (2) oxygen-free, and (3) deoxidized coppers. Another method classifies copper and its alloys into four general categories: (I) those classified as *copper*; (II) the *brasses,* in which zinc is the principal alloy; (III) the *bronzes,* in which tin is the principal alloy (also included as bronzes are those metals which alloy silicon or aluminum with copper); and (IV) the *nickel* alloy. The expanded listing is shown in Table 11-1.

Group I. The *copper group* includes the electrolytic tough pitch (ETP), phosphorized (DHP and DLP), silver, oxygen-gree (OF), free-cutting, and heat-treatable coppers. These are alloys that contain over 99 per cent copper with small amounts of oxygen, phosphorus, silver, lead, tellurium, selenium, nickel, etc.

Cadium is sometimes added to copper to increase its strength without reducing its electrical conductivity.

Beryllium is soluble in copper up to about 2 per cent at 1600°F, as shown in Fig. 11-1(a). Below this temperature the solubility decreases, so that alloy containing up to about 2.25 per cent beryllium in copper will solution heat-treat and age-harden. Solution heat treatment takes place when the material is quenched from approximately 1500°F. Aging takes place when the material is reheated to about 500°F. Cold-working may yield tensile strengths of 200,000 psi and Brinell hardness as high as 400.

Group II. The *brasses,* on the other hand, may be divided into three categories: combinations of copper and zinc; combinations of copper, zinc, and lead; and combinations of copper, zinc, and some element other than lead. The copper-rich end of the copper–zinc equilibrium diagram is shown in Fig. 11-1(b).

The diagram shows a large alpha region below 38 per cent zinc and a beta region between 38 and about 45 per cent zinc. The tensile strength and ductility increase to about 35 per cent zinc. The ductility decreases rapidly as the beta phase appears. Thus most copper–zinc alloys contain less than 40 per cent zinc.

Group III. The *bronzes* are essentially combinations of copper and tin, although combinations of copper and silicon, and copper and aluminum are also called bronzes. Figure 11-1(c) shows the copper-rich end of the copper–tin equilibrium diagram. In this diagram it is seen that, at elevated temperatures, a beta (β) (Cu$_3$Sn) phase exists. As the temperature drops, this beta phase transforms to gamma (γ), delta (δ), and epsilon (ϵ) phases successively. These transformations are so slow that commercially they are nonexistent. The end product, below 10 per cent, generally shows only the alpha phase.

When phosophorus is added, it is used as a deoxidizer. If any phosphorus remains to form Cu$_3$P, a hard compound, it increases the hardness and strength of the bronze. Its presence during the casting process increases the quality of the casting and the fluidity of the melt.

Figure 11-1(d) shows the copper-rich end of the copper–silicon equi-

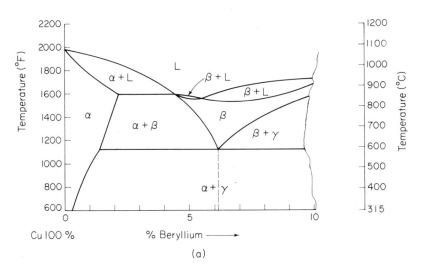

(a)

Figure 11-1

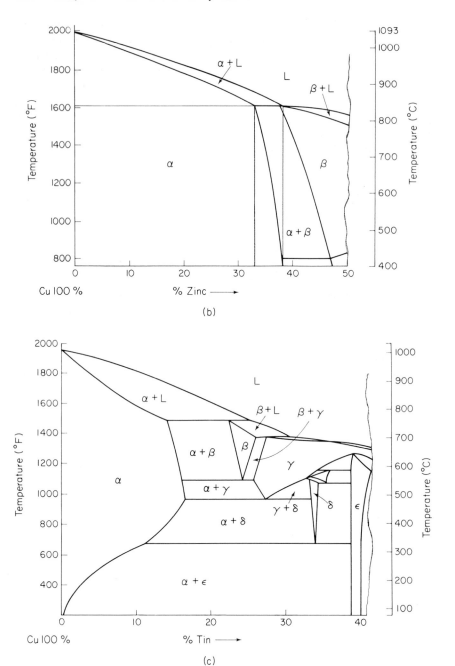

(b)

(c)

Figure 11-1 (continued)

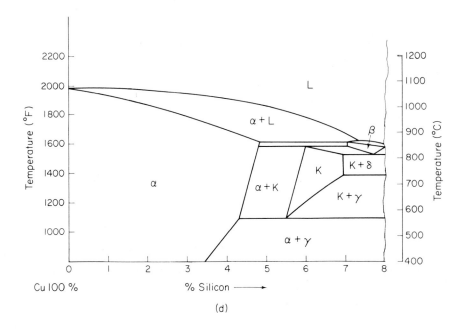

(d)

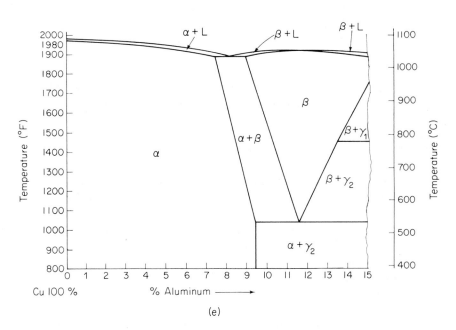

(e)

Figure 11-1 (continued)

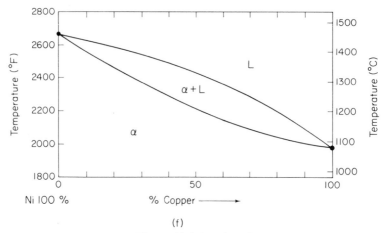

(f)

Figure 11-1 (continued)

librium diagram. These alloys are called the silicon bronzes. The alpha phase has a silicon solubility up to about 5.3 per cent at 1560°F. These alloys will not solution heat-treat but may be cold-worked to give strengths of up to 150,000 psi.

Figure 11-1(e) shows the copper-rich end of the copper–aluminum equilibrium diagram. These alloys are called the aluminum-bronze alloys. The maximum solubility of aluminum in the alpha solution is approximately 9.5 per cent aluminum at 1050°F. The hardness may improve by heating a 10 per cent aluminum bronze to 1700°F and quenching. The alpha phase is transformed to beta phase and trapped on quenching. Subsequent artificial aging by heating to just above 700°F causes the beta phase to transform to alpha and gamma phases, and to harden in the process.

Group IV.   Figure 11-1(f) shows the copper–nickel equilibrium diagram. Since these two elements are completely soluble in each other at room temperature, they can only be work-hardened. The cupronickels are a combination of copper and 10 to 30 per cent nickel as the principal alloy. When 45 per cent nickel in combination with copper is used, the resulting material has high resistivity and a low temperature coefficient of resistivity. This makes it useful for certain resistors and for thermocouples. When used for thermocouples it is called *constantan.*

If the nickel–silver alloys have zinc added to the copper and nickel as a major alloy, they are sometimes called German silvers. They are important because of their resistance to corrosion and their color. Tin is added to nickel–silver to increase its strength and corrosion resistance.

Besides these groups for wrought copper and its alloys, many combinations of copper and its alloys are found in materials used in casting. The broad categories for casting alloys are

1. Tin bronze

2. Leaded tin bronze

3. High-leaded tin bronze

4. Leaded red brass

5. Leaded semi-red brass

6. Leaded yellow brass

7. Manganese bronze

8. Aluminum bronze

9. Nickel brass (nickel silver and bronze)

10. Silicon-bronze and brass

11-2 PROPERTIES OF COPPER

Copper and its alloys have physical properties (see Table 11-1) that are important to many commercial uses. They have excellent electrical and thermal conductivity, corrosion resistance, malleability, formability, and mechanical and fatigue strength, and they are nonmagnetic. These properties make copper very suitable for use as conductors in the electricity field.

In general, wrought copper material is classified into two categories: cold-worked and soft. The cold-worked materials generally are quarter, half, three-quarter, full hardened or spring tempered. The soft, or annealed, state is accomplished after working. Its state is described in terms of grain size (diameter per millimeter).

Cold-work refers to the reduction of size of the material by rolling or drawing. This hardens the material. The amount of cold-working for wrought copper materials is shown in Table 11-2. The percentage of reduction for strip stock is a function of the thickness difference; for wire, the area.

Annealing, as stated, is described after counting the grain or by comparing the sample with a standard. Both operations are accomplished with a microscope. Table 11-3 shows the general classification of grain size.

Electrical tough pitch (ETP), oxygen-free (OF), and the silver-bearing coppers have the highest electrical conductivity. The phosphorized coppers (DLP) rank second, with the electrical conductivity decreasing

Table 11-1　Wrought Copper

	Composition				Mechanical properties				Condition
	Cu	Pb	Ag	$Other$	$T.S. \times 10^3$ psi	$Y.S. \times 10^3$ psi	$Elong.$ %	R_B	
COPPER									
Electro-tough pitch (ETP)	99.9			0.040	50	45	12	50	Hard flat, $\frac{1}{4}$ in.
Phosphorized:									
High residual (DHP)	99.9			0.02 P	55	50	8	60	0.040 in. hard flat
Low residual (DLP)	99.9			0.005 P					
Lake									
Silver bearing			8 oz/$_T$						
Silver bearing			10 oz/$_T$						
			25 oz/$_T$						
Oxygen free (OF)	99.92								
Free cutting	99.0	1			53	50	10	50	Hard, $\frac{1}{4}$ in.
Free cutting	99.5			0.5 Te	53	50	10	50	Hard, $\frac{1}{4}$ in.
Free cutting	99.4			0.6 Se					
Heat-treatable:									
Tellurium	99.0			1.0 Cd	33–53	10–50	45–10	F35–B50	Anneal to hard
Beryllium	97.9			{0.35 Ni, 2.0 Be}	60–200	130–150	35–2	B45–C42	Solution treated to hardened

BRASSES	Cu	Zn	Pb	$Other$					
Plain brasses									
Gilding	95.0	5.0			42–64	32–58	25–4	44–66	$\frac{1}{4}$ to spring, hard
Commercial bronze	90.0	10.0			39–72	14–62	44–3	F60–B78	As rolled to spring
Jewelry	87.5	12.5			47–74	35–62	25–4	47–82	$\frac{1}{2}$ to spring
Red	85.0	15.0			50–84	39–63	25–3	55–86	$\frac{1}{4}$ to spring
Low	80.0	20.0			53–91	40–65	30–3	55–91	$\frac{1}{4}$ to spring
Cartridge	70.0	30.0			54–99	40–65	43–3	55–93	$\frac{1}{4}$ to extra spring
Yellow	65.0	35.0			54–98	40–63	43–3	55–91	$\frac{1}{4}$ to extra spring
Muntz metal	60.0	40.0			54–70	17–50	45–15	80F–B75	Anneal and $\frac{1}{2}$ hard

Table 11-1 (continued)

	Composition				Mechanical properties				
	Cu	Pb	Ag	Other	T.S. $\times 10^3$ psi	Y.S. $\times 10^3$ psi	Elong. %	R_B	Condition
Free-cutting brasses									
Leaded com. bronze	89.0	9.25	1.75		37–52	12–45	45–18	F55–B58	Anneal and ½ hard
Leaded red									
Low (tube)	66.0	33.5	0.5		65–75	50–60	32–7	70–85	Drawn and hard drawn
Medium	65.0	34.0	1.0		54–85	40–62	41–5	55–87	¼ to extra hard
High	65.0	33.0	2.0		54–85	40–62	38–5	55–87	¼ to extra hard
Extra-high	63.0	34.5	2.5		54–74	40–60	35–7	55–80	¼ to hard
High (tube)	66.0	32.4	1.6		65–75	50–60	30–7	70–85	Drawn to hard drawn
Free cutting	61.5	35.5	3.0		49–68	18–52	53–18	F68–B80	Anneal and ½ hard
Forging	59.0	39.0	2.0		52	22	48	42	As extruded
Leaded muntz metal	60.0	39.4	0.6		50	20	35		2 in. max. cross section
Architectural bronze	57.0	40.0	3.0		55	18	35	55	As extruded

	Cu	Zn	Sn	Other	T.S. $\times 10^3$ psi	Y.S. $\times 10^3$ psi	Elong. %	R_B	Condition
Miscellaneous brasses									
Admiralty (strip)	71.0	28.0	1.0		45–88	13–72	69–4	59–109F	Soft to hard
Naval	60.0	39.2	1.1	0.75 Sn	55–70	25–58	50–17	55–75	As rolled to ¼ hard
Leaded naval	60.0	37.5		0.75 Sn	57–75	25–53	40–15	55–82	Soft to ½ hard
Aluminum	76.0	22.0	1.0						
Manganese	70.0	28.0	1.0						
Manganese bronze A	58.5	39.0	1.0	{ 0.1 Mn 1.4 Fe	65–82	30–60	35–25	65–90	Soft to hard
Manganese bronze B	65.5	23.3		{ 4.5 Al 3.7 Mn 3.0 Fe					

Table 11-1 (continued)

	Composition				Mechanical properties				
	Cu	*Pb*	*Sn*	*Other*	*T.S. × 10³ psi*	*Y.S. × 10³ psi*	*Elong. %*	*κ_B*	*Condition*
BRONZES									
Phosphor bronze									
Grade A (5%)	95.0		5.0		49–107	22–105	57–3	33–99	Anneal to extra spring
Grade B	94.0	1.0	5.0						
Grade C (8%)	92.0		8.0		56–119	25–114	73–4	43–102	Anneal to extra spring
Grade D (10%)	90.0		10.0		62–123	26–116	68–5	49–104	Anneal to extra spring
Grade E (1.25)	98.7		1.3		40–76	11–75	47–3	59–80	Anneal to extra spring
444 Bronze	88.0	4.0	4.0	4.0 Zn	44–68	19–57	50–20	14–80	Anneal to hard
	Cu	*Al*	*Si*	*Other*					
Miscellaneous bronzes									
High silicon A	96.0		3.0	1.0 Mn	56–110	21–62	63–4	40–97	Anneal to spring
Low silicon B	97.0		1.7	0.3 Mn	40–75	12–40	30–8	F55–B90	Anneal to extrahard
Aluminum (5%)	95.0	5.0			60–100	25–64	66–8	48–94	Anneal and cold-roll
Aluminum (7%)	91.0	7.0		2 Fe	72–75	34–45	35	74–85	Soft and hard (½ in.)
Aluminum (9%)	91.0	9.0			100	75	8	92	Annealed
Aluminum silicon	91.0	7.0	2.0		95	54	25	84	Extruded cold drawn

Table 11-1 *continued)*

	Composition			Mechanical properties				
	Cu	Zn	Ni	Other	T.S. $\times 10^3$ psi	Y.S. $\times 10^3$ psi	Elong. %	R_B	Condition
NICKEL									
Nickel-containing alloys									
Cupronickel (10%)	88.7		10.0	1.3 Fe	44–60	16–57	42–10	65–100	Anneal and light drawn
Cupronickel (30%)	70.0		30.0		55–84	20–79	45–3	36–86	Hot–cold rolled
Nickel silver (65–18)	65.0	17.0	18.0		58–85	25–74	40–3	40–87	Anneal to hard
Nickel silver (55–18)	55.0	27.0	18.0		60–115	27–93	40–2.5	55–99	Anneal to spring
Nickel silver (German)	65.0	23.0	12.0		56–93	21–79	42–2	37–92	Anneal to extra-hard

Source: Taylor Lyman, ed., *Metals Handbook,* 8th ed. Metal Park, Ohio: American Society for Metals.

Table 11-2

Description	Approx. % reduction by cold-work	
	Strip	Wire
¼ hard	10.9	20.7
½ hard	20.7	37.1
¾ hard	29.4	50.0
Hard	37.1	60.5
Extrahard	50.0	75.0
Spring	60.5	84.4
Extra spring	68.7	90.2

Source: Taylor Lyman, ed., Metals Handbook, Properties and Selection, 8th ed., Vol. I, p. 1006, Table I. Metal Park, Ohio: American Society for Metals.

Table 11-3

Nominal grain size, minimum	Typical use
0.015	Shallow draws for high surface finish; best polishing.
0.025–0.035	Easy drawing, auto wheel covers and hub caps, very good polishing; larger grain size where polishing not necessary.
0.050	Average drawing, will produce roughness or "orange peel" on many parts; can be polished.
0.070	Heavy draws, thick metal gages. Produces rough surfaces or "orange peel"; very difficult to polish.

Source: Taylor Lyman, ed., Metals Handbook, Properties and Selection, 8th ed., Vol. I, p. 1006, Table 3. American Society for Metals.

in the following order: free-cutting (S, Te, Pb), chromium, phosphorized (DHP), cadmium, tellurium–nickel coppers. The bronzes and brasses are low on the list of conducting materials.

In the soft state, ETP copper has a tensile strength of about 32,000 psi; when it is cold-worked, its tensile strength may be increased to 55,000 psi. The ETP coppers will soften at 400°F and embrittle at temperatures above 700°F. The silver-bearing coppers will resist softening up to 650°F. Oxygen-free and phosphor-deoxidized coppers will resist softening at higher temperatures.

Oxygen-free copper (OF) is used when deep drawing and cold-working are required. When tensile strengths of 70,000 psi and good conductivity are required, the chromium coppers are used. When very high tensile strengths of 200,000 psi in combination with high fatigue strengths are required, the beryllium coppers are used. In the latter, electrical conductivity is sacrificed. It may be necessary to machine or form them in the soft state and then spring harden by heat treatment.

The free-machining coppers have high conductivity and high machinability rates. The coppers that have good conductivity and good machinability are tellurium copper (90 per cent conductivity, 85 per cent machinability), leaded copper, and sulfurized copper. Tensile strengths of 75,000 psi, 80 per cent machinability, and low electrical conductivity (50 per cent) are properties of heat-treated and drawn tellurium–nickel copper.

Electrical parts that require good fatigue characteristics but which carry low currents are made from cartridge brass. Where corrosion conditions are severe, the nickel silver, phosphor bronze, or beryllium coppers are used.

Thermal conductivity properties are much the same as the electrical conductivity. The mechanical strength of copper and its alloys is determined essentially by its composition and cold-working. The mechanical properties of some alloys may be achieved by heat treatment. Copper and copper–zinc alloys are stronger than other alloys of copper when used with aluminum, silicon, manganese, tin, or iron. Aluminum and silicon bronzes are used when strength is required for heavy sections of bar, sheet, etc. Of all the brasses, naval brass is not readily cold-formed. Aluminum silicon bronze and tellurium–nickel–phosphorus copper are free machining with high-strength qualities. Leaded naval brass and free-cutting brass are less costly but not as corrosion resistant or as strong as the former materials. Fatigue strengths in descending order are beryllium copper, phosphorus bronze, nickel silver, and silicon bronze.

Under the general heading of fabrication are included machinability, formability, and joining. The machinability of copper alloys relative to free-cutting brass rod is shown in Table 11-4.

In general, half-hard free-cutting brass is used in automatic screw machining, welded torch and soldering tips. Leaded copper alloys provide excellent machining conditions, because lead acts as a lubricant during the cutting operation. This type of copper alloy may be spun, staked, extruded, and forged. Forging brasses and architectural bronze are also suitable for these operations. Low-lead alloys are better than half-hard free-cutting brasses where the operations are severe. Such operations as hammering and similar severe cold-working are examples of severe working.

Formability deals with cold and hot forming. The best results in *cold forming* and *bending* are achieved with the nonleaded alloys such as

Table 11-4

Material	Rating
Free-cutting brass rod	100
Free-cutting copper rod (Te, S, Pb)	90
High-leaded brass	90
Architectural bronze	90
Free-cutting phosphorus bronze	90
Leaded copper	80
Forging brass	80
Tellurium–nickel copper	80
Leaded commercial bronze	80
Medium leaded brass	70
Leaded naval brass	70
Aluminum silicon bronze	60
Low lead brass	60
Copper Plain brasses Phosphorus bronze Naval brass Silicon bronze	under 30

electrolytic tough pitch, oxygen-free, and silver deoxidized coppers; brass with over 63 per cent copper; silicon bronzes; nickel silver; cupronickel; and the solution heat-treatable alloys. Oxygen-free coppers are best used for severe cold-working. Cartridge brass, red brass, and gilding are suitable for cold drawing, bending, and upsetting. Annealed tellurium copper is suitable for severe cold-working operations.

Coppers and brasses with less than 63 per cent copper, leaded brass with less than 63 per cent copper, naval brass, aluminum silicon, the manganese bronzes, and cupronickel materials are suitable for *hot forming.*

Joining of the copper alloys is accomplished successfully by using the following methods: (1) arc welding will successfully join all copper alloys except the zinc-bearing type. The methods used are the inert-gas tungsten arc and consumable-electrode welding; coated metal and carbon arc welding; and resistance seam, spot, and butt welding. (2) Oxyacetylene is used primarily to weld the copper–zinc alloys and, of lesser importance, the phosphorized copper, cupronickel, and copper–silicon alloys. (3) All copper-base metals can be brazed and soldered. The aluminum bronzes can be brazed but not soldered. Silver alloy welding rods are used most frequently for any of the copper alloys, whereas copper–phosphorus welding rods are used primarily on the coppers. Soldering may be done on any of the copper alloys, except the aluminum bronzes, if suitable fluxes are used. Aluminum bronzes may be soldered if they are first plated with a material that is capable of being soldered.

11-3 USES OF COPPER AND ITS ALLOYS

The copper classification referred to in Sec. 11-1 as *group I* (see also Table 11-1) has the following uses: *electrical tough pitch copper* has wide use as building materials, automotive gaskets and radiators, electrical uses of all kinds, household hardware, rivets, soldering coppers, kettles and pans, roadbed expanders, and rolls for printing. *Phosphorized copper* materials are used for gas, oil, and heater lines in household utilities, in industrial lines of all types, and in lines used in the transportation of air, gasoline, and oil. *Oxygen-free copper* is used in wave guides and electrical conductors. *Free-machining copper* is used on screw machine products, motor and switch components, and plumbing fittings. The *silver coppers* are used for commutator parts. *Beryllium coppers* are used in springs, diaphragms, surgical instruments, firing pins, dies, aircraft engine bushings, and for spark-resistant tools.

Group II brasses may be divided into three subcategories: plain brasses, free-cutting brasses, and a miscellaneous grouping (see Table 11-1).

Plain brasses are used for coins, medals, and emblems, costume jewelry, bullet jackets and shells, for architectural application such as etched bronze and screen cloth, for marine hardware, kick plates, screws, and compact and lipstick cases. In addition, red brass is used for heat exchanger and condenser tubing and plumbing tubing. Low brass, besides being used for the purposes listed above, is used for musical instruments, tokens, and clock dials.

Free-cutting brasses are used for screws, screw machine parts, electrical connectors, pump and trap lines for plumbing, gears, nuts, rivets, bearing cages, engraver's plates, hardware butts, hinges, and lock bodies.

In the *miscellaneous* category are materials used for condenser, evaporator, heat exchange, and distiller tubes and for marine hardware, propeller shafts, piston and valve stems, aircraft turnbuckle barrels, balls, nuts, bolts, and rivets. Materials used for clutch plates are also found in this group.

In *group III* are found the phosphor, aluminum, and silicon bronzes.

Phosphor bronzes are used for bridge bearings and expansion plates, bellows, Bourdon tubes, clutch disks, cotter pins, diaphragms, fasteners, lock washers, sleeve bushings, springs, perforated sheets, and as components in textile machinery.

Aluminum bronzes are used for cold- or hot-working forms of sheet, strip, wire, and tubing, for "gold" decorating, corrosion-resisting vessels and tanks, structural components, condenser tubes and piping systems, nuts, bolts, protective sheathing, and fastening.

Silicon bronzes are used in high-pressure lines, marine hardware, pole line hardware, nails, rivets, screws, and nuts, and in heat exchanger tubes, hot water tanks, kettles, screen cloth and wire, and piston rings. The low silicon bronzes are used also in electrical conduit.

Group IV, the nickel-containing alloys, are divided into the cupro-nickel and the nickel silver classes.

The *cupronickel* class is used for condenser, distiller, evaporator, and heat-exchanger tubes. It resists stress, alkali, and saltwater corrosion.

The *nickel silver* class is used for parts that need to be resistant to water, atmosphere, and food products. It may be plated with nickel, chromium, or silver. It is used for rivets, screws, name plates, radio dials, truss wire, and zippers. This material is also used for flatware and hollow-ware, costume jewelry, optical camera products, telephone equipment, electrical controls, surgical and dental equipment, resistance wires, diaphragms, and hardware.

11-4   NICKEL AND ITS ALLOYS

Nickel is one of the metals that has high resistance to corrosion. It is used as an undercoating for parts that are to be chromium plated. Its use as an alloy in cast iron and steel has already been mentioned. It is used in steels to form stainless steel. Nickel, when used with copper, is highly resistant to deterioration by salt water or sulfuric acid. When used with chromium, it is resistant to high-temperature oxidation and corrosion. When alloyed with titanium or aluminum, the alloy will age-harden. It is also used in electroplating. If sulfur is present with nickel, it will form a protective coating around the grains at the grain boundaries, which causes an embrittlement. If small amounts of manganese are added to the melt, the sulfur and manganese combine to form MnS, which leaves the material ductile and noncorrosive.

Table 11-5 shows the compositions of some common alloys of nickel. The pure nickels have 94.0 per cent plus nickel in combination with manganese, aluminum, or silicon. When combined with cobalt, the alloy is heat resistant and may be used in the chemical industry. When combined with manganese, it becomes resistant to oxidation and reduction at elevated temperatures. It is used in spark plugs. When combined with aluminum, it develops high strength as well as becoming corrosion resistant. It is used in springs and shafts that must have both qualities of strength and corrosion resistance.

Monel metals are essentially combinations of nickel and copper. They possess high corrosion resistance and strength. They are, therefore, used to resist corrosion due to sulfuric acids, caustic soda, or salt water in the chemical, electrical, textile, laundry, marine, and pharmaceutical industries. When silicon is added, Monel exhibits good wear-resistant qualities while retaining its corrosion resistance and high strength.

Inconel metals are nickel alloys that contain substantial percentages of chromium and iron. They have good strength and high resistance to corrosion by chemicals at high temperatures.

Hastelloy metals are nickel alloys that contain various combinations of chromium, molybdenum, and iron. They have excellent resistance to oxidizing chemicals even at elevated temperatures. They also have high strength, in many cases retaining this strength even at high temperatures. It is recommended that handbooks be checked before a particular material is selected, since certain precautions must be observed when they are used.

Illium nickels are nickel alloys with the iron component of the Hastelloy metals being replaced by copper. They have good corrosion resistance and strength. They find much of their use in corrosion-resistant castings for bearings, cutting blades, and pump components.

Electrical resistance alloys are nickel alloys that have chromium and iron as alloys. Constantan has a nickel–copper combination. These materials are used in fine resistance wires, heating elements in furnaces or appliances of all types, thermocouples, potentiometers, and rheostats.

The iron–nickel equilibrium diagram is shown in Fig. 11-2.

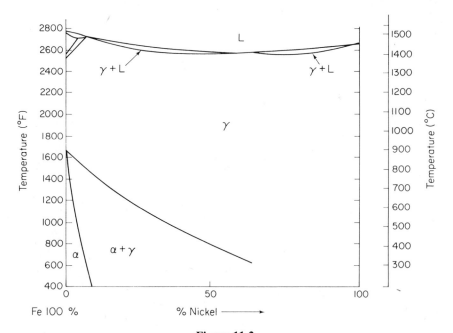

Figure 11-2

Table 11-5 Nickel

Alloys	Ni	Cu	Cr	Fe	Al	Ti	Other	T.S. × 10³	Y.S. × 10³	Elong. %	BHN	Condition
Nickels												
Nickel, pure	99.95							46	8.5	30		A
"A" nickel	99.4							70	20	40	110	A
"D" nickel	95.0						4.5 Mn	86	34	40	147	HR
Duranickel	94.0						4.5 Ti	45	100	40	150	CD
Cast	97.0						1.5 Si	55	25	22	100	SC
Monel Metals												
Monel	67	30						75	30	40	125	A
"K" Monel	66	29					3.0 Al	55	76	21	160	CD
"R" Monel	66	31.5						90	75	25	180	CD
Cast	63	30					1.5 Si	78	36	35	140	SC
"H" Monel	63	30					3.0 Si	115	70	17	265	SC
"S" Monel	63	30					4.0 Si	125	95	25	310	SC
Invar	36	63										
	Ni		*Cr*	*Fe*	*Al*	*Ti*	*Other*					
Inconel Metals												
Inconel	76		16	8.0			3 Al	85	35	25	150	A
702	78		16	8.5			5 Si					
S	68		15.5	7.0			2 Si					
Cast	72		16				3 Mo	100	90	3	350	C
Ni-o-nel	41.4		20	31.6		1	2 Cu	95	37	42	$82R_B$	A
Nimomic–75	77		20					115	55	40	170	A
–80A	74.5		20		1.3	2.5						
–90	57		20.5	9.5	1.6	2.6	17 Co					
Incoloy	31		20.5	45.5				90	40	40	150	A
–901	37.4		13.5	33.7		2.5	5.9 Mo					

Properties

Table 11-5 (continued)

| | Composition | | | | | | Properties* | | | | |
Alloys	Ni	Cr	Mo	Fe	Cu	Other	T.S. × 10³	Y.S. × 10³	Elong. %	BHN	Condition
Hastelloy Nickel											
Alloy B	62		28	5.0			121	57	63	92R_B	SC
Alloy C	54	15	17	5.0		4 W	121	58	47.5	91R_B	SC
Alloy D	85				3	10 Si	115	115	1	35R_C	Cast
Alloy F	47	22	7	17			73	37	20	83R_B	SC
Alloy N	70	7	17	5			87	40	44	—	SC
Alloy W	62	5	24.5	5.5			123	54	55	—	—
Alloy X	47	22	9	18			65	42	22	89R_B	SC
Illium Nickel											
Alloy B	50	28	8.5		5.5		64	56	3	220	3.51 Si
Alloy G	56	22.5	6.5		6.5		68	56	7.5	168	CD
Alloy R	68	21	5.0		3.0		142	95	11.5	238	20% CD
Alloy 98	80	20					54	41	18	160	

| | Composition | | | | | | Properties* | | | | |
Alloys	Ni	Cr	Fe	Mn	Al	Other	T.S. × 10³	Y.S. × 10³	Elong. %	BHN	Condition
Electrical Resist.											
Chromel	90	10					95		32	87R_B	A
Nichrome	80	20					105		30	83R_B	A
60 Nickel	60	16	24				102		30	83R_B	A
35 Nickel	35	20	45				55	21	32	50R_B	Cast
Constantan	45					55 Cu					
Alumel	94.5			2.5	2.0	1.0 Si					

Source: Taylor Lyman, ed., *Metals Handbook*, Properties and Selection, 8th ed., Vol. I, pp. 1118–1130. Metal Park, Ohio: American Society for Metals.

*Average values.

A, anneal; HR, hot rolled; CD, cold drawn; S. spring temper; SC, sand cast.

11-5   LEAD AND ITS ALLOYS

Lead is a heavy, dense material. It has a low melting point, low strength, high coefficient of expansion, and is a very ductile material. It has high corrosion resistance and poor electrical conductivity. It also has lubricating qualities, which make it useful in the drawing operation, bridge bearing plates, and pipe joint compounds.

Lead's greatest use is in storage batteries and as an additive to gasoline. It also finds wide use in red lead and litharge, caulking, soldering, cable covering, type metal, and as a shielding metal in the nuclear industry against gamma radiation. It does not shield against neutron bombardment.

Lead may be alloyed with antimony and tin, which impart greater strength to the metal. Tin as an alloy, however, makes it possible to bond lead with metals such as copper and steel. These alloys are called lead–tin solders. The lead–tin equilibrium diagram is shown in Fig. 11-3. The lead–antimony diagram is shown in Fig. 11-4(a). A photomicrograph of an alloy of lead–tin–antimony is shown in Fig. 11-4(b).

The alloys of lead can be separated into four general categories: (1) plain alloys, (2) solders, (3) antimonial alloys, and (4) babbit metals. These are shown in Table 11-6.

In category 1 the chemical leads are used primarily in the chemical industries; arsenical and calcium leads are used for cable sheathing, and the corroding leads are used in batteries, paints, caulking, type metal, bearing metals, solder, foil, electrical cables, and insecticides.

Solder metals, of course, are used in dip, torch, or furance soldering to join, coat, or as a filler of metal.

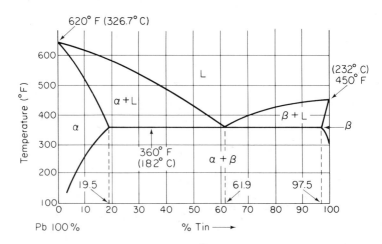

Figure 11-3

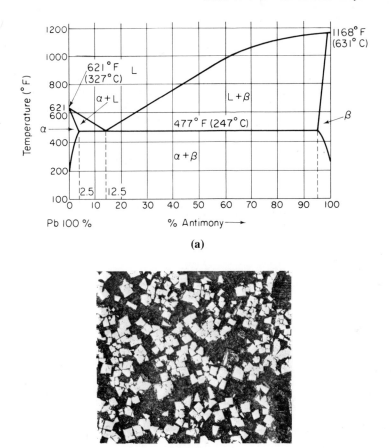

(a)

(b)

Figure 11-4

Antimonial lead alloys are used for cable sheathing, storage battery grids, rolled sheet, and extruded pipe.

Babbit metals are used mainly for bearings in pumps, journals, motors, transmission machinery diesel engines, automobile engines, tractors, trucks, and heavy machinery. The addition of arsenic makes the alloy suitable for heavy-duty action.

11-6 TIN AND ITS ALLOYS

About 40 per cent of the tin produced is used as a coating in cans, 22 per cent in solders, 20 per cent as an alloy in brass and bronze, and the remaining in babbit, white metal, etc. — over 80 per cent of all the tin produced goes into three uses.

Table 11-6 Lead

Alloy	Composition				Properties				Condition
	Pb	Sn	Sb	Other	T.S. × 10²	Y.S. × 10²	Elong. %	BHN	
Plain alloys of lead									
Chemical	99.9+				25	13	27		Rolled sheet
Corroding	99.73+				18	8	30	40	Sand cast
Arsenical	Bal.	0.10		{0.10 Bi, 0.15 As}	23.5		40	4.9	Extrude, air cool
Calcium	Bal.			0.028 Ca	30		40	5	Extruded
Solders									
Silver lead (soft)	97.5	1.0		1.5 Ag				13	
Solder (5–95)	95	5.0			34	15	50	8	
Solder (20–80)	80	20			58	36.5	16	11.3	
Solder (50–50)	50	50			61	48	60	14.5	
Antimonial lead alloy									
Antimonial (1%)	99		1.0		30		50	7	Ext. and aged
Hard lead (4%)	96		4.0		11.7		6.3	24	Aged 1 day
Hard lead (6%)	94		6.0		33		65	10.7	Extruded
Antimonial (8%)	92		8.0		12.4		4.7	26.3	Aged 1 day
Antimonial (9%)	91		9.0		75		17	15.4	Chill cast
Babbitt-lead base									
Alloy 19	85	5	10		100		5.0	19	Chill cast
Alloy 7	75	10	15		105		4.0	22	Chill cast
Alloy 8	80	5	15		100		5.0	20	Chill cast
Alloy 15	83	1.0	15	1.0 As	104		2.0	20	Chill cast
G. babbit	83.5	0.75	12.75	3 As	98		1.5	22	Chill cast

Source: Taylor Lyman, ed., *Metals Handbook,* Properties and Selection, 8th ed., Vol. I, p. 1062. Metal Park, Ohio: American Society for Metals.

Tin is a soft material that does not work-harden permanently. This makes it suitable for use in collapsible tubes such as toothpaste tubes and as tinfoil. The large amount of tin used for coating steel and copper makes the coated surface corrosion resisting. It may be used to coat the inside of cans because of its nontoxicity. It is also used to coat the inside of copper tubing for transporting water. Hot coating is applied to nonferrous materials such as copper, bronze, and aluminum bronze. In some instances, a process known as electrotinning is used.

Tin and its alloys may be separated into four general categories: (1) plain alloys, (2) solders, (3) babbits, and (4) special applications.

The *plain alloys* of tin are as follows: pure tin is used mostly for electrotinning and in chemical compounds; straits tin and hard tin are used for foil and collapsible tubes. Straits tin is also used for tin plate and pipe.

Antimonial and *tin–silver solders* are used to solder electrical apparatus and where high-temperature service is required. Since tin has higher electrical conductivity than lead, it is sometimes advantageous to use a high-tin solder rather than a high-lead solder. The soft solders are used for joining and coating metals. The 61.9 Sn-38.1 Pb soft solder is used extensively. It is called eutectic solder (see Fig. 11-3).

Tin babbit is used as bearings and in die castings. The various combinations of tin, antimony, and copper are used to serve the special-purpose requirements needed in bearing design.

Special applications are the tin alloys used in die casting, tinfoil for food packaging, white metal used for jewelry castings, and pewter used in making bookends, candlesticks, vases, etc.

Table 11-7 shows the composition and average mechanical properties of some tin alloys.

11-7 ZINC AND ITS ALLOYS

Zinc has an electrode potential of -0.76 in the electromotive force series, whereas iron is -0.44. Thus zinc will corrode faster than iron. When coated on iron or steel, it acts as an anode and protects the iron or steel underneath. The processes of galvanizing, hot dipping, electrolysis, metallizing, painting, or packing parts in a fine zinc powder and heating (called *sherardizing*) may be used to apply zinc to surfaces.

Zinc is also used in zinc-base die castings. The two most widely used die-cast materials are known as AG40A and AC41A. Their composition and mechanical properties are shown in Table 11-8. In general, they are used to make die-cast parts for automobiles, building hardware, utensils, toys, and office equipment. The slush metals are used in molding electrical lighting fixtures.

The wrought zinc is used in the manufacture of articles that do not

Table 11-7 Tin

| Classification | Composition | | | | Physical properties | | | | |
	Sn	Sb	Cu	Other	T.S. × 10²	Shear × 10²	Elong. %	BHN	Condition
Plain tin alloys									
Pure tin	99.98+				31		55		Cast
Straits	99.8				21		55 (4 in)	5.3	Cast
Hard	99.6		0.4		40		55 (4 in)		80% red.
							55		
Solder									
Antimonial	95	5.0			142	111			Sold. joint
Silver	95			5.0 Ag	140	106			Sold. joint
Soft	70			30 Pb	142	111			Sold. joint
Eutectic	63			37 Pb	290	80			Sold. joint
Soft	50			50 Pb	61	60	70	14.5	Sold. joint
Babbit						*Comp. st* × 10²			
Alloy 1	91	4.5	4.5		93	44*	2.0	17	Chill cast
Alloy 2	89	7.5	3.5		112	61	2.0	24	Chill cast
Alloy 3	84	8.0	8.0		100	66	1.0	27	Chill cast
Alloy 4	75	12	3.0	10 Pb	80	55	1.5	24	Chill cast
Alloy 5	65	15	2.0	18 Pb	78	50	1.5	22	Chill cast
Special						*Yield st.* × 10²			
Die casting	82	13	5.0		100	60	1.0	29	Die cast
Tinfoil	92			8.0 Zn	87		40		Foil
White metal	92	8.0			65		50	20	Chill cast
Pewter	91	7.0	2.0		86		40	9.5	Sheet anneal

Source: Taylor Lyman, ed., *Metals Handbook, Properties and Selection,* 8th ed., Vol. I, p. 1142. Metal Park, Ohio: American Society for Metals.

*0.125% set.

Table 11-8   Zinc

	Composition					Mechanical properties						Condition
	Zn	Cu	Al	Pb	Other	T.S. × 10³		S.S. × 10³	Elong. %		BHN	
						A*	B		A	B		
Pure zinc alloy												
Pure zinc	99.9+											
Casting materials												
AG40A (ZAMAK-3)	Bal.		4.0		0.04 Mg	41.0		31.0	10		82	Die cast
AC41A (ZAMAK-5)	Bal.	1.0	4.0		0.04 Mg	47.6		38.0	7.0		91	Die cast
Slush alloy	Bal.	0.3	4.7			28.0						Chill cast
Slush alloy	Bal.		5.5			25.0			1.0			Chill cast
Wrought zinc												
Com. rolled (0.08 Pb)	Bal.			0.08		19.5	23.0		65	50	38	Hot rolled strip
						21.0	27.0		50	40		Cold rolled
Com. rolled (0.06 Pb)	Bal.			0.06	0.06 Cd	21.0	25		52	30	43	Hot rolled strip
						22	29		40	30		Cold rolled
Com. rolled (0.3 Pb)	Bal.			0.3	0.3 Cd	23	29		50	32	47	Hot rolled strip
						25	31		45	28		Cold rolled
Copper hard rolled	Bal.	1.0				24	30		50	35	52	Hot rolled
						31	40		40	25	60	Cold rolled
Rolled zinc	Bal.	1.0			0.01 Mg	29	40		20	10	61	Hot rolled
						36	46		25	10	80	Cold rolled
Zn–Cu–Ti alloy	Bal.	0.8			0.15 Ti	32	42		38	21		Hot rolled
						29	37		44	60		Cold rolled

Source: Taylor Lyman, ed., *Metals Handbook*, Properties and Selection, 8th ed., Vol. I, p. 1169. Metal Park, Ohio: American Society for Metals.

*A, parallel to rolling; B, perpendicular to rolling.

require stiffness; when made with 0.08 per cent lead, it is known as commercially rolled zinc, sometimes referred to as "deep drawing zinc." Drawn, spun, or otherwise formed articles such as flashing, address plates, and eyelets are made from commercially rolled zincs. If some stiffness is required, 0.06 per cent cadmium is used as an alloy. When 0.3 per cent lead and 0.3 per cent cadmium are used, the material finds wide use as photoengraver's or lithographer's sheets.

Copper-hardened rolled zinc, sometimes called Zilloy–40, is used for weather stripping, and drawn, spun, or formed articles where stiffness is needed. If magnesium is added to the copper as an alloy, the material develops greater stiffness and may be used for corrugated roofing as well as the other uses mentioned above. If titanium, as an alloy, is substituted for magnesium, maximum creep resistance is developed, and the material is used for corrugated roofing, gutters, and leaders.

11-8   THE PRECIOUS METALS

Precious metals may be separated into four groups: (1) silver and its alloys, (2) gold and its alloys, (3) platinum alloys and the platinum group, (4) palladium and its alloys. In the pure state, precious metals are soft, have good malleability, are corrosion resistant, and have good electrical and thermal conductivity. In the alloyed state they develop a wide range of physical properties. The hardness number ranges from the twenties to more than 600 Brinell. The melting temperature is low enough so that alloys of silver may be used as solders at 1100°F, and high enough so that when alloyed with osmium it melts at temperatures greater than 5400°F.

Table 11-9 shows the alloys of the precious metals, their composition, and physical properties.

Silver.   Silver may be 99.99 per cent pure or commercially pure at 99.9 per cent. Eutectic silver (72 per cent silver, 28 per cent copper) has a low melting point (1435°F) and may be used as a solder. Sometimes nickel, cadmium, palladium, zinc, tin, or phosphorus is added to achieve special purposes. Because of its resistivity to oxidation, it is suitable for electrical connections and contacts. When plated, the base metal underneath may oxidize and cause peeling, especially when used at high temperatures. It has electrical and thermal conductivity almost the same as those of copper. It is not suitable for use as electrical contacts when the voltage is below 20 V, where a 0.2 voltage drop occurs frequently, or in low-level audio circuits because of the noise generation.

Silver-clad copper, iron, nickel, or brass is used in the electrical field as conductors. Glass, mica, and ceramics may be coated with silver and used for electrical conducting parts. In some instances silver coats are used as a conducting base for electroplating. Electroplating of silver on a nickel silver base is used for tableware.

The very good physical properties and the slight change in dimension that takes place when silver–tin–mercury solidifies and cools make it acceptable for use in dental work. Excess moisture, or mercury added during the amalgamation, causes the material to expand during the hardening process. Incomplete amalgamation or improper packing causes corrosion of the alloy.

Sterling silver has high reflectivity, making it suitable for jewelry and cutlery. It is also used in photography because of the photosensitivity of silver crystals. Because of its uniform and high reflectivity, it is also used to coat glass in the manufacture of mirrors. It also finds many uses in the chemical field. Where pure silver is too soft for the service required, coil silver (90 per cent silver, 10 per cent copper) is used. Certain electrical contacts and silver coins are made from this alloy.

Gold.   In its commercially pure state, gold is used for dental and decorative work. It is used in fuses to protect electric furnaces, as a film on glass to filter light, for decorative purposes on glass, and as a high-melting solder. In the electrical field, because of its resistance to corrosion, it is used to line wave guides and as a coating on grid wires, where low-noise contacts are required and where low and stable resistances are required.

Gold–silver–copper alloys are used for jewelry, decorative, and dental purposes. The balancing, or addition, of elements will change the color of the alloy. Eighteen-karat gold (100 per cent gold is 24 karats), which would have a gold content of 75 per cent Au, may have a green, red, or yellow color. If the content is 18K gold and silver (green), it is too soft for general use and may be used for plating. If the content is 18K gold and copper (red), the alloy is too hard to work. Eighteen-karat gold when alloyed with silver and copper has a yellowish color. Ten and 14K gold have their content controlled by the ratio of silver to copper. A range of 14 to 22K gold is used in dental work.

White gold may be of the 18, 14, or 10K variety. It has gold alloyed with nickel, copper, and zinc. The gold content is the same as the green, red, or yellow golds. The control of the copper–nickel–zinc percentages yields the white, greenish white, and pink golds. The latter is a 12K gold.

The combination of 70 per cent gold with 30 per cent platinum produces a high-melting-point solder at 2640°F. Gold–palladium–iron alloys have high resistivity and are used as potentiometer wires. The combination of 49.5 per cent gold, 40.5 per cent palladium, and 10 per cent iron yields the highest resistivity of approximately 1070 microhms per mil-foot after 1 hour annealing.

Table 11-9 Precious Metals*

	Composition					Physical properties				
	Ag	Cu	Ni	Zn	Other	T.S. × 10³	Y.S. × 10³	Elong. %	BHN	Condition
Silver alloys										
Pure	99.99					18.2	7.9	54	27V	(V-Vickers) anneal 1100°F
Comm. pure	99.9					50			90	Quench-aged 17 h 535°F
Sterling	92.5	7.5				52			90	Quench-aged 17 h 535°F
Coining	90.0	10				52			90	Quench-aged 17 h 535°F
Brazing (eut.)	72.0	28				70			68	(30T Rock)
Electrical	Bal.		0.2		0.25 Mg					
Dental amal.	33	2		0.5	{52 Hg, 12.5 Sn}	5	Comp. S 58, 50	15	Knoop 90	

	Au	Cu	Ni	Zn	Other	T.S. × 10³	Y.S. × 10³	Elong. %	BHN	Condition
Gold alloys										
Pure	99.99									
Comm. pure	99.95					32	30	4	58	
Gold–silver–copper									*Vicks*	
Green, 18K	75				{5 Au, 20 Ag}				100	Aged
Yellow, 14K	58.3				{21.7 Au, 20.0 Ag}				275	Aged
Red, 10K	41.7				{33.7 Cu, 20.0 Ag}				310	Aged
Gold–nickel–copper–zinc										
White, 18K	75	2.23	17.3	5.47					380	Aged
White, 14K	58.3	23.5	12.2	6.0					170	Annealed
White, 14K	58.3	22.2	10.8	8.7					165	Annealed
White, 10K	41.7	32.8	17.1	8.4					150	Annealed
White, 10K	41.7	30.8	15.2	12.3					140	Annealed

286

Table 11-9 (continued)

	Composition					Physical properties				
	Ag	Cu	Ni	Zn	Other	T.S. × 10³	Y.S. × 10³	Elong. %	BHN	Condition
Gold–platinum	70				30 Pt	92.7	82.5		169	{ (Hard (66% red) / T.S. at 50% red
Gold–palladium–iron alloys	49.5				{ 40.5 Pd / 10 Fe	200				Peak resistivity alloy, H.T.

	Composition					Physical properties				
	Au	Pt	Pd	Ag	Cu	T.S.	Y.S.	Elong. %	BHN	Condition
Dentistry–gold wire†										
1	30	50	30	11	14	180	150	15	245	
2	60	18	8	8	12	130	102	22	190	
3	50	12	25	16	14	150	120	10	230	
4	64	18	6	15	14	115	80	26	195	Soft-quenched from
5	70	7	5	25	18	120	73	20	200	1300°–1600°F
6	63	5	5	30	18	100	58	28	170	
7	28	25	37	41	21	180	115	20	225	
8		1	44		17	110	87	24	200	

	Composition					Physical properties				
	Au	Ag	Cu	Pd	Pt	T.S. × 10³	P.L. × 10³	Elong. %	BHN	Condition
Dentistry gold cast†										
Yellow, 1	92.5	12	4.5	0.5	0.5	45	15	35	70	Quenched
Yellow, 2	78	14.5	10	4	1	55	25	35	90	Quenched
Yellow, 3	78	26	11	4	3	82	58	20	165	Aged
Yellow, 4	71.5	20	16	5	8.5	120	92	6	235	Aged
White, 5	70	12	10	12	4	75	45	12	170	Aged
White, 6	65	15	12	10	8	120	83	3	260	Aged
White, 7	30	30	25	20	7	130	100	3	280	Aged
Zn (0.5–2%)†										

Table 11-9 (continued)

	Composition					Physical properties				
	Pt	Other	Cu	Ag	Au	T.S.	Y.S.	Elong.	BHN	Condition
Platinum alloys										
Comm. pure	99.85					24	2	40	40	Annealed
Pt–ruthenium		5–14				60		34	130	Annealed, 5% Ru
Pt–rhodium		3.5–40				30		35	70	Annealed, 5% Rh
Pt–palladium		10–25				48			100	Annealed, 40% Pd
Pt–iridium		5–35				40			90	Annealed, 5% Ir
Pt–nickel		5–20				93			130	Annealed, 5% Ni
									Vick	
Pt–Tungsten	99.85	2–8				112		5	133	Annealed, 4% W
Cobalt		40–60								
Palladium alloys										
Pure	99.85					42	30		105	Annealed, 500°F
Pd–silver	60			40		46			180	Annealed
Pd–copper	60	40				75				Annealed
									Vick	
Pd–silver–copper	40		30	40					450	Aged
Pd–silver–gold	30			40	30				80	Aged
Pd–ruthenium	95.5	4.5				80	39	25	152	1830°F air cooled
Iridium	99.+					90		34	170	Annealed, 75°F
Osmium	99.+									
Rhodium	99.+					138			122	Annealed
Ruthenium	99.+					78			220	Hot rolled, 50%

Source: Taylor Lyman, ed., *Metals Handbook, Properties and Selection.* 8th ed., Vol. I, p. 1181. Metal Park, Ohio: American Society for Metals.

*All values maximum.
†Ni (1 to 2%), Zn (0.5 to 2%)

Platinum. Pure platinum is used in resistance thermometers and thermocouples. It is alloyed with other elements to develop properties such as high corrosion resistance and hardness so that it may be used in precision electrical contacts, brushes, and potentiometers. It is also used in the pure, alloyed, or clad state for jewelry and decorative purposes. It is an outstanding catalyst and therefore finds wide use in the chemical industry.

When alloyed with palladium, it is used for jewelry and electrical contacts. When alloyed with rhodium, it is used in high-temperature applications where oxidation is liable to take place. Platinum–rhodium thermocouples are used to measure and control temperatures up to 3500°F. Standard combinations of rhodium and platinum are a platinum–10 per cent rhodium wire in combination with a platinum wire, or a platinum–13 per cent rhodium wire and a platinum wire.

Other thermocouple combinations of rhodium and platinum that have been used are 5 per cent in combination with 20 per cent, 6 with 30 per cent, and 20 with 40 per cent. Platinum–rhodium is also used in the glass industry and for windings in electric furnaces.

Other elements used as alloys with platinum are iridium, ruthenium, nickel, tungsten, and cobalt. Combined with iridium, platinum is used for high-temperature applications in the chemical and electrical fields. It is also used for jewelry. Because of its stability, the platinum–10 per cent iridium bar is the material used in the measuring standard of length and weight stored by the Bureau of Standards. The platinum–ruthenium alloy has the same application as the platinum–iridium alloy. Both iridium and ruthenium when alloyed with platinum have wide use as hypodermic needles.

The platinum–nickel alloy develops good high-temperature strength. Oxide-coated wires of platinum–5 per cent nickel find their greatest use in electron tubes. Where oxidation is objectionable at high temperature, this alloy should not be used.

When alloyed with tungsten, it is used in potentiometers where wear resistance and low noise are required, such as in power tube grids and spark plug contacts. It is an excellent alloy for making strain gages because of its good high-temperature physical properties. This alloy should not be used where oxidation can take place.

The platinum–cobalt alloy makes excellent permanent magnets because of the disordered lattice structure at elevated temperatures and the ordered lattice structure that occurs at room temperatures. The strain develops as a result of the lattice change from face-centered cubic to a face-centered tetragonal structure, and the ordering of the lattices hardens the material.

Palladium.   This metal has about the same physical property characteristics as platinum. It is lighter and less dense than platinum. Because it
remains relatively free from tarnish, it is used in electrical contacts for
audio transmission where freedom from noise is required. It may be hot-
or cold-worked or fabricated into very thin leaf. Cold-working will increase its tensile strength by as much as 40,000 psi for 80 per cent reduction.

   When alloyed with silver, it is used for electrical contacts. All the
advantages of palladium and few of the disadvantages of silver are pre-
served. Because of its 252-ohm/circular-mil-foot resistivity and its coefficient of thermal expansion of $6.8 \times 10^{-6}/°F$ between 30 and 212°F,
when alloyed with 90 per cent silver it is used as a brazing alloy for joining Inconel, nickel, and other of the heat-resisting alloys. A combination
of 60 per cent Pd–40 per cent Ag yields an alloy that is used in precision
resistance wires.

   When alloyed with copper, it is used primarily for electrical contacts
that are to operate in circuits in the milliampere range, and for slip rings
where a hard material is used for brushes.

   Palladium–silver–copper and the palladium–silver–gold alloys are
used for electrical contacts that are subject to wear. Both are subject to
age-hardening and are used in dental work. Platinum–silver–gold alloys
are used for high-strength brazing. The latter alloy may also be clad to
other metals. It is easily worked and is very corrosion resistant. When alloyed with ruthenium, it is used in jewelry and for electrical contacts.

Iridium.   This element has a very high specific gravity of 22.5. Up to this
point, iridium has been referred to only when alloyed with one of the other
elements. Commercially pure iridium is used in extrusion dies in the processing of high-temperature-melting glass. It is a very highly corrosion
resistant material and possesses very high temperature tensile strength.
Radioactive iridium–192 is used in gamma photography in the nonde-
structive testing of materials.

Osmium.   This element has the second highest specific gravity of 22.57.
It is a very hard metal and has high wear and corrosion resistance. Because of this high wear resistance, it is used in phonograph needles, pens,
electrical contacts, and bearings in instruments. It is very hard (600 BHN)
and has the highest melting point (4900°F) of all the precious metals.

Rhodium.   Since this metal maintains its high uniform reflectivity, it is
used for mirrors. It is also used on jewelry articles because of its non-
tarnish characteristic. It is sometimes applied to electrical contact surfaces.

Ruthenium. This is a soft material that is used primarily as an alloy. This aspect of ruthenium was discussed earlier.

11-9 SPECIAL-PURPOSE METALS

Materials that are used as solid-state semiconductors are germanium, silicon, gallium, selenium, and tellurium.

Germanium. This is a grayish-white metal with a diamond cubic crystal structure that makes it useful as transistor material. As such, it is used to amplify electric currents, to rectify alternating to direct current, or as amplifier or detector material in the field of radio and radar. When used with arsenic or phosphorus, the crystal becomes more negative and more conducting. When used with gallium or aluminum, it becomes more positive. Junction-type transistors use a combination of these materials. When alloyed with gold, it is used as a jewelry solder. When impregnated with helium, the inoculated germanium material will rectify an electric current, allowing it to pass in one direction but not in the other. It cannot be used as a semiconductor at temperatures much above 175°F.

Selenium. High-purity selenium is a gray powder. It exists in several allotropic forms. Amorphous selenium, a red powder, is used in electrical instruments and in photoelectric cells. It is also used as a coating on nickel-plated steel or aluminum for rectification of electric currents. One main use is as a photoconducting cell, which is sensitive to light. Selenium is also used for coloring glass and ceramics and plays an important role in metals when alloyed with steel, copper, and stainless steel by increasing their machinability.

Silicon. Another of the semiconductor materials, in its pure state it may be used in rectifiers, transistors, and other electronic devices. It is a better semiconductor material than germanium, because it will operate at higher temperatures (300°F) and transmit more electricity. It also makes better rectifiers than selenium for the same reasons. Its use as a deoxidizing and reducing agent when alloyed with steel, cast iron, and nonferrous metals has been discussed already.

Tellurium. A silver-gray material that behaves like a semiconductor, its chief use is as an alloy with copper, to form tellurium copper or bronze, and steel, to increase its machinability. When used as an alloy with lead, it improves the work-hardening and increases the recrystallization temperature. It is also used in the manufacture of thermoelectric elements.

Gallium. A silver-white material, which, when alloyed with arsenic (gallium arsenide), becomes a very effective semiconductor material. It may be used in amplifiers, oscillators, diodes, switching circuits, and FM transmitters.

Problems

11-1.   List the four classifications of copper and state the chief characteristics for each. What are the principal alloying elements in each class?

11-2.   (a) What effect does beryllium have when it is added to copper? (b) Cadmium?

11-3.   When present, what effect does the β phase have on the ductility of brass and bronze?

11-4.   (a) What effect does the addition of phosphorus have on bronze? (b) On the casting process?

11-5.   (a) How is silicon bronze hardened? (b) How is aluminum bronze hardened? Explain fully.

11-6.   Is it possible to solution heat treat or precipitation harden alloys of copper and nickel?

11-7.   What combination of copper and nickel constitute the cupronickel alloys?

11-8.   What is German silver? What are its characteristics?

11-9.   What is the material called constantan? What are its characteristics?

11-10.   List the 10 copper alloys frequently used in casting. Select any two. Refer to Table 11-1, and discuss their composition and mechanical properties.

11-11.   (a) How are cold-worked copper materials classified? (b) How are the soft materials classified?

11-12.   How is the degree of annealing of copper or its alloys described? Explain.

11-13.   Discuss the physical properties of (a) ETP, (b) OF, (c) DLP, and (d) DHP copper.

11-14.   Discuss the physical properties of (a) the free-machining and (b) the heat-treatable coppers.

11-15.   (a) How is the mechanical strength of copper and its alloys determined? (b) How may the mechanical properties of some copper alloys be achieved?

11-16.   (a) What effect does the addition of aluminum have on copper and copper–zinc alloys? (b) Which other metals have similar effects?

11-17. Which of the copper alloys have acceptable machinability? Which have the poorest machinability rating?

11-18. List some of the uses of free-cutting brass.

11-19. (a) Which of the copper alloys are suitable for cold forming? (b) Which are suitable for severe cold-working?

11-20. Which of the copper alloys are suitable for hot-working?

11-21. All the copper alloys, except one, may be arc welded. (a) Which one cannot be welded? (b) List the types of arc-welding processes that may be used.

11-22. Which of the copper alloys may be gas welded?

11-23. Which of the copper alloys may be brazed or soldered?

11-24. List some uses of (a) electrical tough pitch copper, (b) phosphorized copper, (c) oxygen-free copper, (d) free-machining copper, (e) silver copper, and (f) beryllium copper.

11-25. List some uses of (a) plain brasses and (b) free-cutting brasses. (c) List some uses of the brasses in the general category.

11-26. List some uses of (a) phosphor bronzes, (b) aluminum bronzes, and (c) silicon bronzes.

11-27. List some uses of (a) cupronickel alloys of copper and (b) nickel silver coppers.

11-28. List some uses of nickel alloys and state the alloys used.

11-29. (a) What are the principal elements used in Monel metals? Discuss some of their properties. (b) List their uses.

11-30. What are the principal elements used in the Inconel metals? Discuss some of their properties.

11-31. What are the principal alloys that are characteristic of the Hastelloy metals? Discuss some of their properties.

11-32. List the alloys that are used in the Illium metals. What are some of the properties of these metals?

11-33. What are the principal alloys of the electrical resistance alloys? List some of their uses.

11-34. List some of the uses of lead. How do antimony and tin alter the properties of lead when they are used as alloys?

11-35. What are the uses of the plain alloys of lead?

11-36. Which of the alloying elements are added to lead to convert it to solder?

11-37. What are some uses of antimonial lead?

11-38. Which alloys are added to lead to create the babbits?

11-39. List the three major uses of tin.

11-40. List the uses of the plain tin alloys.

11-41. What are some of the advantages of tin as a solder material over lead?

11-42. What elements are alloyed to form tin babbit?

11-43. Discuss the special alloys of tin. List the properties that these alloys must exhibit to be useful.

11-44. What are some uses of commercially pure zinc?

11-45. List some of the alloys and the characteristics that these alloys impart to zinc.

11-46. Refer to the position of zinc in the electromotive series and explain how it protects iron when coated with it.

11-47. (a) What is wrought zinc? (b) When it is used?

11-48. (a) What is copper-hardened rolled zinc? (b) What effect does the addition have on the copper when magnesium is added? (c) When titanium is added?

11-49. List the precious metals. What are some common characteristics of these metals?

11-50. List the properties and uses of silver and its alloys.

11-51. Discuss silver and its alloys as a dental material.

11-52. Discuss the use of silver alloys as (a) an electrical material, (b) sterling silver, and (c) coin silver.

11-53. What are some of the uses of commercially pure gold?

11-54. (a) List the karat rating of green, red, and yellow gold. (b) Which element gives the material its characteristic color? (c) How is the karat rating determined? Illustrate.

11-55. How are the characteristic white, greenish-white, and pink colors controlled in the "white" golds?

11-56. Which element when alloyed with gold yields (a) a high-temperature solder? (b) A high-resistivity material?

11-57. List the uses of commercially pure platinum.

11-58. List several of the alloys of platinum and state their major uses and properties.

11-59. State the use of platinum when it is alloyed with (a) palladium, (b) rhodium, (c) iridium, and (d) ruthenium.

11-60.　List some of the effects of alloying platinum with (a) nickel, (b) tungsten, and (c) cobalt.

11-61.　What are some of the chief advantages of palladium?

11-62.　Discuss the characteristics of an alloy of palladium and (a) silver, (b) copper, (c) silver–gold, and (d) ruthenium.

11-63.　List the principal properties of the element (a) iridium, (b) osmium, (c) rhodium, and (d) ruthenium.

11-64.　Discuss the use of germanium as a semiconductor when it is alloyed with (a) arsenic or phosphorus, and (b) gallium or aluminum.

11-65.　Discuss some uses of (a) selenium, (b) silicon as a semiconductor material, (c) tellurium, and (d) gallium.

12 | Corrosion

12-1 TYPES OF CORROSION

The chemical attack of metals with the resultant deterioration, destruction, and loss of material is known as corrosion. The attack may be internal as well as external. The rate at which corrosion takes place is a function of factors such as the composition of the metal, the degree of exposure to humidity and temperature, and the presence of voids, stresses, and types of surface films.

One method of classifying corrosion is by describing the mechanism by which corrosion progresses. Such mechanisms are galvanic cell, direct chemical corrosion, or oxidation.

As stated, one form of corrosion is generated by the presence of a galvanic cell as a result of electrochemical action. Galvanic cells are present when two dissimilar metals are placed in contact, or near contact, with each other. If they are in near contact, they need a transporting medium. The result is that one metal is more negative than the other. Movement of electrons from the more negative metal to the less negative metal will take place. Thus, using a copper fitting on a steel pipe will create a galvanic corrosion cell.

To form a galvanic cell there must be an anode and an electrode present, and there must be a flow of a direct current. Corrosion takes place at, or in the vicinity of, the anode. The anode is the place where current leaves the material, causing the metal to disintegrate. Disintegration is used in the opposite sense from plating. The student should be aware that *electron* and *current "flow"* are opposite in direction. Thus, if current

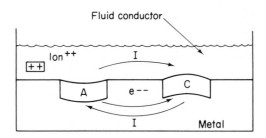

Figure 12-1

flows clockwise, the electrons flow counterclockwise. This is shown in Fig. 12-1.

The current enters the metal at the cathode. Corrosion does not take place at the cathode. However, plating does take place at this point.

Anodes and cathodes may form at the surface of a material when there are local differences of composition in the metal. Local differences between metals and their surrounding environment may also create an anode–cathode couple. These may be created by impurities; microstructure differences; stresses, scratches, hole, or notch differences; or grain boundary differences.

At the anode the material loses electrons, which migrate toward the cathode. The loss of electrons creates a positively charged ion, which leaves the surface of the metal. The surface of the metal at the anode is said to be *oxidized*. At the cathode the electrons combine with other ions, generally hydrogen, and form hydrogen molecules. This process is called *reduction*. If a piece of iron is immersed in water, the corrosion reactions are as follows:

Oxidation at anode: $Fe \rightarrow Fe^{++} + 2e^-$
Reduction at cathode: $2H^+ + 2e^- = H_2$

The equation is

$$Fe + 2H^+ = Fe^{++} + H_2$$

The reduction process is referred to as *polarization*. The polarization mechanism creates a resistance to electron flow at the cathode. This blocking effect results from the accumulation of hydrogen on and around the cathode. It always retards current flow, and thus reduces the rate at which materials corrode. Any mechanism that reduces polarization increases corrosion.

The electromotive series shown in Table 12-1 lists the hydrogen ion as the reference electrode. Any material that is listed as more negative than hydrogen will corrode. The more negative, the greater the tendency to corrosion. Any differences in size between the anode and cathode will increase the corrosion process. Thus a small anode, with reference to the cathode, will accelerate the corrosion process. In some instances, metal is more negative than iron and should therefore create a cell conducive to corrosion, and yet they do not corrode. Such is the case of aluminum when added to iron. When chromium is added to iron, the alloy is also corrosion resistant. In fact, the latter combination is called stainless steel. In both cases a thin layer of a gas or corrosive material forms at the surface of these materials to prevent corrosion. This is a process called *passivity*.

Even though elements have a certain position in the hydrogen galvanic series (Table 12-1), this may change with the electrolytic solution. Thus tin will increase the corrosion of steel when the electrolyte is salt water. It will protect the steel when the electrolyte is fruit juice. The same effect is perceived when different alloys of the same base metal are considered.

Table 12-1   Galvanic Series — Salt Water

Metal	Ion	Potential
Lithium	$Li^+ + e^-$	−3.02 anode end (highest corrosion possible)
Potassium	$K^+ + e^-$	−2.92
Calcium	$Ca^+ + 2e^-$	−2.87
Sodium	$Na^+ + e^-$	−2.71
Magnesium	$Mg^{++} + 2e^-$	−2.34
Beryllium	$Be^{++} + 2e^-$	−1.70
Aluminum	$Al^{+++} + 3e^-$	−1.67
Uranium	$U^{++++} + 4e^-$	−1.40
Manganese	$Mn^{++} + 2e^-$	−1.05
Zinc	$Zn^{++} + 2e^-$	−0.76
Chromium	$Cr^{+++} + 3e^-$	−0.71
Iron	$Fe^{++} + 2e^-$	−0.44
Cadmium	$Cd^{++} + 2e^-$	−0.40
Cobalt	$Co^{++} + 2e^-$	−0.28
Nickel	$Ni^{++} + 2e^-$	−0.25
Tin	$Sn^{++} + 2e^-$	−0.14
Lead	$Pb^{++} + 2e^-$	−0.13
Hydrogen	$H^+ + e^-$	0.00
Antimony	$Sb^{+++} + 3e^-$	−0.10
Bismuth	$Bi^{+++} + 3e^-$	+0.23
Copper	$Cu^{++} + 2e^-$	+0.34
Copper	$Cu^+ + e$	+0.52
Silver	$Ag^+ + e^-$	+0.80
Lead	$Pb^{++++} + 4e^-$	+0.83
Mercury	Hg^{++}	+0.85
Platinum	$Pt^{++++} + 4e^-$	+1.20
Gold	$Au^{+++} + 3e^-$	+1.42 cathode end (lowest corrosion possible)

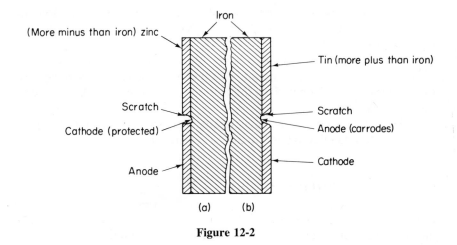

Figure 12-2

Figure 12-2(a) and (b) show iron that has been coated with zinc and tin, respectively. Reference to Table 12-1 will show that the iron is protected by the zinc, since it is closer to the high-corrosion end of the electromotive series. Zinc is the anode and serves to protect the cathodic iron. When tin is plated, the iron becomes the anode and the tin plate is the cathode. The iron will corrode more rapidly. The iron when exposed receives no protection from the tin.

12-2 CHEMICAL FACTORS

Moisture increases the electrical conductivity of the carrier between the anode and the cathode. In general, corrosion is a function of the rate of electrical conductivity. An increase in the conductivity of the environment increases the corrosion rate. Studies show that atmospheres with less than 30 per cent humidity generally will not increase the corrosion of materials.

Nitric *acid,* which is oxidizing, may increase the corrosion rate if it does not form a protective coating on the metal surface. The nonoxidizing acids, such as hydrochloric acid, will increase the rate of corrosion of the surface of metal because of the production of hydrogen. In both cases, if a protective film is present on the surface of the metal, the metal will not corrode. If the acid removes the coating, the metal will corrode. As long as the metal is in contact with this acid solution, no new film will form to protect the metallic surface, and the metal will corrode.

Alkalides when concentrated in solution will cause localized corrosion. This is especially true at elevated temperatures. In general, alkalides have very little effect upon the rate of the overall corrosion of a metal. Localized corrosion may take place.

Natural salts in the presence of moisture increase corrosion. In many instances the localized corrosion effect is more damaging than the overall corrosion effect.

Oxygen in solution causes severe localized corrosion of steel. In some cases where a large amount of oxygen is present, such as in the case of stainless steel, it depolarizes the cathode and forms an insulating coating on the anode.

Sulfur, as a compound, accelerates corrosion. It renders the protective scale less effective to corrosion resistance.

12-3 CORROSION MECHANISM

High temperatures, in general, increase the rate of corrosion. If increasing the temperature of a solution increases the conductivity of the solution, the rate of corrosion of the metal increases. It also decreases the polarity of the cathode. If increasing the temperature of the solution decreases the oxygen concentration that has been dissolved in the solution, corrosion will be retarded because of the continuous protective films that are formed.

In general, increasing the temperature of an electrolyte exposed to air results in an increase in corrosion to the metal up to about 180°F.* Between 180 and 212°F the rate of corrosion may decrease. It may increase again above the boiling point.

There is a type of corrosion that generally takes place where high velocity turbulence is present. This happens on propellers in agitators, pumps, or valves. Since the corrosion takes place as a result of the wearing away of the protective film, it is called *erosion corrosion. Cavitation* and *fretting corrosion* are forms of erosion corrosion. The former results from turbulence, the latter from stress concentration. These corrosion mechanisms result from high velocity impingement of fluids on the metal surfaces. This may result in a cavitation effect due to the rapid breakdown of gas or vapor bubbles at the metal fluid contact point. Such action increases the corrosion damage to the metal.

Direct chemical attack takes place over the entire surface of the metal. This occurs when the anode keeps moving around over the surface of the metal. This is sometimes referred to as *uniform corrosion.* It is a surface phenomenon.

Corrosion may occur *intergranularly.* This is an internal mechanism. It occurs when chromium carbides precipitate into the grain boundaries, thus lowering the chromium content adjacent to the grain bounda-

The Making and Shaping of Steel, p. 932. U.S. Steel (1964).

ries. A galvanic cell is created between the large grains (cathode) and boundary materials (anode). See Fig. 12-3. Austenitic stainless steel at temperatures of 1500°F is subject to intergranular corrosion. Fine-grained steels are more subject to corrosion than coarse-grained steel.

If the carbon content of the steel is low (below 0.03 per cent), the possibility of chromium carbides forming is very small. If the carbon percentages are greater than 0.03 per cent, chromium carbides form instead of iron carbides. The chromium, which acts to passivate the steel, is precipitated into the grain boundaries as chromium carbide. Intergranular corrosion increases. Steels that have had tantalum, titanium, or columbium added form carbides at elevated temperatures and eliminate carbon precipitation at the grain boundaries, thereby eliminating the possibility of the formation of chromium carbides. Slow cooling from elevated temperatures will also give carbon time to form chromium carbides, which precipitate into the grain boundaries. Fast cooling does not allow these carbides to form. Therefore, controlled cooling can affect the corrosion qualities of steel.

Corrosion resistance is also important when making cast-iron castings. High-silicon cast alloys, austenitic nickel alloys, and chromium casting alloys are the three main types of corrosion-resisting cast irons.

The high-silicon irons alloy about 0.75 per cent carbon with approximately 15 per cent silicon and about 0.75 per cent manganese. These castings are very hard and brittle and have a low tensile strength at about 20,000 psi. They are very resistant to oxidizing and nonoxidizing acids such as phosphoric, sulfuric, and nitric acid. The other acids require special alloying with iron to achieve acid corrosion resistance.

Austenitic nickel alloys may also produce a corrosion-resisting casting. The alloys that enter into this type of material are silicon, 2 per cent; tempered carbon, 3 per cent; nickel, 4 to 20 per cent; chromium, 2 per cent; copper, 6 per cent; and manganese, about 1 per cent. These materials expand very little when heated because they remain austenitic at all times. They are corrosion resistant to alkalies, acids, and salts. Tensile strength for this material is about 30,000 to 40,000 psi.

When differences in the environment create an *anode–cathode cou-*

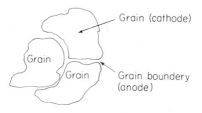

Figure 12-3

ple, cells may form in a localized area and a current is caused to flow. Thus two electrodes of the same material placed into different concentrations of an acid will cause a current to flow. This is called *cell concentration corrosion.* Figure 12-4(a) shows corrosion due to ion cell concentration. Figure 12-4(b) shows corrosion due to oxygen ion concentration.

Pit corrosion is another type of corrosion that is localized and results in holes or pitting. Pitting results from the accumulation of small concentrations of cells at a point or in an existing pit. If the material underneath this concentration develops an increase in ions, as contrasted with a constant ion supply at a distance from the concentration, corrosion will proceed.

Stress corrosion results from any mechanism that will generate high stress concentrations in the presence of conditions conducive to corrosion. Thus the effect of the stress is to accelerate the corrosion process, rather than create it. The corrosion that takes place is a localized phenomenon and therefore results in failures in the form of cracking.

Season cracking and *caustic embrittlement* are two types of stress corrosions. The former takes place when certain materials, such as severely worked brass, crack in the presence of ammonia. Welds that have not been stress relieved will fail in the presence of sodium hydroxide at elevated temperatures. Caustic embrittlement occurs in steel when it is

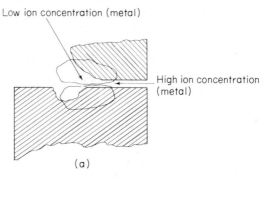

(a)

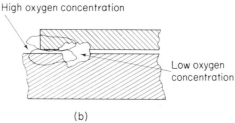

(b)

Figure 12-4

under high stress and high temperature in the presence of sodium hydroxide. In steam boilers, at joints or seams that leak, the evaporation of the high-temperature steam in the presence of water softeners will cause failures. The steam provides the stress and the temperature. The softener provides the sodium hydroxide. These conditions are ideal for supporting corrosion.

Dezincification is a mechanism wherein a material loses one of its components in such a way that a galvanic cell is created. Once started, the process continues until the part fails. The removal of zinc from brass, cobalt from high-cobalt alloys, iron from cast iron, or silicon from copper–silicon alloys causes corrosion by this process.

When dezincification takes place in brass, it corrodes by the removal of the zinc from the brass. The zinc remains in the solution. The copper plates back on the brass. The structure that remains is pitted, porous, and brittle. It consists almost entirely of copper oxide and copper.

12-4   PROTECTION AGAINST CORROSION

Several procedures may be used to avoid failure that results from corrosion. The careful selection and manufacture of materials used, surface coatings, the use of inhibitors, and the use of cathode protection are some.

Selection of the appropriate material for a specific use, in a specific environment, is of the utmost importance. The careful selection of a corrosion-resisting material will reduce the cost of subsequent maintenance and may result in long-term savings, even though the initial cost of the material selected may be high. Alloys or heat-treated materials must be considered along with other factors when one is selecting corrosion-resistant materials. The selection of appropriate material for a particular use must be made based on previous experience with the material or upon tests conducted to justify its use.

In many instances the design of the part or assembly may affect the rate of corrosion. Where crevices, joints, and overlaps occur there is likely to be a concentration of moisture, or electrolyte. Such concentration will cause the base metal to corrode. Changing the design will usually correct this condition. Scale or dirt will polarize a metal surface. In many instances the surface under the dirt or scale will become anodic, whereas the clean metal will become cathodic.

Cathode protection is achieved if the material that is protected is capable of being converted into a cathode. It should be remembered that it is the anode that corrodes. Thus a material may be protected by the process of generating a direct current and controlling its direction of flow. The flow should be into the part to be protected. The introduction of a more negative material than the object being protected will cause the object to become cathodic with reference to the negative material. Thus, if zinc is placed in the vicinity of iron, it becomes the anode and the steel the cathode. The zinc is sacrificed in order to keep the steel from corroding. Aluminum and magnesium may also be used as polarizing materials to prevent corrosion of base materials used in ships' hulls, propellers, and hot-water tanks. Direct current from a battery, generator, or rectifier will also produce the desired protection of the cathode.

Inhibitors may be used in the corrosive medium for the purpose of making them inert. They may be used to "coat" the anode and prevent its action. The cathode may be tendered inactive by coating it with a passive film. An example of the former is a passive film of ferrous oxide (Fe_2O_3) that is formed on steel by sodium phosphate in water. Cathodes may be made passive with films that block the flow of electrons or oxygen through them. Oils, in some instances, create effective films that will combat corrosion. Inhibitors may also be used to retard the action of the electrolyte. The chromates are the most widely used inhibitors. Others are phosphates, silicates, and some organic compounds.

One of the most important methods used to prevent corrosion is the use of *coatings* that present a barrier to the movement of electrons. There are several methods of classifying the coatings. They may be classified as organic, inorganic, or nonmetallic coatings.

Organic coatings operate by preventing moisture, air, or other materials from attacking the base material. Since they will disintegrate in the presence of heat, they should be used at room temperatures. Organic coating materials are paints, varnishes, adhesive plastic tapes, tar, grease, and oil. One important consideration is thoroughly to clean the surface to be coated so that the impurities are removed. Washing, electrolytic cleaning, and sand-blasting are but a few of the methods used to clean castings.

Inorganic coatings are typified by glass, resins, or enamel. These also create a barrier between the environment and its surroundings. Nickel, tin, or other metals may also be applied to the surface to be protected. The inorganic coatings are brittle and hard. All are subject to failure due to shock. They are applied by plating, spraying, baking, and dipping. These materials become fused when they are applied, or are processed after they are applied to create a homogeneous protective coat.

In some instances, *nonmetallic materials* are used because they are noncorrosive. In one instance, the entire design could be made from these materials. In another, they could be used as protective coatings. Such materials are the ceramics, plastics, rubber, wood, or carbon.

Metal purification, alteration of the environment, shape of the part designed, or avoidance of stress situations will help considerably in the prevention of corrosion. Thus the use of pure metals helps to forestall corrosion because of the lack of nuclei or cells where it can start. The shape of a part could generate stress concentration or turbulent conditions.

Problems

12-1. List the factors that affect the rate at which the corrosion of metals takes place.

12-2. In a galvanic cell, describe the function of (a) the anode and (b) the cathode.

12-3. Describe the operation of a galvanic cell.

12-4. In how many ways may an anode–cathode couple be created?

12-5. Explain the oxidation–reduction of iron.

12-6. Discuss the term polarization as it refers to the cathode.

12-7. Which material will corrode when coupled with iron and immersed in salt water: (a) magnesium, (b) aluminum, (c) zinc, (d) nickel, (e) copper?

12-8. (a) Which metal is said to corrode—one that is more or less negative than hydrogen? (b) List two such metals.

12-9. If the anode is smaller than the cathode, will it accelerate or retard corrosion?

12-10. What is passivity?

12-11. What effect does the electrolyte have on the galvanic series?

12-12. Explain Fig. 12-2(a) and (b).

12-13. What effect does moisture have on the rate of corrosion? Explain.

12-14. Discuss the effect of nitric and hydrochloric acids on the rate of corrosion of metal surfaces.

12-15. Repeat Problem 12-14 for (a) the alkalides, (b) the natural salts, (c) oxygen, (d) sulfur.

12-16. Discuss the effects of increasing the temperature of the cell system.

12-17. Describe the corrosion mechanism referred to as erosion corrosion.

12-18. What are cavitation and fretting corrosion?

12-19. What is uniform corrosion?

12-20. Describe the mechanism that results in intergranular corrosion.

12-21. Discuss the effect of carbon precipitation into the grain boundaries of steel when in the presence of chromium.

12-22. Discuss the corrosion of high-silicon cast iron and austenitic nickel alloys when cast.

12-23. Describe cell concentration corrosion and how it operates.

12-24. What is pit corrosion?

12-25. What effect does stress have on corrosion? Why?

12-26. Season cracking and embrittlement are two forms of stress corrosion. Describe them.

12-27. Describe the corrosion process known as dezincification.

12-28. List the various ways in which corrosion may be combated.

12-29. What effect does the design of a part or assembly have on the rate of corrosion? Give examples to support your discussion.

12-30. Can a direct-current source be used to combat corrosion? Explain.

12-31. Explain the creation of an electrolytic cell when dirt or scale are present on steel.

12-32. How may inhibitors be used to prevent corrosion?

12-33. Describe the method for combating corrosion called cathode protection.

12-34. Describe the use of organic coating to combat corrosion. List several such organic coatings.

12-35. List several inorganic materials used to prevent corrosion.

12-36. Which are the materials listed as nonmetallic and which are used to prevent corrosion?

12-37. How can metal purification prevent corrosion?

12-38. Explain why the shape of a machine part is an important consideration in preventing corrosion.

13 | Inorganic Materials

13-1 CERAMIC MATERIALS

In general, ceramic materials have structures that are a combination of nonmetallic and metallic elements. But the properties of ceramic materials differ from those materials that are metallic or nonmetallic. Because there are many combinations of metallic and nonmetallic materials, there are many ceramic phases.

The ceramic materials considered in this chapter are stones, clays, refractories, glass, and asphalt. In general, their structure is complex, with but few free electrons. Like steel, the complex structure is crystalline. The lack of many free electrons at room temperature makes most of these materials electrically nonconducting and poor thermal conductors. In addition, they are brittle, hard, and stiff. They are generally stronger in compression than in tension. They are always totally elastic and as such exhibit practically no evidence of plastic flow when a load is applied. Thus, removing the load results in the material immediately returning to its original dimensions. Also, deformation of the material is very little prior to fracture.

The resistance of ceramics to chemical attack is greater than that of metals and organic materials. They usually have high melting points. Their stability results from electron sharing, which produces strong ionic bonding, or from covalent bonding.

Ceramics, as a class, have the highest melting points. They range as high as 7000°F and as low as 3500°F, with an average of about 4200°F. Table 13-1 shows the physical properties of some ceramic materials. Many of these materials will be referred to throughout this chapter.

Table 13-1   Ceramic Materials

Chem. symbol	Material	Melting point °F	Mod. of elast. × 10⁶ psi	Tensile strength × 10³ psi	Compr. strength × 10³ psi
Al_2O_3	Alumin	3700	50	40	400
BeO	Beryllia	4450	50	14	185
B_4C	Boron carbide	4400	42	22	420
Cr_4C	Chromium carbide	4000	36		450
	Electric porcelain	3200	10	5	60
General	Graphite		0.5–1.7	0.5–1.5	2–6
Muscovite	Mica	2500	25	45	150
SiC	Silicon carbide	4950	13	2	15
Dense	Silicon carbide	5000	68	25	150
TiC	Titanium carbide	5700	50	65	300
WC	Tungsten carbide	5025	80	130	650
ZrO_2	Zirconia	4900	20	10	70

13-2   STONE

Stone is one of the oldest engineering materials known to man. It is used in the form of limestone, sandstone, slate, marble, and granite. Very little of these materials is imported already processed. The processing of these materials into a form so that they may be used as an engineering material takes place almost entirely within the borders of the United States.

Rocks are the solid room-temperature state of the molten material. The original cooling of the earth took place and resulted in a hard crust. Volcanic eruptions and cracks in this crust continued to produce molten materials that cooled into rocks. The solid state of these molten materials is called *igneous rock.*

When the melt cools very rapidly, such as if quenched, the resulting structure has a crystalline appearance. If the melt is allowed to cool slowly, the visible crystals may orient themselves in such a manner that they take a form that we call *granite. Basalt,* or *felsite,* has a structure in which the individual crystals are not visible. Basalt rock that has a dark color is called *traprock.*

Igneous rock that has disintegrated and, as a result of weather conditions, been moved in layers into a new rock formation is called *sedimentary rock.* Limestone, sandstone, and shale are all sedimentary rocks.

Marble, slate, asbestos, mica, and soapstone are all classes of rocks that have resulted from the breaking up of igneous and sedimentary rocks. They are called *metamorphic rocks.*

The properties of stone are related to their chemical and physical structure. The three chemical structures that are most commonly found to make up stone are the following:

1. *Silica,* which is silicon dioxide (SiO_2).

2. *Silicate of alumina.*

3. *Calcium carbonate* ($CaCO_3$).

Silica is durable and will withstand attack from any material except hydrofluoric acid, which dissolves it. Silicate of alumina, which is the basis for shale, slate, feldspar, and clay, is much less durable than silica. Calcium carbonate is soluble in dilute acid solutions. It therefore will disintegrate in polluted moist atmospheres, because sulfur dioxide when mixed with water forms the sulfurous acids. Limestone and marble fall in this category. The physical properties of these materials are shown in Table 13-2.

Limestone.   Limestone is a sedimentary stone composed largely of calcium carbonate ($CaCO_3$) and calcium magnesium carbonate ($MgCO_3$). Upon calcination, limestone yields lime (CaO). Nearly 70 per cent of all stone used in construction is limestone. It has a specific weight of about 155 lb/ft^3 and a crushing strength of about 9000 psi. It also has a fair resistance to temperatures below 1650°F. Its low porosity and its workability makes it suitable for use in masonry, bridges, and building construction. Limestone is also used in the manufacture of acetylene. When burned with coke, it forms calcium carbide, which is then converted into acetylene.

Sandstone.   Sandstone is an aggregate of many sand particles bound together into a natural state by limestone, hydrated silica, iron oxide, or clay. If the grain sizes are small and tightly bound together, the material approaches shale in appearance and physical properties. If the grains are large, the end product approaches a unified mass, or conglomerate.

Calcium carbonate, the binder in sandstone, is easily worked, but will disintegrate rapidly under severe weather conditions. It is homogeneous and may split in any direction. A material called "itacolumite" is a flexible sandstone that contains talc or mica. Chert, known also as hearthstone, is a silicate sedimentary stone used in buildings and paving. It splinters when it disintegrates. The various colors found in this material are the result of the impurities in the stone. The pure stone has a white or cream color. Iron oxides produce a range of colors from yellow to red. Glauconite yields green. Manganese oxide produces a black-colored sandstone. Some sandstones are streaked with combinations of the above colors.

The average compressive strength of sandstone used in building materials is about 10,000 psi. Its specific weight is about 148 lb/ft^3. It has

Table 13-2 Physical Properties of Stone*

Material	Compres. st. ×1000 psi	Porosity %	Weight lb/ft³	Therm. coef. 10⁻⁶/°F	Mod. rupt. ×1000 psi	Sh. st. ×1000 psi	T.S. ×1000	Elast. mod. ×10⁶ psi	Wear resist.
Limestone	14	13.0	155	2.5	2.5	2.0	0.6	7.0	8.5
Sandstone	13	18.0	148	5.0	1.5	1.7	0.4	4.8	13.0
Shale			172						
Slate	18	0.4	176	5.1	8.0	2.7	3.6	14.0	8.0
Marble	16	0.4	165	3.8	2.0	3.2	1.5	11.0	19.0
Granite	26	1.0	166	3.6	3.2	3.0	0.8	7.0	60.0
Asbestos	15		153				1.2	7.2	48.0
Mica	25		175		7.0		6.0		
Soapstone			170						

*Average values.

a modulus of elasticity of about 3.5×10^6 psi. Temperatures of up to 1500°F have very little effect on this material.

Building materials are called by names that refer to the localities from which they come. Bluestone (N.Y.) has an even grain and up to 20,000 psi crushing strength. Kemrock is the trade name for sandstone that has been impregnated with a black resin material and baked into a hard finish. It is used for laboratory tables because it is acid- and alkali-resistant. Holystone is the trade name of a very fine grained sandstone used for rubbing down furniture and concrete. Macstone is the trade name for concrete slabs, faced with sandstone, and used as building blocks.

Shale.   Shale is composed of fine-grained silica and alumina. It is formed under pressure into a usable material. Shale is not porous like sandstone. It is hard and has the properties of slate. In its calcinate form, it is used with lime in cements. Another type of shale (Swedish) contains uranium oxide, molybdenum, and vanadium. In one form, oil shale, it is impregnated with a black oil substance which, when heated, breaks down into oil, gas, and coke. In Scotland and England oil shale produces from 15 to 100 gal of oil/ton of shale. In the United States (Colorado) the yield is from 15 to 30 gal of oil/ton of shale.

Slate.   Slate is a form of shale. It is a metamorphosed material that formed from mud and clay deposits. It forms into layers or very flat cleavage planes. This makes it possible to cut and grind flat slabs usable as blackboards, fireproof roofing shingles, flagstone, and electrical insulated panels.

It is a low-porosity material that has high strength and high weather and abrasion resistance. The average compressive strength of slate is 15,000 psi. It is heavy, weighing about 176 lb/ft³. When used for roofing shingles the roof members must be much stronger than normal in order to support the weight. The *black slate* used in roofing comes from coal beds. It is fine-grained, but does not have the weather resistance or hardness of the other slate materials. Slate is also ground into a fine powder, which is then used in caulking compounds and linoleum. A lime–slate powdered mixture (60 per cent lime to 40 per cent slate) when mixed with cement produces an insulating porous concrete.

Marble.   As a result of metamorphic processes, marble has a crystalline structure, is hard, and may be polished. Because of its resistance to wear and abrasion and its decorative acceptability, it is used extensively as a building material for walls, steps, floors, etc. It is also used for electric panels, statues, and ornaments.

Marble has the same chemical structure as limestone. However, the presence of mica or other materials causes imperfections in marble. The term *marble* refers to any of the forms of limestone that can be polished. Pure limestone polishes white, but marble is streaked by many colors. Vermont marble is available in many different colors, such as mottled, black, gray, red, green, or white. Georgia marble is white, pink, gray, or blue. Alabama marble has pure colors and low porosity. Marble from Italy is white and very hard. Marble from Uruguay comes in many colors.

Marble weighs about 165 lb/ft³ with a compressive strength of about 16,000 psi. It has a tensile strength of 500 psi and a modulus of elasticity of 11×10^6 psi. Marble can be purchased in matched slabs for walls, translucent or semitransparent, in block form for statues, as chips to be used in the manufacture of artificial stone, or as marble dust to be used as a filler in soaps and rubber.

Granite.   Granite is an igneous rock whose composition is feldspar and quartz. Small quantities of mica will not harm granite. Large quantities create a line of stress concentrations, which causes the granite to disintegrate. It is a hard material with a compressive strength of about 26,000 psi, a shear strength of 3000 psi, and a tensile strength of 800 psi. Its modulus of elasticity is about 7×10^6 psi. It has a specific weight of approximately 166 lb/ft³ and a specific gravity of about 2.7.

It is a dense, hard, strong, material. It also has good abrasion- and moisture-resistant qualities. It may be finished with a very high polish, and its resistance to abrasion makes it desirable for use as steps. It is used for surface plates in machine-shop part layout. It is available in a variety of colors and color combinations, such as gray, green, or red hues. It may have black or dark green background threaded into the red hues, or pink background threaded into the dark grays and greens. North Carolina granite is light gray, pink, or black. Virginia granite may be green, red, or multicolored. New England granite is mined in many colors.

Mica.   Mica is a metamorphic material. Its cleavage planes permit it to be split off larger blocks into very thin flexible sheets. These sheets are processed into electrical parts. Much of the waste mica is processed, by bonding, into sheets.

Mica is processed in many varieties. The most common variety has an aluminum component, which is used in radio parts. Another variety of mica has a magnesium component — called "amber mica" — which disintegrates when in contact with sulfuric acid. It has a light brown tint and very good heat-resistant qualities. Chromium and vanadium mica has a green tint. Still another type of mica has a calcium component, which gives it a yellow and purple tint. A ruby- and a green-tint mica are mined in India. Other mica is mined in Argentina. It should be noted that synthetic mica has been made successfully.

Carbon.   Carbon is one of the most widely prevalent materials in existence. In the natural state it exists in several different allotropic forms. It forms or may be processed into more compounds than any other element. Amorphous and crystalline black carbons have specific gravities of 1.9 and 2.25, respectively. Diamond has a specific gravity of 3.5. It has a melting point of about 6330°F, a density of 0.078 psi, and a thermal conductivity of 90 Btu/ft/h°F.

Carbon is used in steel and cast iron; it is found in petroleum, coal, charcoal, carbon black, graphite, coke, and plant life. It may be pressed or molded into bricks, blocks, or other shapes for use as electrical components. When processed into a very dense material, it is made into bricks and used as liners in the chemical industry. Because it may be machined and is acid resistant, it may be used in pumps, valves, seals and piston rings in the chemical industry. In this form, called *impervious carbon,* it has a tensile strength of about 2500 psi, a compressive strength of 10,000 psi, and a thermal conductivity of about 100 Btu/ft/hr/°F.

When used for electric light arc electrodes, carbon is usually impregnated with other elements to intensify the current-carrying capacity. Such materials are fluorides of the rare earths to increase the brilliance of the arc; cerium metal to introduce blue into the radiation; and iron or a combination of iron, nickel, and aluminum to produce 3000-angstrom wavelengths for medical purposes. These short wavelengths cannot be seen by the eye. By introducing strontium, powerful infrared radiation wavelengths are produced.

Asbestos.   Asbestos fibers are long, thin, rock crystals formed from old rock by metamorphism. It is mined chiefly in Canada, Vermont, Arizona, Turkey, and South Africa. This material is heat and chemical resistant. It has long fibers, which can be bonded, compacted, or spun. They are used in automobile lining, aircraft air ducts, fireproof gloves and clothing, insulating board, fireproof cloth, shingles, tiles, siding, and pipe covering.

Asbestos board is manufactured by molding fibrous asbestos with cement under pressure. With sodium silicate, it is processed into thin paper sheets. These sheets have good heat-insulating qualities and are fire resistant; they are used to cover pipes, as gaskets, and for electrical insulation. When impregnated with rubber, asbestos is used for packaging chemicals. When clay is used as a binder for asbestos, the sheets are used to insulate against high voltage.

Jade is a granular-type asbestos. It is mined in Burma in green, white, very dark green, or mauve colors.

Soapstone.   This is a soft impure talc material that can be cut easily. Upon heating, it loses its water and hardens. It has a low talc component and is used for lining tanks and heating equipment, electrical panels, and as an abrasive material.

Steatite, a high-talc soapstone, is cut and used for insulators in the electrical and electronic industry. It may be powdered and then molded into desired shapes.

One variety of soapstone, which is blue-gray and is mined in Virginia, has good electrical and thermal resistance. It is used in laboratory table tops and sinks. A green-gray variety is used as a building stone.

13-3　CLAY

The most widely used clay is a hydrated silicate of alumina. Impurities that may be found in clay are ferric oxide, lime, potash, and magnesium. These impurities impart aesthetic qualities to clay. Actually, clay is the end result of the disintegration of igneous rock into very fine particles. These particles when mixed with water form a plastic mass that can be formed or molded to desired shapes, and then fired to a hard material.

Shale and *slate* are two materials that are actually forms of clay. Various combinations of the three major ingredients that enter into the clay product may be extracted from the clay-composition diagram, Fig. 13-1(a). The diagram shows the various combinations of alumina (Al_2O_3), silica (SiO_2), impurities, and water. The impurities, as pointed out, could be ferric oxide, potash, magnesium, or lime.

Point A in Fig. 13-1(a) indicates a percentage combination of Al_2O_3, SiO_2, and impurities for a particular shale. The lower right vertex of the triangle is 100 per cent Al_2O_3; the lower left vertex of the triangle is 100 per cent SiO_2; the upper vertex is 100 per cent impurities. If a dotted line is drawn from the desired 100 per cent vertex perpendicular to the opposite side of the triangle, it will also be perpendicular to the 20 per cent Al_2O_3 line, which passes through point A. This is shown in Fig. 13-1(b). The side opposite the SiO_2 (100 per cent) vertex is the 70 per cent line. The side opposite the impurities (100 per cent) vertex is the 10 per cent line. The composition of this clay (shale) at point A is

$$Al_2O_3 = 20\%$$

$$SiO_2 = 70\%$$

$$\text{Impurities} = 10\%$$

Example 1

State the percentages by weight of the ingredients of the clay designated as B in Fig. 13-1(a).

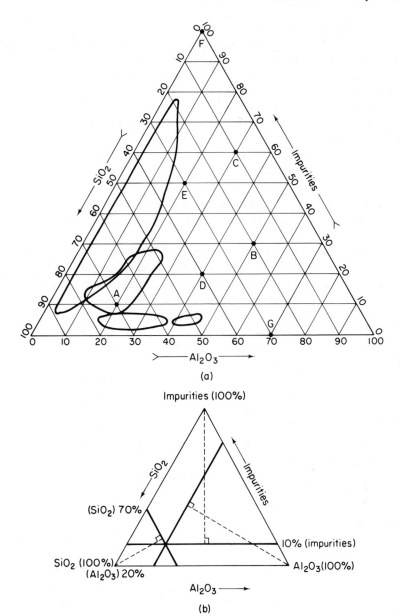

Figure 13-1

solution:

(a) $Al_2O_3 = 50$ per cent.

(b) $SiO_2 = 20$ per cent.

(c) Impurities $= 30$ per cent.

The following clays may be classified as *sedimentary:*

Glacial clays are sediments that have been brought to their location over a long period of time by glaciers. Since they were brought to their location over the earth's surface, they picked up much sand, stone, and gravel while making their trip. In most instances, this type of clay is not suitable to be used in the manufacture of clay products.

Sedimentary clays are those which have been transported by water and deposited by sedimentation. This is one of the best kind of clays for use in the manufacture of clay products. The depositions may take place in the ocean, lakes, or swamps.

Residual clays are clays that have formed by the disintegration of rocks as a result of weather conditions or internal structural collapse. This type of clay generally contains impurities, among which are iron oxide and lime.

Loess clay is a type of clay deposited by the action of the wind.

Fire clay is a clay with about 7.5 per cent impurities. It is a sedimentary clay and is very suitable for use in the manufacturing of clay products. It may be used for temperatures of from 2800 to 3000°F.

If clays are classified according to their silicate structure, the following types of clay are available for engineering applications:

Aluminous clay: bauxite ($Al_2O_3 \cdot 3H_2O$) and diaspore ($Al_2O_3 \cdot H_2O$) are two materials that yield clay.

Kaolin ($Al_2O_3 \cdot 2SiO_2 \cdot 2H_2O$): a white clay in its pure state. It is used in the manufacture of porcelain, firebrick, chinaware, paper, rubber, pigment in paints, and insulation materials. In its pure state its melting point is 3200°F.

Montmorillonites: a common variety of clay, sometimes called steatite. The talc clays are in this group.

Flint clay: these clays are not plastic and are used to reduce plasticity in some clays.

Ball clays: a very plastic type of clay used in electrical porcelain and white ware products.

Feldspar: this material is found in clay. It is, however, classified as rock, not clay.

Vermiculite: a material in the mica class. When heated to about 1700°F, it expands at right angles to its cleavage planes into fibers, which may be ground or processed for use as insulation that will withstand 2000°F temperatures.

Figure 13-2 shows the equilibrium diagram for combinations of silicon oxide, SiO_2, and aluminum oxide, Al_2O_3. These are the principal ingredients that, combined with water, produce clay.

Cristobalite is pure silicon oxide (SiO_2) above 2678°F. It is a quartz material used as a refractory. It differs from quartz in its crystal structures. Below 2678°F, SiO_2 is called *tridymite*.

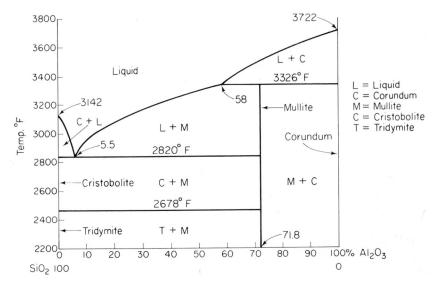

Figure 13-2

Mullite ($3Al_2O_3\cdot2SiO_2$) is a refractory material found in the mineral buckite, or it may be manufactured by prolonged heating, at very high temperatures, in an electric furnace, which fuses silica sand and bauxite. The resulting material, mullite, is used in crucibles, extruding dies, and spark plugs. Mullite mined in California is used in spark plugs; that mined in Brazil is used mainly as gemstones.

Corundum (Al_2O_3) is an aluminum oxide mined in South Africa, India, Burma, Brazil, and Georgia, U.S.A. It has a specific gravity of 4 and a melting point of about 3722°F. In its natural state it is mined and used as gemstone. When it contains chromic acid, it is called ruby; iron oxide and titanic oxide content produces sapphire; ferric oxide, yellow topaz. It may also be produced synthetically. In its synthetic sapphire and ruby forms, it is used as bearings in watches and precision instruments, gages, wear surfaces, and wear guides.

Aluminum oxide, after it is fused, may be crushed and graded, and used as an abrasive material in grinding, lapping powders, emery cloth, and in grinding wheels. When in brick form, it is used as a refractory insulating material to line furnaces.

13-4 CLAY MATERIAL PROCESSING

Clays are mined, graded to a uniform size, blended, and stored. If the mined lumps are too large, they are crushed to the desired size and then stored. As needed, the clay is removed from storage and sent through the crushers, which remove the rocks and break the lumps into small pieces. From the crushers, the clay moves through the grinders, which grind it into fine particles. The entire aggregate passes over a screen, which allows the small particles to pass through so that they may be stored. The material that remains on the screen is returned to the crushers for reprocessing.

The next step is to mix the powdered clay with water. This is called *tempering* and is done in a *pug mill.* Four processes are used: (1) slip, (2) dry-press, (3) soft-mud, and (4) stiff-mud processes.

Slip Process.   Slip is a water–clay mixture of 10 to 50 per cent water with clay. It is poured into molds of plaster of Paris. This plaster absorbs the moisture and causes the outer layer of clay to solidfy. The center liquid is poured off, and the mold is removed and dried.

Dry-Press Process.   This uses a relatively dry mix of less than 10 per cent water. The material is placed into molds, and under high pressure the end product is produced. Rather dense and durable bricks are made in this manner.

Soft-Mud Process.   Used when the clay has a high component of water, this process is used to make only brick. The clay, containing about 30 per cent water, is pressed into and formed in molds. Clay may be worked into the mold by hand or with tools. These molds are moistened and dusted with sand so that the clay will not stick to the mold walls. When sand is used to dust the mold, the brick manufactured by this process is called *sand struck. Water-struck* brick refers to brick that is forced into a very wet mold.

Stiff-Mud Process.   The most widely used method to process clay is the *stiff-mud* process. Clay is first processed in a *deairing* machine. This machine removes air and air bubbles from the clay so that few, if any, inclusions develop. From the deairing process the clay is moved by an auger conveyer to the extrusion dies. As the clay is pushed through the dies, a ribbon of clay of the desired shape is forced out the other end of these dies. These extruded ribbons of clay are cut into desired lengths. Because of the low water content of this clay, cracks due to shrinkage are less apt to develop.

Once extruded in dies or shaped in molds, the material is ready for drying. Prior to firing, the clay must be *dried,* either naturally or artificially. To avoid cracking as a result of shrinkage, temperatures of about 300°F and proper humidity levels must be maintained. Heating is slow and extends for as long as 2 days.

In this process the control of temperature is a function of the type of clay and kiln used, the type of glaze used, and the conditions under which the glaze must be heated. Fast heating may cause the moisture to vaporize and produce voids or imperfections in the finished product.

At the end of the drying period, coatings of ceramic glazing materials are sprayed on the dry clay. These glazing materials soften at temperatures that are lower than the drying temperatures. This material softens, fuses to the clay surface, and produces a glassy surface that is very hard and colorful.

The next step in the procedure is the burning of the clay. The process takes place in four stages: dehydration, oxidation, vitrification, and cooling. *Dehydration* takes place at about 400 to 1200°F. That is, at the selected temperature, the excess moisture that has not been removed during the drying cycle is driven off. The temperature is again increased to a range of between 1000 to 1800°F. During this *oxidation* stage, the carbonaceous materials, sulfur, and other impurities are burned off. *Vitrification* is the stage at which the material is hardened so that it will retain its shape. It takes place at elevated temperatures, sometimes as high as 2400°F. The temperature depends upon the conditions, as indicated earlier.

The time during which these three stages take place is a slow process requiring close monitoring. Temperature monitoring is accomplished with thermocouples, or with small cones that, when heated, bend at different preestablished temperatures. They are called "pyrometric cone equivalents" and are shown in Table 13-3. The time required to complete these stages ranges from 24 hours to as much as 6 days.

Table 13-3

Cone no.	°F	Cone no.	°F
6	2246	20	2786
7	2282	23	2876
8	2300	26	2903
9	2345	27	2921
10	2381	28	2939
11	2417	29	2984
12	2435	30	3002
13	2462	31	3056
14	2552	32	3092
15	2605	33	3173
16	2669	34	3200
17	2687	35	3245
18	2714	36	3290
19	2768	37	3308
		38	3335

Source: Kent, *Mechanical Engineering Handbook*, pp. 5–54. New York: John Wiley & Sons.

The final stage, that of *cooling,* must be very carefully controlled. If the clay is cooled too fast, it becomes brittle and cracks or warps. Also, color changes can take place, which might be objectionable. The time required to cool clay objects could range from 1 to 7 days.

When one is selecting clay materials, certain precautions must be observed. The clay, when mixed with water, must be plastic so that it may be shaped. It must be capable of heat-treating so that it will produce the physical characteristics desirable. It must also be able to develop certain desired coloration and other aesthetic appearances suitable for the application for which it has been selected.

13-5   BRICK

Bricks are building materials made from sand and clay or sand and shale. The physical characteristics that can be expected depend upon the proportion and kinds of the various materials used, the method used to manufacture the product, and the time and temperature of firing.

As the temperature of firing is increased, the color darkens, the porosity decreases, and the strength of the brick increases. This darker brick is generally used for *face brick. Common brick* is used in the basic building of structures. This brick absorbs about 15 per cent by weight of water and face brick about 8 per cent by weight of water. The face brick is more dense, more resistant to damage from frost, and stronger. It is used to protect the outer surfaces of buildings or walls.

Brick is *coursed* as shown in Fig. 13-3(a). A *course* of brick is defined as the vertical brick dimensions and the thickness of the mortar between two courses.

Figure 13-3(c) shows a common brick that is produced by the dry-press process. The indentation shown (frog) is easily produced in a mold. The brick shown in Fig. 13-3(d) is produced by the stiff-mud process.

Bricks are generally laid so that the $2\frac{1}{4}$- by 8-in. face is vertical, as shown in Fig. 13-3(a). If the mortar is $\frac{3}{4}$ in. thick, each course of brick is 3 in. and permits four courses for every foot of height. This course of laid brick is called a *running course.* Common brick may be laid so that the $3\frac{3}{4}$- by 8-in. face is exposed, in which case it is called a *stretcher course.* When a course of brick is laid so that the $2\frac{1}{4}$- by $3\frac{3}{4}$-in. face is exposed, as shown in Fig. 13-3(a), the layer is called a *header course.* If the brick is laid as shown in Fig. 13-3(b), the layer is called a *Flemish course.*

The method of laying up courses to form a pattern in a brick wall is called a *bond.* Staggering stretcher courses is called a *running bond.* Layers of five stretcher courses and one header course are called a *common bond.* Alternating header and stretcher courses constitute an *English*

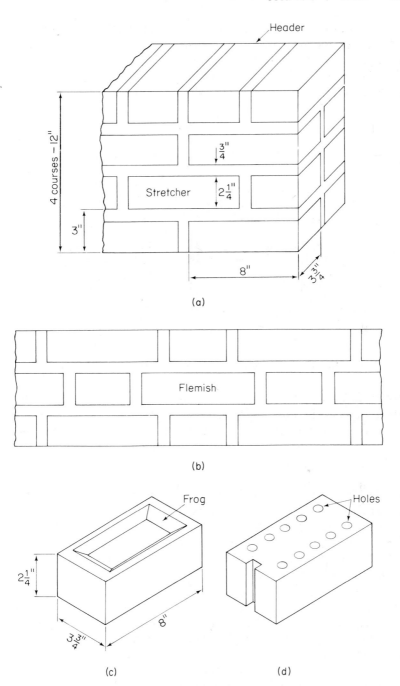

Figure 13-3

bond structure. Layers of Flemish courses are called a *Flemish bond*. Other types of bonds are cross Flemish, garden wall, stack, double Flemish, and one-third diagonal.

Table 13-4 shows different types of brick that may be used as structural materials.

Table 13-4

| | Course dimensions, in. | | | | |
Type	Thick	Height	Length	No. C	In.
Double	4	$4\frac{1}{3}$	8	3	16
Economy	4	4	8	1	4
Engineering	4	$3\frac{1}{5}$	8	5	16
King Norman	4	4	12	1	4
Modular	4	$2\frac{2}{3}$	8	3	8
Norman	4	$2\frac{2}{3}$	12	3	8
Norwegian	4	$3\frac{1}{5}$	12	5	16
Roman	4	2	12	2	4
St. Clay Res.	6	$2\frac{2}{3}$	12	3	8
Triple	4	$5\frac{1}{3}$	12	3	16

There are three grades of common brick:

Grade SW.   The compressive strength of this brick is about 3000 psi when it is used in a flat position. It has a structure that has a high resistance to frost action and wet environment such as exist underground. It has a water absorption percentage of about 18, a maximum saturation coefficient of 0.80, and a maximum water absorption of 20 per cent.

Grade MW.   This grade is intended for use in a relatively dry location exposed to temperatures that may be below freezing. This type of brick is generally used above ground. It has a minimum compressive strength of about 2500 psi. It has a maximum water absorption percentage of about 22, a maximum saturation coefficient of about 0.90, and a maximum water absorption of 25 per cent.

Grade NW.   This grade of common brick is used where no freezing occurs. It is used as an interior brick or backup brick. Its compressive strength is about 1500 psi. Its maximum water absorption percentage and maximum saturation coefficient of water absorption have no limit.

Common brick is generally made with a smooth surface. The surface is generally treated with a color or glaze. Ceramic, or salt glaze, is sometimes added to protect the surface or provide a special effect. Textures

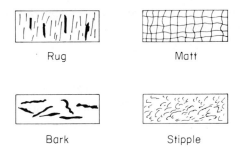

Rug   Matt

Bark   Stipple

Figure 13-4

are applied to the surfaces to produce a desired effect. Figure 13-4 shows several textures of brick.

Something should be said at this point about the mechanical properties of common brick. The physical properties of a few types are shown in Table 13-5. The compressive strengths listed are values per brick, which means that the weight-supporting ability of a flat-laid brick course could be in excess of 100,000 lb if the load is distributed evenly.

Table 13-5

	Comp. st. $\times 10^3$ psi	Mod. rupt. $\times 10^3$ psi	Porosity %
Common brick			
SW	3.0		
MW	2.5		
NW	1.5		
Fireclay brick			
Super duty	2.0–5.0	0.6–3.0	8–16
High duty	1.0–6.0	0.5–2.0	10–25
Intermediate	1.2–6.0	0.8–2.5	10–25
Low	2.0–6.0	1.0–2.5	10–25
High alumina brick			
50%	2.5–5.0	1.0–1.5	18–25
99%	5.0–9.0	2.0–3.5	20–29
Mullite	3.0–8.0	0.8–2.5	20–30
Silica brick	1.5–4.0	0.5–1.2	20–30
Basic brick			
Magnesite	5.0–10	1.5–3.0	16–24
Mag–chrome	2.0–4.0	0.7–1.2	20–25
Chrome	3.0–6.0	1.2–3.0	18–27
Chrome–mag.	2.0–3.5	0.9–1.5	18–27
Forstenite	1.5–4.0	0.7–1.2	18–27

Porosity ranges from those bricks which exhibit a high degree of porosity to those which permit almost no penetration of moisture or gas.

Porosity is a function of the amount of water absorbed by the brick. Bonding depends upon the ability of a brick to absorb some of the bonding material from the mortar. The equation for calculating the percentage of absorption is

$$A_c = \frac{W_c - W_d}{W_d} \times 100$$

W_c = cold saturated weight
W_d = dry weight
A_c = cold per cent absorption

The *weight* of the saturated brick is determined after a dry brick has been immersed in cold water for 24 hours. Some air remains trapped.

The *saturation coefficient* is determined by submerging a dry brick in boiling water for 5 hours and allowing both to cool to room temperature. The hot-water percentage absorption is

$$A_h = \frac{W_h - W_d}{W_d} \times 100$$

W_h = hot saturated weight
W_d = dry weight
A_h = hot per cent absorption

The coefficient of saturation is the ratio of the cold saturation to the hot saturation percentages. Thus

$$C_s = \frac{A_c}{A_h}$$

A_c = cold-water coefficient
A_h = hot-water coefficient
C_s = saturation coefficient

Example 2

Given the following data for a common brick: the weight of dry brick is 4 lb, 12 oz; the weight after 24 hours submerged in cold water is 5 lb, 8 oz; after 5 hours submerged in boiling water it is 5 lb, 10 oz. Calculate (a) the cold-water percentage of absorption, (b) the hot-water percentage of absorption, and (c) the coefficient of saturation.

solution:

(a)   The cold-water percentage of absorption:

$$A_c = \frac{W_c - W_d}{W_d} \times 100 = \frac{5\frac{1}{2} - 4\frac{3}{4}}{4\frac{3}{4}} \times 100$$

$$= 15.8\%$$

$W_d = 4$ lb, 12 oz $= 4\frac{3}{4}$ lb
$W_c = 5$ lb, 8 oz $= 5\frac{1}{2}$ lb
$W_h = 5$ lb, 10 oz $= 5\frac{5}{8}$ lb
$A_c = ?$
$A_h = ?$
$C_s = ?$

(b) The hot-water percentage of absorption:

$$A_h = \frac{W_h - W_d}{W_d} \times 100 = \frac{5\frac{5}{8} - 4\frac{3}{4}}{4\frac{3}{4}} \times 100$$

$$= 18.4\%$$

(c) The coefficient of absorption:

$$C_s = \frac{A_c}{A_h} = \frac{15.8}{18.4} = 0.86$$

The durability of brick is far better than one would suspect. It has a long life, some brick being as much as 50 centuries old. If structures collapse, it is generally because the joints deteriorate. Brick is fired at higher temperatures than the melting point of most metals. Thus, if the structure collapses as a result of high temperatures, it is not because the temperatures have caused the brick to disintegrate.

Also, moisture absorbed by brick expands when frozen. The coefficient of saturation for brick is not likely to be over 90 per cent. Thus 10 per cent or more of the inner structure of a brick is air space. Since water expands about 9 per cent upon freezing, it has room to expand without increasing the stress on the brick. Thus, if the specifications call for a coefficient of absorption of 0.80 (0.78 for an average of five bricks), the structure will be able to withstand much moisture and freezing conditions.

It has been seen that *building brick* is a general building material used for structures. *Face brick* is a dense brick that results when the clay material is pressed before burning to achieve density. *Glazed brick* is the result of coating the clay with a thin coating of slip and then burning the brick to achieve a smooth, hard, transparent coating. *Paving brick* is made from *shale* that is pressed and fired.

Terra-cotta is a building material made from a mixture of clay and shale. It is strong, has a low coefficient of expansion, is plastic in its pre-fired state, and has a fine texture. It is often hand-shaped or carved, glazed, and fired. It is manufactured as *architectural* terra-cotta, *ceramic* terra-cotta, and *lumber* or *block* terra-cotta. Whereas the former two kinds of terra-cotta have a closed structure, the latter two types have a porous structure.

Tile is another building material. It is a clay material, defined as having not over 75 per cent of its volume solid. All clay tile is made by the extrusion stiff-mud process. There are many types of tile produced. They are roofing, floor, wall, facing, backup, or fireproofing tile.

Conduit is another tile form. It is made from a clay and shale combination with salt, shaped by extrusion, and fired. It must be resistant to all types of materials, nonporous, and absorption free.

13-6   REFRACTORY MATERIALS

Refractory materials are used for high-temperature applications, such as linings for heat-treating furnaces. They consist of the fireclays, high alumina, silica, magnesia, magnesite and chrome combinations, or insulating firebrick. Some of the physical properties are shown in Table 13-5.

Fireclay brick contains mainly *kaolinite* ($Al_2O_3 \cdot 2SiO_2 \cdot 2H_2O$), which is roughly 40 per cent alumina, 46 per cent silica, and 14 per cent water, before it is subjected to elevated temperatures. At elevated temperatures the percentages are 54 per cent silica and 46 per cent alumina with small quantities of other impurities, which are called glass-forming substances. This material is made from preshrunk and prefired clay to ensure stability. These bricks will not carry heavy loads. They are used to line chambers for the purpose of confining high temperatures. They are made to fit one another without the use of very much mortar. The mortar used is made from finely ground fireclay and water.

Another clay used as a refractory is *flint* clay. This hard material, when mixed with water and fired, possesses low shrinkage and plasticity. It has high refractory qualities.

Kaolins are white clays that have good shrinkage qualities during drying and firing. If they contain appreciable amounts of silica, their drying and shrinkage qualities are low.

High-alumina firebrick contains more than 48 per cent alumina. Alumina may be used at high temperatures when the alumina content ranges from 50 to 99 per cent. Increasing the alumina content increases its resistance to higher temperatures. This material is resistant to slag and fume attack.

Silica brick retains its strength at high temperatures. It is classed chemically as an acid, and it is used in linings for melting furnaces, when the lining is to be in contact with metals that are acid by nature. Since it spalls (flakes or chips) when used at temperatures below 1000°F, it is rarely used in furnaces which operate below that temperature.

Magnesite bricks are chemically basic materials. They are used in basic open-hearth or Bessemer furnaces because they will not react chemically with basic slag, but will react with and remove acid impurities from steel. The mortar used between the brick is thin sheets of steel, which oxidize. Ferric oxide then reacts with the magnesia, bonding the brick.

Chrome brick is a refractory material containing up to 50 per cent chromic oxide, with varying amounts of alumina, magnesia, silica, and ferrous oxide. These are sometimes referred to as *neutral refractories.*

Combinations of *chrome–magnesia* are also manufactured into bricks that are suitable for use at high temperatures without spalling.

Silicon carbide is made in the electric furnace by fusing silica and coke. It has a high resistance to spalling at elevated temperatures.

In some cases *graphite and clay* are bonded to produce a lining for the hearths of iron-melting furnaces. Since its strength increases as the temperature increases, it is possible to use this material at temperatures as high as 6000°F.

Both the *silicate* and the *oxide* of *zirconium* are refractory materials. They are used chiefly in the melting of glass, because they are resistant to alkaline fluxes.

Insulating firebrick is used to conserve the heat required to bring furnaces up to temperature. These bricks are lightweight and porous. The porosity is achieved by mixing fine coke or sawdust in the fireclay mix. The additive burns out, producing gas voids, which make the brick porous. Since the brick weighs about one-third that of a regular firebrick, they require one-third the Btu to bring them up to heat. They are, however, used chiefly in intermediate-temperature brazing and heat-treating furnaces. They have also found use on the outside of high-temperature furnaces to reduce heat loss through the walls of the furnace.

13-7 GLASS

Glass dates back to at least 3000 B.C. It is made by fusing silica with an oxide. It is an amorphous substance, which is cooled slowly to avoid crystallization. Some materials that have crystalline structure are now classified as glass because of their transparency. Whether the final structure is amorphous or crystalline depends upon the heat treatment. Generally, crystal formation in glass has an adverse effect on its physical properties. Glass is actually a solid liquid. The characteristics that make it desirable and useful are its hardness and transparency, its rigidity at room temperature, its plasticity at elevated temperatures, and its resistance to all types of weather and temperature. It is subject to attack by hydrofluoric acid.

The compressive strength of glass is very high. Its tensile strength is low compared to most metals. It has poor electrical and thermal conductivity. It has poor resistance to thermal shock and mechanical impact. When used as a conductor, it must be coated with a material that supplies and acts as a vehicle for ion motion.

Glass is manufactured by melting the ingredients in a pot when small batches are required or in regenerative furnaces when high production is needed. The material is mixed, fed into one end of a furnace, melted, and extruded, or drawn into sheets, rods, or tubes at the other end of the furnace. In some instances an appropriate mass of molten glass is blown into the desired shape. In other instances, the molten glass is press forged, rolled, or cast into the desired shape.

The *annealing* process for glass consists of reheating it to a temperature below the melting point and slowly cooling it to room temperature.

It was pointed out that glass is not very strong in tension. If it fails, it usually fails in tension. If glass is heated to a temperature slightly below its melting point, about 1000°F, and cooled with an air blast or in an oil bath, the surface of the glass will be placed in compression while the inside of the glass is in tension. With the surface in compression, the glass is less likely to break when tensile forces are applied. This process is called *tempering.*

There are many types of glass manufactured. They are soda lime, lead alkali, borosilicate, aluminosilicate, high silicate, and fused silicate. Glasses are combinations of silicate, which is the glass-forming component, lime or soda ash, which acts as a fluxing agent, and an oxide, which acts as a modifier. These oxides are listed in Table 13-6.

Chemicals are added to glass to satisfy special requirements. Table 13-7 shows some of these materials and the effects produced when they are used.

Soda-Lime Glass ($NA_2O \cdot CaO \cdot 6SiO_2$).   Used to make sheet or plate. Increasing the ratio of sodium to silicate oxide decreases the resistance of the glass to chemical attack and lowers its melting temperature. *Waterglass* is that glass which loses its solubility to water when the Na_2O is considerably increased. If both CaO and Na_2O are increased at the expense of the silicate oxide, a hard glass results that has improved chemical resisting qualities. Soda-lime glass has a coefficient of thermal expansion of about $5 \times 10^{-6}/°F$.

Table 13-6

	Types of glass				
Oxide	Soda lime, %	Lead, %	Borosilicate, %	96% Silica	Fused quartz, %
SiO_2	70–75	53–67	73–82	96	99.8
Na_2O	12–18	5–10	3–10		
K_2O	0–1	1–10	0.4–1		
CaO	5–14	0–6	0–1		
PbO		20–40	0–10		
B_2O_3			5–20	3+	
Al_2O_3	0.5–1.5		2–3		
MgO	0–4				

Source: Kent, *Mechanical Engineering Handbook,* 12th ed., pp. 5–26. New York: John Wiley & Sons.

Table 13-7

Materials added	Effect
Iron	Green
Barium	Brilliance
Soda ash	Fogging
Manganese oxide	Violet or blue
–small amounts	Neutralize green
Cobalt oxide	Blue
Copper chloride	Red
Sulfur and iron oxide	Yellow to ruby amber
Neodymium oxide	Yellow sunglass
Fluorides, sulfides, or metal oxides	Opalescence or alabaster glass
Selenium and cadmium sulfide, or copper chloride	Ruby glass
Gold chloride and tin oxide	Ruby glass
Tourmaline or peridate	Polarized glass

Lead-Alkali Glass. These glasses are produced by substituting lead oxide for the lime in the soda-lime glass mixture. Lead oxide increases the light-dispersive power of glass. High values for dispersive power produce brilliance in glass. Cut glass and other ornamental glass make use of this quality. Because lead absorbs gamma and X-rays so readily, this glass is used for shielding from high-energy rays. The heavier lead ions also produce electrical properties in this type of glass, which increase its resistance to electric flow. Combinations of K_2O, PbO, BaO, and ZnO produce glass suitable for optical applications. They are called *flint* glasses.

Borosilicate Glass. When boric acid (B_2O_3) is substituted for the alkali and the lime in soda-lime glass, the coefficient of expansion of the glass is reduced by about one-third or about $1.7 \times 10^{-6}/°F$. This decreases the possibility of cracking from thermal shock, such as sudden changes in temperature. The reduced alkali content makes the glass more resistant to chemical attack. Pyrex, astronomical telescope lenses, and chemical piping are examples of the use of this type of glass.

96 Per Cent SiO_2 Glass. This glass is produced from borosilicate glass. A special heat-treating procedure produces a borax–boron oxide in combination with a silicon oxide phase in the glass. An acid bleaching process removes all the soda and produces a 96 per cent silica in combination with about 3 per cent B_2O_3. This glass has a very low coefficient of thermal expansion ($0.4 \times 10^{-6}/°F$) and a high melting temperature. It therefore combines high temperature strength with high resistance to thermal shock. It also has very good corrosion resistance and low absorption of ultraviolet light.

Fused Silica or Quartz Glass.   This glass is made from commercially pure SiO_2. It is very difficult to work because of its high melting temperature, which is over 3000°F. Because of the high cost of manufacture, this glass is used for special purposes. It has a coefficient of thermal expansion of $0.3 \times 10^{-6}/°F$.

Aluminosilicate Glass.   This glass has a high silica and alumina content, which gives it a high melting temperature. The coefficient of expansion of this glass is about one-half that of soda-lime glass. It has good high strength and thermal shock characteristics.

13-8   PROPERTIES OF GLASS

Because of the importance of glass as an engineering material, the properties of glass are reviewed here.

Glass may be rolled, drawn, blown, molded, pressed, or built up out of larger sections by fusing, welding, or soldering two halves. These operations are used in the making of *glass brick*. Many of the physical properties of glass may be controlled by the addition of chemicals and heat treatment. Tempering creates a glass that is stronger in tension than ordinary glass.

Safety glass is wired or laminated. Laminated glass is a sandwich of a layer of tough transparent material, such as plastic, between two sheets of glass. The sandwich is rolled at about 250°F into a single unit. Transparency, resistance to ultraviolet light, and resistance to shattering of this type of glass make it useful in automobiles.

Fiber glass is produced when molten glass, under steam pressure, is forced through very small holes. When cooled, it forms a network of thin glass fibers and air. The air acts as an insulator and therefore makes this an excellent material for insulating walls, ceilings, refrigerators, etc.

Glass is a liquid, which at room temperature is in a solid state. It is called a brittle amorphous solid. Its optical properties show an index of refraction of about 1.5, which can be varied to accommodate special needs. It becomes ductile at elevated temperatures. Its importance lies in the fact that it is chemically stable. It is insoluble in water, although, over long periods of time, water will erode and fog glass. In general, it has greater resistance to wear and abrasion than steel. It also has a coefficient of thermal expansion that is lower than steel.

The tensile strength of ordinary glass increases inversely with its size. Thus $\frac{1}{2}$-in.-diameter glass rods have ultimate tensile strengths of about 8000 psi, whereas fibers, which have diameters of 5×10^{-5} in., may have ultimate tensile strengths of 3,000,000 psi. It was pointed out that tempering may increase ultimate strength by a factor of 5, and even by factors of 10. The compressive strength also ranges as high as 150,000 psi.

The specific gravity of glass averages about 2.5. The thermal expansion of ordinary glass is about $5.0 \times 10^{-6}/°F$, and for fused silicate glass it may be as low as $0.3 \times 10^{-6}/°F$. The modulus of elasticity is about 10×10^6 psi for most glass. Fused silicate is about 12×10^6 and aluminosilicate about 18×10^6 psi. Leaded glass softens at a temperature as low as 1160°F and fused silicate glass as high as 3000°F.

Table 13-8*

Material	Sp. gr.	Ther. ex. $\times 10^{-6}/°F$	Mod. elast. $\times 10^6$ psi	Index refract.	Soften temp. (°F)	T.S. (anneal) $\times 10^3$ psi
Soda lime	2.47	5.0	9.8	1.512	1280	4–10
Lead alkaline	2.85	5.0	9.0	1.542	1160	4–10
Borosilicate	2.23	1.8	9.8	1.474	1510	4–10
96 per cent SiO_2	2.18	0.4	9.7	1.458	2730	4–10
Fused silicate	2.20	0.3	12.0	1.459	3000	4–10
Aluminosilicate		1.5	18.0		2000	

Source: Kent, *Mechanical Engineering Handbook,* 12th ed., pp. 5–27. New York: John Wiley & Sons.

Table 13-8 shows some physical properties of some of the glass discussed in this section.

13-9   ASPHALTS

Asphalt is a mixture of hydrocarbons that is soluble in turpentine, petroleum solvents, and carbon disulfide. It has an animal origin rather than a vegetation origin. It ranges from a brown to black color, depending upon the region from which it came. Asphalt mined in this country contains a large amount of mineral matter. Asphalt from Venezuela and Trinidad, where it is found in great asphalt "lakes," has about 50 per cent bitumen and equal parts of clay and water in its crude state.

Sandstone that contains asphalt is found in Oklahoma, Kentucky, Utah, and other states. This type of sandstone also occurs in Canada, Europe, Asia, and South America. In most instances, these materials are used directly for paving. When a particular consistency is needed, it is mixed with sand.

Asphalt is also a by-product of petroleum. The three types of asphalt are straight run, air blown, and cracked.

Straight-run asphalt represents about three-quarters of the total asphalt produced from petroleum. It is a viscous material and is used in runways and roads. When used this way, it is mixed hot with hot stone, and poured on a rolled stone base. It is then rolled when hot to produce a smooth surface.

Air-blown asphalt is resilient and viscous. It is less susceptible to temperature change than straight-run asphalt. It is used in paper laminates, paints, pipe covering, roofing, and shingles.

Cracked asphalt also is a viscous material, but it is more susceptible to temperature change than straight-run asphalt. It is used for impregnating insulation boards and as a spray for dust control.

Artificial asphalt is a bituminous residue from coal that is purified and then mixed with sand or limestone. It is used for roofing, cold-molded products, and paints and varnishes. It produces an acid-resisting paint and an insulating varnish.

Cutback asphalt is asphalt that has been mixed with petroleum distillates. It is used as a base for paints that, when applied, form protective coatings against nonoxidizing acids, alkalies, and salts. It is used as the cementing material for floor covering and for waterproofing walls, wood, masonry, or metal. When combined with rosin it becomes a tacky adhesive, which is used either for laminating paper or for impregnating floor felts.

Emulsified asphalt is a water solution used for waterproofing walls and painting pipes and other surfaces. When mixed with rubber latex, it is used for road filling. It is mixed as a powder or as an additive preparation. When mixed with asbestos fiber, it is used for painting steam pipes. When mixed with pitch, it is used as a roof covering.

Asphalt, mixed with fillers and pigments, is used as flooring tile in factories. As a binder with asbestos fibers and pigments, it is used as a chemical-resisting floor covering.

Problems

13-1. Discuss the structure and properties of ceramic materials in general.

13-2. List the materials classified as stone.

13-3. What is (a) igneous rock? (b) granite? (c) basalt? (d) sedimentary rock?

13-4. List the rocks that are classified as metamorphic.

13-5. List the three chemical structures most commonly found in stone. Discuss these three structures in terms of durability. List examples of each structure.

13-6. Note the compressive strengths of the materials listed in Table 13-2. State your own reasons for the various values.

13-7. Repeat Problem 1-6 for the "wear resistance" values in Table 13-2.

13-8. Do you detect any relationships between the columns marked "modulus of rupture," "shear strength," and "tensile strength"? If so, discuss them.

13-9. Discuss limestone as an engineering material.

13-10. Discuss sandstone in terms of grain size, binders, colors, and physical properties.

13-11. List some sandstones in common use. What are these uses?

13-12. Discuss the material shale.

13-13. Discuss slate and its physical properties.

13-14. Discuss marble as a building material.

13-15. Discuss the color of marble and its various uses.

13-16. Discuss the composition and physical properties of granite.

13-17. List the chemicals and the color of the mica produced by them.

13-18. List some physical properties of carbon.

13-19. List the uses of carbon and state the related physical form.

13-20. List the elements that may be used in combination with carbon to produce special needs. What effects do the addition of these elements have?

13-21. What is asbestos? List some of its uses.

13-22. What is jade?

13-23. What is the effect of the talc content on soapstone?

13-24. What is clay?

13-25. From Fig. 13-1(a) determine the percentage composition by weight of the clay at point C.

13-26. Repeat Problem 13-25 for point D.

13-27. Repeat Problem 13-25 for point E.

13-28. Repeat Problem 13-25 for point F.

13-29. List the sedimentary clays and the methods by which they were formed.

13-30. List the various clays according to their silicate structures.

13-31. What are cristobolite, tridymite, mullite and corundum? Explain their production and uses.

13-32. Figure 13-2 is an equilibrium diagram of Al_2O_3 and SiO_2. Select a composition between 0 and 70 per cent Al_2O_3 and calculate the percentage of each phase at 3600, 3000, 2700, and 2400°F.

13-33. Discuss the processing of clay.

13-34. Describe the slip process of making clay parts.

13-35. (a) Describe the dry-mud process for using clay. (b) What is meant by sand-struck molding? (c) Water-struck molding?

13-36. Discuss the use of the stiff-mud process in the manufacture of clay parts.

13-37. Discuss the following stages of the stiff-mud process: (a) deairing, (b) extrusion, (c) drying, (d) glazing, (e) burning.

13-38. Discuss the four stages that are used during the burning process in the stiff-mud procedure.

13-39. Discuss the use and making of common and face brick.

13-40. What is a *course* of brick?

13-41. What is a running course? A stretcher course?

13-42. What is a header course? A Flemish course?

13-43. Define the term "bond" as related to brick laying. Name at least five of the eight bonds listed in this chapter. Describe and make a freehand sketch of the bonds listed.

13-44. List the three grades of brick and explain the meaning of each grade.

13-45. (a) How is the weight of a saturated brick determined? (b) What is the hot-water absorption percentage? (c) How is the saturation coefficient determined?

13-46. The weight of a dry brick is 4 lb, 8 oz. After being submerged in cold water, it weighs 5 lb, 6 oz. After being submerged in hot water, it weighs 5 lb, 12 oz. Calculate (a) the cold-water percentage of absorption, (b) the hot-water percentage of absorption, and (c) the coefficient of saturation.

13-47. Given the cold-water weight of a brick of 5 lb, 8 oz, a hot-water weight of 5 lb, 14 oz, and a coefficient of saturation of 0.705, calculate (a) the dry weight of the brick, (b) the cold-water percentage absorption, and (c) the hot-water percentage absorption.

13-48. Explain the capability of brick to withstand moisture and freezing temperatures.

13-49. Describe in general terms the following types of brick: (a) common, (b) face, (c) glazed, (d) paving.

13-50. What is terra-cotta? How many varieties of terra-cotta are there?

13-51. What is tile? Give examples of tile.

13-52. What is kaolinite?

13-53. Describe the various types of refractory brick and materials discussed in this chapter.

13-54. Explain the use of silica and magnesite bricks in melting furnaces.

13-55. List some characteristics and physical properties that make glass useful as an engineering material.

13-56. How is glass manufactured? Discuss the annealing process.

13-57. Explain the effect of tempering glass.

13-58. List the six types of glass discussed in this chapter. What are the three ingredients that make up all types of glass? What is the purpose of each ingredient?

13-59. Describe soda-lime glass and state some of its physical properties.

13-60. Describe lead-alkali glass. What are some of its uses?

13-61. Explain the advantages and list the uses of borosilicate glass.

13-62. How is 96 per cent SiO_2 glass produced? What are its advantages over other types of glass?

13-63. What are the characteristics of fused silica glass?

13-64. What are some characteristics of aluminosilicate glass?

13-65. List the processes that may be used to join or fuse glass.

13-66. How is safety glass produced?

13-67. How is fiber glass produced?

13-68. Discuss the physical properties of glass.

13-69. List three types of materials that will dissolve asphalt, a hydrocarbon material.

13-70. List the three materials that are generally found in the Trinidad asphalt.

13-71. Which type of asphalt is found in Utah or Oklahoma?

13-72. List the three types of asphalt that are processed from petroleum. Discuss each type.

13-73. Describe artificial asphalt and its use.

13-74. Discuss cutback asphalt. How is it used?

13-75. What is emulsified asphalt? How is it used?

13-76. List a use for asphalt that is mixed with asbestos fibers and pigments.

14 | Cement and Concrete

14-1 MORTARS AND CEMENTS

Cements have been used for centuries. In 1824 the first powder was made from limestone and clay to form what we now know as portland cement.

Cements are essentially calcium based. One calcium cement is made from *lime* ($CaCO_3$). Another is made from *gypsum* ($CaSO_4 \cdot 2H_2O$). Lime cements include portland cement and natural cements. The gypsums include plaster of paris, hard-finish plaster, and gypsum wall board. These cements are used in thin layers to protect masonry, or as a binder of stone, brick, or aggregate.

In the broadest sense, cementing materials are those which bind other materials. Mortar is such a material. It is used for bedding and bonding brick, stone, tile, etc. It should contain as much water as is consistent with the troweling operation. Since the water content determines the quality of the bond achieved, the water content should be high.

Four types of mortar are recognized:

Type M mortar consists of one part portland cement, three parts sand, and one-fourth part hydrated lime. Also acceptable is a mixture of one part portland cement, one part masonry cement, and six parts sand. This type of mortar is recommended for use where it is to be in contact with the earth. It is also recommended for general use.

Type S mortar consists of one part portland cement, four and one-half parts of sand, and one-half part of hydrated lime. Also acceptable is a mixture of one-half part portland cement, one part masonry cement, and

336

four and one-half parts sand. This mortar is recommended where high resistance to shearing forces is needed. It is also recommended for general use.

Type N mortar consists of one part portland cement, six parts sand, and one part hydrated lime. Also acceptable is a mixture of one part masonry cement and three parts sand. This cement is recommended for use above ground in exterior walls where the exposure to weather is severe.

Type O mortar consists of one part portland cement, nine parts sand, and two parts hydrated lime. Also acceptable is a mixture of one part masonry cement and three parts sand. This type of mortar should not be used in places where it will be subjected to freezing and thawing in moist environments. It is essentially an indoor mortar. It also should not be used in places where it is likely to be subjected to compressive stresses that are greater than 100 psi.

14-2 LIME

Limestone is calcium carbonate (calcite) with the chemical symbol $CaCO_3$. In another raw state, magnesium carbonate is also present in substantial amounts. In general, magnesium aids in the purification of lime. Lesser amounts of alumina, silica, and iron oxide are also present as impurities. These impurities may be harmful to the production of purified lime.

The purification process consists of controlled heating of the raw materials to between 1650 and 1800°F, after it has been crushed by grinding and fed into furnaces. This process, known as *calcining,* causes the dissociation of the calcium carbonate into quicklime (CaO) and carbon dioxide (CO_2). If magnesium carbonate is present, it will also dissociate into magnesia (MgO) and carbon dioxide (CO_2). The equation may be written

$$CaCO_3 + heat \rightarrow CaO + CO_2$$

$$MgCO_3 + heat \rightarrow MgO + CO_2$$

The resulting lime is sold as pulverized, lump, or screened lime.

Sometimes lime is mixed with controlled amounts of water, pulverized, and sold as *slake* or *hydrated lime.* The calcined lime when slaked increases in volume and generates much heat. The equation is

$$CaO + H_2O \rightarrow Ca(OH)_2 + heat$$

This type of lime is marketed in the form of a fine white powder.

Also present is the calcined magnesia (MgO) that was not slaked. It must be removed; otherwise, it will cause cracks, since it will slake and expand after the cement has set.

When lime hardens, it does so by absorbing carbon dioxide from the air. In the presence of water and CO_2, calcium carbonate reforms. As the calcium carbonate crystals reform, they interlock, causing solidification. The equation is

$$Ca(OH)_2 + CO_2 \rightarrow CaO_3 + H_2O$$

If water is added to the quicklime in greater amounts than needed, $Ca(OH)_2$ forms as a paste. The same thing happens to slaked lime. The exposure of the paste to the air causes the $Ca(OH)_2$ to revert to calcium carbonate according to the above formula. If too much water is used, particles of water will remain in the structure after solidification. This water will evaporate and leave voids in the structure.

If quicklime is to be used as a cementing material, it must be hydrated by slaking. After several weeks the slaked lime is ready for use. Hydrated lime may be used immediately.

Lime may be used as a plastering material. It is mixed four parts sand to one part lime and used as a plaster. It is slow setting. It is also used in agriculture as a soil reconditioner. It has a compressive strength of about 400 psi.

Besides use as an ingredient in plaster and mortar, lime may be mixed and used as a *lime–cement mortar,* which has better strength than *lime mortar,* but not the strength of cement mortar. The mixture for lime–cement mortar is generally two parts lime, one part cement, and six parts sand. The mixture is brought to a thick consistency by adding water. Lime plaster produces a rough-finish plaster. The ingredients are one part lime and three parts sand. A hard plaster may be produced by mixing two parts plaster of paris with one part lime.

14-3 GYPSUM

Gypsum in its raw, pure state is white. In its impure state, it contains some limestone, silica, and alumina and is light gray. Its chemical symbol is $CaSO_4 \cdot 2H_2O$. The process of refining raw gypsum consists of heating it to drive off the two parts of water, as indicated in the equations

$$CaSO_4 \cdot 2H_2O \underset{\text{harden}}{\overset{\text{calcine}}{\rightleftharpoons}} CaSO_4 \cdot \tfrac{1}{2}H_2O + 1\tfrac{1}{2}H_2O \qquad \text{at about } 240°F$$

$$CaSO_4 \cdot \tfrac{1}{2}H_2O \rightarrow CaSO_4 + \tfrac{1}{2}H_2O \qquad \text{at about } 390°F$$

This process is the calcination of raw gypsum. The material is control heated in large kettles or in almost horizontal kilns. When the horizontal kiln is used, the material is fed into the end of the kiln and rotated. The gypsum works its way down the kiln and out the other end.

The end product of the 390°F stage yields $2CaSO_4 \cdot H_2O$, which is called *plaster of paris*. This material is used in wallboard, rock lath, sheathing, plastering, and molds for casting dental and medical materials. It is used with lime to produce a finished plaster.

Sometimes other materials, such as lime or clay, are present in plaster of paris. If mixed with water, the end product is called *wall plaster* or *cement plaster*. When mixed with sawdust or general cinders, the aggregate is cast into blocks and used as a building material. Firm setting is caused by interlocking crystallization.

When the temperatures of calcination exceed 390°F, all the water is driven off, and the $CaSO_4$ remains. This is called *flooring plaster*.

When alum or borax is added, the workability of the plaster is increased. This is called *hard-finish plaster*.

Keene's cement results when alum is added to calcinated gypsum and recalcinated. After it is ground, it is ready for use. This is a very hard gypsum product which sets in about 3 hours.

When mixed with water, the half-hydrate and the anhydrite of *gypsum* will reform into a hard interlocking phase of the original $CaSO_4 \cdot 2H_2O$. The taking on of water and subsequent hardening, or setting, classifies plaster of paris as a hydraulic cement. Impurities, such as glue and borax will slow down the setting process. Warm water will increase the setting time.

The strength of plaster is increased by adding hair or wood fibers to the product. Its ultimate strength is a function of the water-to-plaster ratio. Excess water weakens the final product. Figure 14-1 shows a curve

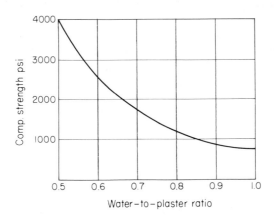

Figure 14-1

that relates the mixing ratio of water to plaster and the compressive strength of the plaster.

14-4 CEMENT

A kind of *portland cement* was used by the Romans, who discovered that by mixing volcanic ash and sand with slake lime, they could make a stronger building material than that which they made from lime alone. This cement is called *Pozzolan cement* because it was first made from the ash collected at Pozzuoli, Italy.

Natural cement is made from limestone, alumina, and silica. When calcinated, calcium, silicate, and alumina form clinkers. These are ground into a fine powder. When mixed with water, this fine powder forms a cement material that is about 50 per cent as strong as portland cement.

When limestone, which contains a fair amount of silica but very little alumina, is calcined with water, the lime will slake. Its setting time is very slow, and its strength does not compare with portland cement. It is referred to as *hydraulic lime.*

A low-strength, but economical, cement is sometimes made from blast-furnace slag mixed with an accelerator and hydrated lime. It is called *slag cement.*

Magnesium oxychloride cement, called Sorel's cement, hardens as a result of hydration. It is very strong and hard. It is used with many different kinds of filler materials, ranging from copper filings, to sand asbestos, to wood and wood fibers. It is used for floors and as a facing material and steam pipe covering.

Portland cement is made from lime- and clay-bearing materials. These materials are ground or pulverized and stacked in separate areas so that they may be mixed in appropriate proportions. After the materials have been carefully analyzed, they are moved to drying kilns, where they are mixed in dry form. This is called the *dry process* of moving the ingredients. Another, more accurate, method is to process the materials as a wet slurry. In both processes the materials must be ground to a fine powder, blended, mixed, and stored.

The material is then fed into one end of a long cylindrical furnace. This furnace is inclined about 5 degrees with the horizontal. The input end is the high end. This cylinder is lined with firebrick and fired with gas, coal, or oil. The temperature at the output end is about 2700°F. As the furnace rotates, the slope of the furnace causes the materials to move toward the output end. The materials fuse into clinkers, which are cooled in a second furnace. This air is used to preheat the first furnace. As the clinkers leave the cooling furnace, they are mixed with gypsum and small quantities of water, ground to a fine powder, and pumped into storage tanks ready for packaging and shipping.

The various components for the many types of cement used are reflected on the composition diagram, Fig. 14-2(a).

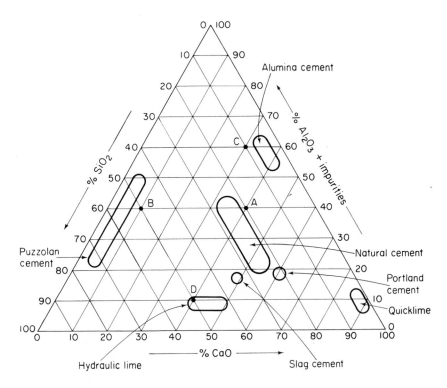

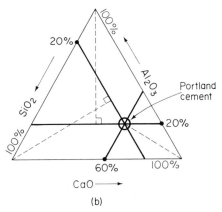

(b)

Figure 14-2

Example 1

From Fig. 14-2, indicate one possible composition of major ingredients of portland cement.

solution:

(a) CaO: in Fig. 14-2(b) a dotted line is drawn from the 100 per cent CaO apex of the triangle to the perpendicular which passes through the region marked "portland cement." This line terminates at the 60 per cent side of the triangle.

(b) SiO_2: in Fig. 14-2(b) the solid line terminates at 20 per cent SiO_2.

(c) Al_2O_3: in Fig. 14-2(b) the solid perpendicular line terminates at 20 per cent Al_2O_3.

(d) Thus, the ingredients for this portland cement are 60 per cent CaO, 20 per cent SiO_2, and 20 per cent Al_2O_3.

There are four major compounds used in portland cement, as shown in Table 14-1. When cement reacts with water, a paste is formed that solidifies into a solid mass, or acts as a binder for other materials in the mix.

Table 14-1

Compound	Formula	Abbr.
Tricalcium silicate	$3CaO \cdot SiO_2$	C_3S
Dicalcium silicate	$2CaO \cdot SiO_2$	C_2S
Tricalcium aluminate	$3CaO \cdot Al_2O_3$	C_3A
Tricalcium aluminoferrite	$4CaO \cdot Al_2O_3 \cdot Fe_2O_3$	C_4AF

Five types of cement are shown in Table 14-2.

Table 14-2

Type	Cement	Percentages				
		C_3S	C_2S	C_3A	C_4AF	Impur.
I	Normal	45	27	11	8	6
II	Modified low heat	44	31	5	13	5
III	Quick setting	53	19	10	9	7
IV	Low heat	28	49	4	12	5
V	Sulfate resisting	38	43	4	9	4

Source: American Concrete Institute.

Types I, II, and III cements are classified as *air-entraining portland cements*. These cements contain about 4 to 7 per cent very small air bubbles. They resist damage from freezing temperatures and damage from salt used to melt ice and snow.

Type I cement is a general-purpose cement used in sidewalks, walls, buildings, sewers, and in cases where heat generated during hydration is not damaging. It is also used in reinforced concrete. This cement should not be used where there is danger of sulfate attack.

Type II cement generates heat of hydration much slower than does normal portland cement. This cement will withstand sulfate attacks. It may be used where large masses of concrete are needed.

Type III cement produces structures that possess high strength a short time after pouring. They are used when early setting is required, when it is necessary to remove forms early, or when the possibility of frost or cold weather exists.

Type IV cement generates the lowest amount of heat of hydration of all types of cement. It is used in dams or other large concrete structures, where the heat of hydration must be kept at a minimum.

Type V cement is affected the least by sulfate attacks. Structures that may be attacked by sulfates are those subjected to seawater, effluents from manufacturing plants, and some ground waters.

When lime reacts with water (hydration), it does so with a release of heat. The end products of the hydration process of tricalcium silicate, C_3S, or dicalcium silicate, C_2S, are calcium hydroxide, $Ca(OH)_2$, and calcium silicate hydrate, $3CaO \cdot SiO_2 \cdot 3H_2O$ with a release of heat. When large masses of concrete are poured, this heat may generate cracking. The release of heat by C_3S is twice as great as that released by C_2S; the release of heat by C_3A is three times greater than that released by C_2S. C_3A and C_3AF react and release very little lime during hydration. Table 14-2 shows the low-heat cements (IV and V), which have greater amounts of C_2S and smaller amounts of C_3A and C_4AF than do the air-entraining cements.

A typical reaction is

$$2(3CaO \cdot SiO_2) + 6H_2O \rightarrow 3CaO \cdot 2SiO_2 \cdot 3H_2O + 3Ca(OH)_2$$

There is some delay before cement sets after water is added. The setting of cement is completed in a matter of hours and depends upon the composition of the material and the amount of water added. The hardening of cement takes months. It should be noted that setting and hardening of cement are not the same. Finer cement particles, the use of hot water, the temperature at which aging takes place, increasing the amount of C_3S, and the generation of great amounts of heat are all accelerators or indicators of rapid setting. Gypsum is a retarder of setting in cement. Fast-setting cement is not as strong as slow-setting cement.

It was noted that the temperature at which curing takes place is important. To attain maximum strength, curing should take place at temperatures between 50 and 90°F. This takes about 28 days. When curing takes place at temperatures higher than 90°F, cement (concrete) loses about three-quarters of its strength. When curing takes place below 40°F, the strength of cement is reduced about 15 per cent.

Curing must also take place in a moist atmosphere. Dry atmospheres stop the hardening processes. Moist temperatures permit hardening to continue for very long periods of time, and in some instances the strength of the cement binder will double.

The graphs in Fig. 14-3(a) show the compressive strength of cement as a function of the curing time of the various components.

Table 14-3 shows the effect of water and aging time on the compressive strength of concrete.

Table 14-3   Effect of Water–Cement Ratio and Aging Time on Compressive Strength of Concrete

Gal. of water per sack of cement	Compressive strength after aging (max. value) × 10^3 psi							
	Type I (days)				Type III (days)			
	1	3	7	28	1	3	7	28
4	2.0	4.3	5.8	7.5	4.2	6.0	7.3	8.5
5	1.4	3.3	4.7	6.3	3.3	5.0	6.0	7.2
6	1.0	2.5	3.4	5.2	2.5	4.0	5.0	6.2
7	0.7	1.8	3.0	4.4	1.8	3.1	4.1	5.2
8	0.4	1.5	2.4	3.6	1.2	2.3	3.3	4.2

Source: E. R. Parker, Materials Data Book. New York: McGraw-Hill Book Company.

Figure 14-3(b) shows an idealized visual picture of the mechanism of the setting and hardening of cement. As seen in the upper left segment of the figure, grains of cement disperse in water. In the upper right segment of Fig. 14-3(b), the gel begins to form, thicken, and disperse from the cement grains. If the water content is high enough, the gel expands. If it is not sufficient, the gel contracts. The water permeates the surface gel to form more gel at the surface of the cement. As the gel thickens, it forms the network shown in the lower left of Fig. 14-3(b). This network is called "setting." As the gel gets thicker, the time required for hardening increases,

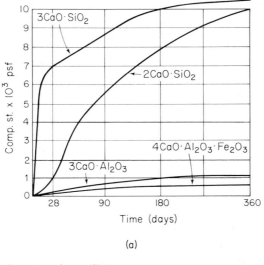

(a)

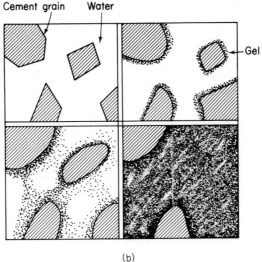

(b)

Figure 14-3

because now it takes longer for the water to penetrate the gel. The hardening is idealized in the lower right of Fig. 14-3(b).

Any unreacted water evaporates and leaves voids in the cement. The more unreacted water present, the greater the number of voids. It is also true that, if any unreacted cement remains, the mass will harden when water or moisture becomes available.

Table 14-4   Water Leakage Due to 20-Psi Pressure Drop Across a 1-Inch-Thick Section

Gallons of water per sack of cement	Leakage after aging, in.³/ft²/h days			
	1	3	7	28
5	24	12	0	0
7	144	55	32	0
9	–	–	70	7

Source: E. R. Parker, *Materials Data Book*. New York: Mc-Graw-Hill Book Company.

Table 14-4 shows the leakage through cement at 20-psi pressure when related to the water used. As the volume of water used for each bag of cement increases, the leakage through the cement increases. This leakage decreases as the aging proceeds. After 7 days of aging, when 5 gallons of water per bag of cement is used, the cement is waterproof; if 7 gallons is used, 20-psi pressure will force 32 in.³ of water through each square foot of cement per hour. After 28 days of curing both 5 and 7 gallons of water mixed with one bag of cement produce a waterproof cement.

14-5 CONCRETE

Concrete is a combination of an aggregate, a cementing material, and water. The aggregate may be different proportions of fine and coarse gravel, portland cement, and water. By introducing other agents, the concrete could become air entraining. About three-quarters of the volume of a concrete mix is aggregate. The rest is a paste made from the cementing material and water. Ideally, each piece of aggregate should be totally surrounded by the paste, and any spaces between the pieces of stone should be full of paste. Various sizes of crushed rock and gravel are used. The idea is that smaller rocks will fill spaces between larger rocks, thus decreasing the need for as much cement paste.

As indicated, the aggregates can be graded roughly into fine and coarse. Fine aggregate is a combination of materials under $\frac{1}{4}$ in. in diameter. These materials may be crushed stone or sand of various sizes. They must be free from impurities, which could react with the cement or which could form voids if they disintegrate after hardening. Coarse aggregate is also graded crushed stone, or rounded stone, up to about 3 in. in diameter. Whether the aggregate is fine or coarse, the stone should be clean, hard, and more or less round.

A *silt test* will reveal the amount of silt or clay in sand. One-quarter of a quart jar is filled with sand. Three-quarters of the jar is filled with clear water. With the cover on the jar, it is shaken vigorously. The jar is allowed to stand until the water is clear once again. The heavy particles will settle out first. The silt settles out next. A layer of fine silt that exceeds $\frac{1}{8}$ in. at the top of the deposit would prohibit the use of the sand from which the sample was taken.

Sand may be tested for *organic matter*. A sample of sand (about 4 oz) is placed in a glass bottle (12 oz), which is then filled about half full with a 3 per cent solution of sodium hydroxide. The bottle is shaken and allowed to stand for about 24 hours. A color of the liquid ranging from clear to light straw indicates that the sand may be used. A color of straw to dark brown indicates too much organic matter in the sand and that the sand should not be used.

14-6 AGGREGATE CLASSIFICATION

As indicated, aggregate needs to be sorted and classified according to size. These sizes are sifted through sieves. The last sieve through which the material passes yields the size of that material. Standard sieves* for fine aggregate are numbers 4, 8, 16, 30, 50, and 100, and pan. For coarse aggregate the screen sizes used in the sieves are number 4, $\frac{3}{8}$ in., $\frac{3}{4}$ in., $1\frac{1}{2}$ in., 3 in., 6 in., and pan. The numbered sieves indicate the openings per linear inch. Thus a number 8 sieve has eight openings per linear inch of screen, or 64 openings per square inch. If there are eight openings per linear inch, then each opening is $\frac{1}{8}$ in. square.

The procedure is to place a 1000-g sample of *fine* aggregate into the top sieve. A more rigorous test uses a 500-g sample. A 5000-g sample is used for *coarse* aggregate. The sieves are shaken for about 3 min. The material that remains in each sieve is weighed. The percentage of the whole that this weight represents is then calculated and recorded. Then *cumulative* percentages are calculated and recorded. Cumulative percentages are calculated by adding each percentage retained to the preceding total. The cumulative percentage totals should correspond to Table 14-5.

If the *fineness particle number* is desired, the cumulative percentages are added and divided by 100. Allowable fineness numbers range from 2.5 to 3.0. Fine sands yield a fineness number of approximately 2.50; medium sands, 2.75; coarse sands, about 3.00

Example 2 illustrates the calculation of the fineness modulus number.

---

*Note the expanded sieve sizes listed in Table 14-9 of this chapter.

Table 14-5

Sieve no.	Cumulative % retained
4	0–5
8	10–20
16	20–40
30	40–70
50	70–88
100	92–98

Example 2

A 500-g sample of fine aggregate is placed in standard sieves and shaken. The following weights are determined: sieve number 4, 12.4 g; number 8, 62.6 g; number 16, 92.4 g; number 30, 101.5 g; number 50, 122.6 g; number 100, 105.8 g; pan, 8.2 g. Calculate (a) the percentage of aggregate retained in each sieve, (b) the cumulative percentage retained, and (c) the fineness modulus number.

solution:

Sieve no.	Grams retained	% Retained	Cumulative % retained
4	12.4	2.45	2.45
8	62.6	12.38	14.83
16	92.4	18.28	33.11
30	101.5	20.08	53.19
50	122.6	24.25	77.44
100	105.8	20.93	98.37
Pan	8.2	1.62	
	Σ 505.5 g	Σ 99.99%	Σ 279.39%

(a) The percentage retained in each sieve is

$$\text{No. 4 sieve} = \frac{12.4}{505.5} \times 100 = 2.45\%$$

The results of each calculation are documented above.

(b) The cumulative percentages retained are shown above and are

No. 4   = 2.45

No. 8 = 2.45+12.38 = 14.83

No. 16 = 14.83+18.28 = 33.11 . . .

(c) The fineness modulus number is the summation of the cumulative percentages retained divided by 100.

$$\text{Fineness mod.} = \frac{279.39}{100} = 2.79$$

14-7 SLUMP TEST AND SPECIFIC GRAVITY

The slump test, which yields the consistency of cement or concrete, is carried out by selecting the desired water–cement ratio of aggregate, cement, and water to make up about 30 lb of concrete, or about 0.2 ft.^{3} This is mixed, and a cone, shown in Fig. 14-4, is filled about one-third full of this mixture and rammed with a $\frac{5}{8}$-in.-diameter round rod that has a spherically shaped end. The cone is then filled to the two-thirds mark and rammed

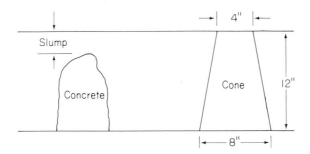

Figure 14-4

again. It is then filled to the top, rammed, and the excess is struck off. When the cone is lifted vertically off the mixture, the concrete will slump an amount depending on its consistency. An *inches-of-slump* reading is taken when the slumped sample is compared with the erect cone, as shown in Fig. 14-4.

Table 14-6 shows some values of acceptable minimums for the slump of concrete.

Table 14-6

| | *Slump* | |
Construction	*Min. in.*	*Max. in.*
Reinforced foundation walls and footings	2	4
Plain footings, caissons, substructure walls	1	3
Slabs, beams, reinforced walls	2	5
Building columns	3	5
Pavement	1	2
Sidewalks, driveways, ground slabs	2	4
Heavy mass construction	1	2

Source: Design and Control of Concrete Mixtures, 10th ed. Portland Cement Association.

The *bulk specific gravity* is a determination that is used to calculate the amount of aggregate, cement, and water needed for a quantity of concrete.

Two specific gravity calculations are made. One is based on oven-dried material, and the second on saturated surface-dried material. *Oven-dried material* is defined as that material whose surface and pores are free from moisture. *Surface-dried material* is defined as that material whose pores are saturated with moisture, but whose surface is dry.

A coarse aggregate specific gravity value (saturated-surface, dried-bulk specific gravity) is calculated by taking a 5000-g sample of coarse aggregate that is washed and dried. It is then immersed in water for 24 hours. This saturates the sample for a fixed period of time. The sample is surface dried, placed in a wire basket, immersed in water, and weighed while in the water. The specific gravity is determined from the equation

$$S_g = \frac{W_a}{W_a - W_s}$$

W_a = saturated weight in air
W_s = weight submerged
S_g = specific gravity coarse aggr.

The percentage absorption of moisture may be determined by oven drying the sample until its weight, upon repeated weighing, remains constant and then by calculating this percentage from the equation

$$A\% = \frac{W_a - W_o}{W_o} \times 100$$

W_a = saturated weight in air
W_o = oven-dried weight
A = percentage absorption, *coarse aggr.*

Example 3

A coarse aggregate that weighs 5005 g and has a specific gravity of 2.68 is immersed for 24 h in water. The sample surface is dried and weighed again. It is found to weigh 5140 g. (a) How much does it weigh when suspended in water? (b) What is the oven-dry weight of the moisture if the percentage of moisture absorption is 1.4 per cent?

solution:

(a) The equation for specific gravity is

$$S_g = \frac{W_a}{W_a - W_s}$$

Therefore,

$$W_s = \frac{S_g W_a - W_a}{S_g} = \frac{2.68(5140) - 5140}{2.68}$$

$$= 3222 \text{ g}$$

$W_a = 5140$ g
$W_s = ?$
$S_g = 2.68$
$A = 1.4$ per cent

(b) From the percentage of absorption equation

$$A = \frac{W_a - W_o}{W_o} \times 100$$

the value of the oven-dried weight is

$$W_o = \frac{W_a}{\dfrac{A}{100} + 1} = \frac{5140}{0.014 + 1}$$

$$= 5069 \text{ g}$$

A *fine aggregate* specific gravity is determined by using a 1000-g sample which is oven dried and saturated for 24 hours in water. This saturated sample is warmed and dried until it slumps when tested with the slump test. Once it slumps, the sample is in the saturated surface-dried condition. *Five hundred grams* of this sample are placed in a flask that has a known volume. The remaining space in the flask is filled with water. The saturated bulk specific gravity of the fine aggregate is calculated from

$$S_g = \frac{500}{V_f - V_w} \qquad \begin{aligned} V_f &= \text{volume of flask} \\ V_w &= \text{volume of water} \\ S_g &= \text{sp. gr. fine aggr.} \end{aligned}$$

The percentage of moisture absorption is calculated after the sample has achieved constant weight when oven dried.

$$a\% = \frac{500 - W_o}{W_o} \times 100 \qquad \begin{aligned} a &= \text{percentage of absorption, } \textit{fine aggr.} \\ W_o &= \text{oven-dried weight} \end{aligned}$$

Example 4

A fine aggregate of 500 g of saturated surface-dried material is placed into a 280-cm³ flask. The volume of water required to fill the flask is 92 cm.³ Calculate (a) the bulk specific gravity of the sample, and (b) the percentage moisture absorption if the oven-dried weight of the sample is 480 g.

solution:

(a) The bulk specific gravity is

$$S_g = \frac{500}{V_f - V_w} = \frac{500}{280 - 92} \qquad \begin{aligned} V_f &= 280 \text{ cm}^3 \\ V_w &= 92 \text{ cm}^3 \\ W_o &= 480 \text{ g} \end{aligned}$$

$$= 2.66$$

(b) The percentage of absorption is

$$a = \frac{500 - W_o}{W_o} \times 100 = \frac{500 - 480}{480} \times 100$$

$$= 4.17\%$$

14-8   MIXING CALCULATIONS

The problem now is to determine the amount of aggregate, cement, and water needed to produce a desired quantity of concrete. Any calculation is an estimated value. The volume of the concrete produced is directly related to the water that rises as the concrete sets, the voids in the concrete that result from the evaporation of water, and the volume of the aggregate, cement, and water used in making the cement. These volumes may be taken from Fig. 14-5.

The values for specific gravities and weights shown in Table 14-7 may be used to find the specific weights of the ingredients of concrete.

Table 14-7

Material	*Specific gravity*
Water	1.00
Weight	8.33 lb/gal
Cement	3.25
Weight	94 lb/bag
Sand	2.65
Gravel	2.60

One method for calculating the various amounts of ingredients needed to produce concrete follows. Fig. 14-5 shows all possible combinations of cement, aggregate, and water percentages. The trapezoid in the lower left yields the more usual combinations for making concrete. An array of straight lines is projected from the 100 per cent aggregate corner of the triangle, through the trapezoid, to the cement–water graph. The use of Fig. 14-5 is best explained with the illustrated problem that follows.

Example 5

Using point A in Fig. 14-5, (a) calculate the water–cement ratio. (b) Calculate the ultimate compressive strength of a standard and high-strength cement for this cement–water ratio. (c) Is the slump for this mixture acceptable? Explain. (d) Calculate the quantity of water, aggregate, and cement needed to produce 1 yd³ of concrete. (e) Assume that 2 per cent needs to be added for voids and lost water. What is the weight of the concrete? (f) How many bags of cement are needed? (g) How many gallons of water are needed?

solution:

(a)   One of the lines extends from the 100 per cent aggregate apex of the triangle through point A to the 0.7 cement–water ratio.

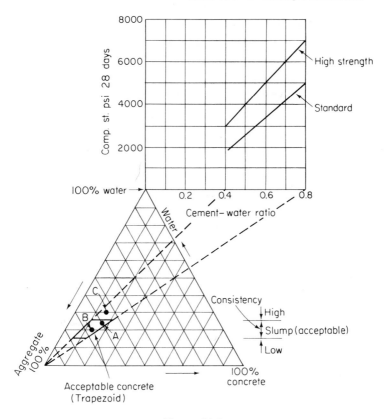

Figure 14-5

(b) If a vertical line is drawn from the 0.7 cement–water ratio point, it will cross the standard cement curve and yield a 28-day ultimate strength of about 4200 psi. The vertical line will also intersect the high-strength cement curve and yield an ultimate strength of 6000 psi.

(c) If a line is drawn parallel to the base of the triangle from point A to the right, it will intersect the slump graph in the acceptable consistency region. A 4-in. slump concrete may be workable for some combinations, and not for others. See Table 14-6.

(d) The ingredients needed to produce 1 yd^3 of concrete are calculated in the following manner:

   (1) From the triangle diagram in Fig. 14-5, the amounts of each of the ingredients are

   Water = 22%

   Cement = 18%

   Aggregate = 60%

(2) Therefore, in terms of the fractional part of *1 yd³ of concrete* for each of the ingredients, the following may be used:

Water $= 22\% = 0.22$ yd³

Cement $= 18\% = 0.18$ yd³

Aggregate $= 60\% = 0.60$ yd³

(3) The pounds of water required for *each cubic yard of concrete are*

$W_w = 0.22$ yd³ $\times 27$ ft³/yd³ $\times 62.4$ lb/ft³

$= 371$ lb

where W_w is the weight of water.

(4) The cement, specific gravity 3.25, for each cubic yard of concrete has a weight of

$W_c = 0.18$ yd³ $\times 27$ ft³/yd³ $\times 3.25 \times 62.4$ lb/ft³

$= 986$ lb

where W_c is the weight of cement.

(5) The weight of the sand–stone aggregate (assume a specific gravity of 2.62) required for each cubic yard of concrete is

$W_a = 0.60$ yd³ $\times 27$ ft³/yd³ $\times 2.62 \times 62.4$ lb/ft³

$= 2648$ lb

where W_a is the weight of the aggregate.

(e) The total quantity of concrete to this point is

$W = W_w + W_c + W_a = 371 + 986 + 2648$

$= 4005$ lb

If 2 per cent is added because of voids and water losses, the weight is

4005 lb $\times 1.02 = 4085$ lb of concrete

(f) The number of bags (cement factor) of cement needed for 1 yd³ of concrete is

$$N = \frac{986 \text{ lb}}{94 \text{ lb/bag}} = 10.5 \text{ bags} \sim 11 \text{ bags}$$

where N is the number of bags.

(g) The gallons of water needed for 1 yd³ of concrete is

$$\text{Gallons of water} = \frac{371 \text{ lb}}{8.33 \text{ lb/gal}} = 44.54 \text{ gal}$$

A quick and *approximate method* is to use the arbitrary ratios of volume shown in Table 14-8. This method is not as accurate as the method just discussed, but it is widely used.

Table 14-8

Concrete	Ratios
Reinforced	1:2:4
Large sections	1:3:5
Pavement	1:2:3
Sidewalks	1:2:3

Thus a ratio of volumes for each ingredient for sidewalks could be

$$\text{Cement} = 1 \text{ volume}$$

$$\text{Fine aggregate} = 2 \text{ volumes}$$

$$\text{Coarse aggregate} = 3 \text{ volumes}$$

Water must be added to produce the desired slump.

Example 6

Assume a concrete sidewalk 6 ft wide by 100 ft long by 8 in. thick. Using the approximate volume method, how much of each ingredient is needed?

solution:

(a) The volume of concrete needed is

$$V = 6 \times 100 \times \tfrac{8}{12} = 400 \text{ ft}^3$$

$$= \frac{400 \text{ ft}^3}{27 \text{ ft}^3/\text{yd}^3} = 14.8 \text{ yd}^3$$

(b) From Table 14-8, the ratio needed is 1:2:3.

Cement	= 1 volume
Fine aggregate	= 2 volumes
Coarse aggregate	= 3 volumes
Total	6 volumes

(c) The cubic yards of each ingredient needed is

Cement	$= \tfrac{1}{6} \times 14.8 =$	2.5 yd³
Fine aggregate	$= \tfrac{2}{6} \times 14.8 =$	4.9 yd³
Coarse aggregate	$= \tfrac{3}{6} \times 14.8 =$	7.4 yd³
	Total	14.8 yd³ *check*

Another method for determining concrete mix follows.

We have seen that measuring particle sizes is important when constructing an aggregate in preparation for mixing concrete. As indicated, the raw material is placed in a series of sieves, largest openings on top, and agitated. Openings $\frac{1}{4}$ in. and larger are given in inches. Smaller meshes are given as the number of openings per linear inch. Table 14-9 gives some standard screen meshes.

Table 14-9

Mesh*	Opening	Mesh	Opening	Mesh	Opening
	1.050	10	0.065	150	0.0041
	0.742	14	0.046	200	0.0029
	0.525	20	0.0328	270	0.0021
	0.371	28	0.0232	400	0.0015
3	0.263	35	0.0164		
4	0.185	48	0.0116		
6	0.131	65	0.0082		
8	0.093	100	0.0058		

*Taylor meshes.

Figure 14-6 shows an idealized pack of uniformly arranged spheres, all of which have the same size. Under these conditions, the spheres occupy 74 per cent of the space available, and air occupies the remaining 26 per cent. Bricks, which have perfect size, will, theoretically, stack and occupy 100 per cent of the space available.

Both conditions described above are ideal. Irregularly shaped aggregate, if packed with matching surfaces, will reduce the percentage of volume occupied by the air. If different-sized particles are used, the percentage of volume occupied by the air spaces will also be reduced. When concrete is mixed, the available air space is filled with the cement paste.

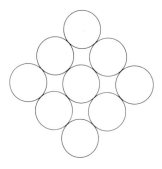

Figure 14-6

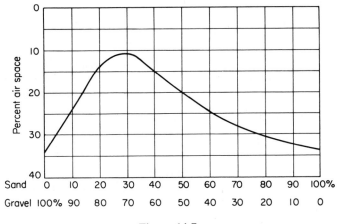

Figure 14-7

From Fig. 14-7 it can be seen that a mixture of 60 per cent sand and 40 per cent gravel will have about 24 per cent of the total volume as air space. This space, ideally, would be occupied by the cement paste.

Example 7

What percentage of the available space is occupied by a mixture of aggregate that has a ratio of 40 per cent sand to 60 per cent gravel?

solution:

From Fig. 14-7 about 15 per cent of the space is available for cement. Therefore, 85 per cent of the space is occupied by the aggregate.

To find the weight of the sand and gravel, the procedure in Example 4 may be used. To find the weight of each in the aggregate, the method that follows may be used.

(a) Volume of the bulk gravel:

$$V_g = \frac{W_g}{D_g}$$

W_g = weight gravel, lb/ft³
D_g = bulk density gravel, lb/ft³
V_g = volume of gravel, ft³

(b) Actual volume of the gravel:

$$V_{ag} = \frac{W_g}{S_g \times 62.4 \text{ lb/ft}^3}$$

V_{ag} = actual volume gravel, ft³
S_g = sp. gravity gravel

$$= \frac{W_g}{2.60 \times 62.4} = \frac{W_g}{162.24 \text{ lb/ft}^3}$$

(c)   Volume of the pores:

$$V_p = V_g - V_{ag}$$

V_p = volume of pores in gravel, ft³

(d)   Volume of the bulk sand:

$$V_s = \frac{W_s}{D_s}$$

W_s = weight sand, lb/ft³
D_s = bulk density sand, lb/ft³
V_s = volume bulk sand, lb/ft³

(e)   Bulk sand needed:

$$V_{as} = V_s - V_p$$

V_s = volume of sand, ft³
V_{as} = additional bulk sand, ft³

(f)   Total mixture of the bulk aggregate:

$$V_a = V_g + V_{as}$$

V_a = volume of aggregate per 100 lb, ft³

(g)   Weight of the total bulk aggregate:

$$W_a = \frac{100 \text{ lb} \times 27 \text{ ft}^3/\text{yd}^3}{V_a \text{ft}^3}$$

W_a = weight bulk aggregate, lb/yd³

Example 8

Using the ratio from Example 7, and assuming a bulk density of gravel of 108 lb/ft³ and of sand of 105 lb/ft³, calculate (a) the bulk volume of the gravel, (b) of the pores, (c) of the additional sand, (d) the total volume per 100 lb of mixture, and (e) the weight of the total mixture per cubic yard. The specific gravity of sand is assumed to be 2.65 and of gravel, 2.60.

solution:

(a)   Volume of the bulk gravel is

$$V_g = \frac{W_g}{D_g} = \frac{60 \text{ lb}}{108 \text{ lb/ft}^3}$$

$W_s = 40 \text{ lb}$
$W_g = 60 \text{ lb}$
$D_g = 108 \text{ lb/ft}^3$

$$= 0.555 \text{ ft}^3$$

$$V_{ag} = \frac{W_g}{2.60 \times 62.4} = \frac{60 \text{ lb}}{2.60 \times 62.4 \text{ lb/ft}^3}$$

$$= 0.370 \text{ ft}^3$$

(b)   The volume of pores is

$$V_p = V_g - V_{ag} = 0.555 - 0.370$$

$$= 0.185 \text{ ft}^3$$

(c)   The bulk volume of the sand is

$$V_s = \frac{W_s}{D_s} = \frac{40\ lb}{105\ lb/ft^3}$$

$$D_s = 105\ lb/ft^3$$
$$S_g = 2.60$$
$$S_s = 2.65$$

$$= 0.381\ ft^3$$

Additional sand needed is

$$V_{as} = V_s - V_p = 0.381 - 0.185$$
$$= 0.196\ ft^3$$

(d) The total bulk volume of gravel and sand aggregate per 100 lb is

$$V_a = V_g + V_{as} = 0.555 + 0.196$$
$$= 0.750\ ft^3/100\ lb$$

(e) The weight of the aggregate per cubic yard is

$$W_a = \frac{100 \times 27}{V_a} = \frac{100\ lb \times 27\ ft^3 yd}{0.750\ ft^3}$$
$$= 3600\ lb/yd^3$$

To find the number of sacks of cement needed to pour 100 ft³ of concrete, the following procedure may be used.

The following equations are based on *one* bag of cement:

(1) The volume of gravel is

$$V_{ag} = \frac{V_g D_g}{2.60 \times 62.4}$$

(2) The volume of sand deposited in the gravel pores is

$$V_{as} = \frac{V_s D_s}{2.60 \times 62.4}$$

(3) The volume of the cement that will form the paste and deposit in the pores is

$$V_c = \frac{D_c}{3.25 \times 62.4}$$

D_c = Bulk density of cement lb/ft³

Sp. gr. cement = 3.25
Sp. wt. cement = 94 lb/ft³
V_c = vol. cement, ft³
V_w = vol. water, ft³
D_w = unit volume of water, gal
V = vol./bag of cement
C = sacks of cement

(4) The volume of water that will also deposit in the pores is

$$V_w = \frac{D_w(8.33\ lb/gal)}{62.4\ lb/ft^3}$$

(5) The total volume V for *every bag of cement* is

$$V = V_{ag} + V_{as} + V_c + V_w$$

(6) The sacks of cement required per 100 ft³ of concrete is

$$C = \frac{100\ ft^3}{V}$$

Example 9

Assume that a mix will have a bulk volume of 2.40 ft³ of gravel and a bulk volume of 1.70 ft³ of sand. Assume that 5 gal of water are needed per bag of cement. How many bags of cement will be needed to pour 350 ft³ of concrete?

solution:

(a) The volume of gravel is

$$V_{ag} = \frac{V_g D_g}{2.60 \times 62.4} = \frac{2.40 \times 108}{2.60 \times 62.4}$$

$$= 1.60 \text{ ft}^3 \text{ of gravel}$$

$V_g = 2.40$ ft³
$V_s = 1.70$ ft³
$D_g = 108$ lb/ft³
$D_s = 105$ lb/ft³
Sg of cement $= 3.25$
$D_c = 94$ lb/ft³
$D_w = 5$ gal/bag

(b) The volume of sand in pores is

$$V_{as} = \frac{V_s D_s}{2.65 \times 62.4} = \frac{1.70 \times 105}{2.65 \times 62.4}$$

$$= 1.08 \text{ ft}^3 \text{ of sand}$$

(c) The volume of the cement that forms in pores is

$$V_c = \frac{D_c}{3.25 \times 62.4} = \frac{94^*}{3.25 \times 62.4}$$

$$= 0.46 \text{ ft}^3 \text{ of cement}$$

(d) The volume of the water in the pores is

$$V_w = \frac{D_w \times 8.33}{62.4} = \frac{5 \times 8.33}{62.4}$$

$$= 0.67 \text{ ft}^3 \text{ of water}$$

(e) The total volume of concrete per bag of cement is

$$V = V_{ag} + V_{as} + V_c + V_w = 1.60 + 1.08 + 0.46 + 0.67$$

$$= 3.81 \text{ ft}^3 \text{ per sack of cement}$$

(f) The total number of bags needed to pour 350 ft³ of concrete is

$$C = \frac{350 \text{ ft}^3}{3.81 \text{ ft}^3/\text{sack}} = 91.9 \simeq 92 \text{ sacks}$$

14-9 PHYSICAL PROPERTIES OF CONCRETE

It was pointed out in this chapter that there are many factors which determine the physical properties of concrete. The physical properties of the aggregate, the ratios of the various components of the mix, the curing rate, moisture, temperature, hydration, etc. — all have an effect upon the physi-

*One cubic foot of loose cement weighs 94 lb/ft³.

cal properties of concrete. In almost all instances, a decrease of the water-to-cement ratio will increase the physical properties of concrete. In general, the strength of the concrete increases as the cement content increases and as the void spaces in the concrete decrease.

Probably the most important physical property of concrete is its *comprehensive strength.* The ability of concrete to support loads in compression is of the order of 4000 to 5000 psi after 28 days of curing. If failure occurs, it generally takes place in the cement rather than in the aggregate. Any fracture plane will take place at between 30 and 40 degrees with the cross-sectional area of the test specimen.

The *tensile strength* of concrete is low. It is about one-eighth its compressive strength. For this reason it is rarely designed to load in tension except when prestressed.

The *modulus of elasticity* is generally about 4×10^6 psi and increases as curing proceeds. Its *linear coefficient of expansion* averages about $6 \times 10^6/°F$.

The *flexure strength* is generally about one-tenth its compressive strength. This property of concrete is important when it is to be used in beams, slabs, floors, or highways where heavy loads may be applied, subjecting the concrete to dynamic loading. Since the tensile strength of concrete is less than its compressive strength, it will fail at the surface when subject to oscillating loads.

Its *fatigue strength* is about 0.4 its ultimate strength. It has a specific gravity of about 1.9 to 2.5 and a specific weight of 144 lb/ft³. Table 14-10 summarizes suggested trial mixes of concrete.

14-10 STRESSED CONCRETE

Normally reinforced concrete, when manufactured as precast concrete or poured on the job, has the concrete poured around a network of steel reinforcing rods. This creates a high-strength concrete, because the embedded steel takes the tensile forces rather than the concrete.

Since concrete is comparatively weak in tension, various methods are used to place concrete in compression initially, so that in use any tensile forces that operate on the concrete will be balanced by these built-in compressive forces. Cold-worked steel rods are used. The rods are cold-bent. These bends act as anchors. The surface of the rods have radial ribs or are scored so that a positive bond is created between the rod and the cured concrete.

The process is to place a steel rod in tension as concrete is being poured around it [Fig. 14-8(a)]. The concrete solidifies and bonds to the rod. The ends of the rod are released. This places the concrete in compres-

Table 14-10   Suggested Trial Mixes for Non-Air-Entrained Concrete of Medium Consistency (3- to 4-in. slump)

Water gal/bag of cement	Max. size aggr. in.	Air content %	Water gal/yd³ of concrete	Cement bags per yd³ of concrete	With fine sand Fineness no. = 2.50			With coarse sand Fineness no. = 2.90		
					Fine ag. % total	Fine ag. lb/yd³	Coarse ag. lb/yd³	Fine ag. % total	Fine ag. lb/yd³	Coarse ag. lb/yd³
4.5	3/8	3.0	46	10.2	53	1320	1190	57	1430	1080
	1/2	2.5	44	9.8	44	1150	1460	48	1260	1350
	3/4	2.0	41	9.1	36	1000	1760	40	1110	1650
	1	1.5	39	8.7	32	910	1940	36	1030	1820
	1½	1.0	36	8.0	28	830	2150	32	970	2030
5.0	3/8	3.0	46	9.2	54	1400	1190	58	1510	"
	1/2	2.5	44	8.8	46	1230	1460	50	1340	
	3/4	2.0	41	8.2	38	1070	1760	42	1180	
	1	1.5	39	7.8	34	990	1940	38	1100	
	1½	1.0	36	7.2	30	910	2150	34	1030	
5.5	3/8	3.0	46	8.4	55	"	"	59	1570	"
	1/2	2.5	44	8.0	47			51	1400	
	3/4	2.0	41	7.5	39			43	1230	
	1	1.5	39	7.1	35			39	1160	
	1½	1.0	36	6.6	31			35	1080	
6.0	3/8	3.0	46	7.7	56	"	"	60	1630	"
	1/2	2.5	44	7.4	48			52	1450	
	3/4	2.0	41	6.9	40			44	1280	
	1	1.5	39	6.5	36			40	1210	
	1½	1.0	36	6.0	32			36	1120	
6.5	3/8	3.0	46	7.1	57	"	"	61	1660	"
	1/2	2.5	44	6.8	49			53	1500	
	3/4	2.0	41	6.3	41			45	1330	
	1	1.5	39	6.0	37			41	1250	

Table 14-10 (continued)

Water gal/bag of cement	Max. size aggr. in.	Air content %	Water gal/yd³ concrete	Cement bags per yd³ if concrete	With fine sand Fineness no. = 2.50			With coarse sand Fineness no. = 2.90		
					Fine ag. % total	Fine ag. lb/yd³	Coarse ag. lb/yd³	Fine ag. % total	Fine ag. lb/yd³	Coarse ag. lb/yd³
7.0	3/8	3.0	46	6.6	58	"	"	61	1720	"
	1/2	2.5	44	6.3	49			53	1540	
	3/4	2.0	41	5.9	42			45	1360	
	1	1.5	39	5.6	37			41	1280	
	1½	1.0	36	5.2	33			37	1190	
7.5	3/8	3.0	46	6.2	58	"	"	62	1750	"
	1/2	2.5	44	5.9	50			54	1570	
	3/4	2.0	41	5.5	42			46	1390	
	1	1.5	39	5.2	38			42	1310	
	1½	1.0	36	4.8	34			38	1220	
8.0	3/8	3.0	46	5.8	58	"	"	62	1780	"
	1/2	2.5	44	5.5	51			54	1600	
	3/4	2.0	41	5.2	42			46	1410	
	1	1.5	39	4.9	38			42	1330	
	1½	1.0	36	4.5	34			38	1240	

Increase or decrease water per cubic yard by 3 per cent for each increase or decrease of 1 in. in slump, then calculate quantities by the absolute volume method. For manufactured fine aggregate, increase percentage of fine aggregate by 3 and water by 2 gal/yd³ of concrete. For less workable concrete, as in pavements, decrease percentage of fine aggregate by 3 and water by 1 gal/yd³ of concrete.

Source: Design and Control of Concrete Mixtures, 11th ed. Portland Cement Association (1968).

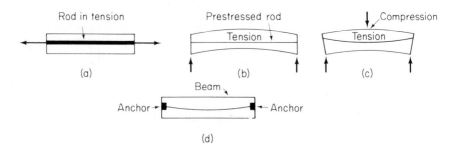

Figure 14-8

sion, which makes it bow as shown in Fig. 14-8(b), generating built-in resistance to tensile forces. If loaded as shown in Fig. 14-8(c), it will resist efforts to straighten. This is called *prestressed concrete*.

Post-stressed concrete is cured concrete that has been reinforced normally with steel rods. Also embedded in the cured concrete are tubes or channels. Cables or rods are inserted into these tubes, anchored at one end, stressed, and anchored at the other end, as shown in Fig. 14-8(d). This type of stressing is generally used for very heavy concrete sections or very long members.

Problems

14-1. What is the difference between calcium and gypsum cement?

14-2. (a) List the four types of mortar discussed in this chapter. (b) List the proportional parts of cement, sand, and lime for each. (c) State the recommended uses of each.

14-3. (a) Give the chemical symbol for limestone. (b) What is calcining? (c) What are the end products of calcining?

14-4. (a) What is pulverized lime? (b) What is hydrated lime? (c) What is slaked lime? (d) Why is it necessary to remove the magnesium oxide from the lime?

14-5. How does the lime harden? Write the equation.

14-6. Discuss the effect of water when it is added to lime.

14-7. Describe the differences and uses of (a) lime–cement mortar and (b) lime mortar.

14-8. What is the content of gypsum in its raw state? How is it refined?

14-9. (a) What is plaster of paris? (b) What are some of its uses? (c) What effect does the addition of water have on its compressive strength? (d) What is interlocking crystallization?

14-10.   (a) What is wall plaster? (b) How is plaster of paris processed into building blocks? (c) What is flooring plaster?

14-11.   How does hard-finish plaster differ from Keene's plaster?

14-12.   Referring to Fig. 14-1, what water-to-plaster ratio should be used to achieve a plaster compressive strength of 2000 psi?

14-13.   What is the effect on the compressive strength of plaster as the water-to-plaster ratio increases from 0.6 to 0.9?

14-14.   What is Pozzolan cement?

14-15.   What is natural cement? How is it made?

14-16.   (a) Describe hydraulic lime. (b) What is slag cement?

14-17.   What is Sorel's cement? What are its uses?

14-18.   Describe the process for making portland cement.

14-19.   What is the composition of the cement in Fig. 14-2(a) at point *A?*

14-20.   Repeat Problem 14-19 for point *B*.

14-21.   Repeat Problem 14-19 for point *C*.

14-22.   Repeat Problem 14-19 for point *D*.

14-23.   List the four major compounds used in portland cement.

14-24.   List the three types of portland cement classified as air-entraining cement. State the uses of each.

14-25.   What are the special characteristics of types IV and V portland cements?

14-26.   Discuss the hydration of lime in cement and the effects of the heat released.

14-27.   Discuss the factors that accelerate the setting of cement. What effect does gypsum have on the setting rate of cement?

14-28.   (a) What is the ideal temperature for curing cement? (b) What are the effects of too high or too low a temperature on the strength of cement?

14-29.   What are the effects of moisture on the strength of cement?

14-30.   Discuss the effect of water and aging time on the compressive strength of concrete.

14-31.   Assume that 5 gal of water are used for each bag of cement; what is the compressive strength of the resulting concrete (see Table 14-3) for the following days of aging for a Type I cement: (a) 3 days, (b) 28 days.

14-32.   Assume that a type III cement is aged for 7 days; (see Table 14-3) what compressive strength can be expected from the concrete (a) if 4 gal of water is used per bag of cement? (b) 6 gal? (c) 8 gal?

14-33. Describe the formation of gel and its effect on setting and hardening of cement.

14-34. Referring to Table 14-4, if zero leakage of water is required through cement, how many days should the cement age if (a) 5 gal of water per bag of cement is used? (b) 7 gal? (c) 9 gal?

14-35. List the three essential components of concrete. Discuss them.

14-36. Describe the silt test for sand.

14-37. Describe the test used to determine the organic matter content in sand.

14-38. How are the various aggregate sizes determined? What are the standard sieve sizes?

14-39. How many openings per inch are there in a 50-mesh screen? How many openings are there in 1 in.^{2} of mesh? What is the size of each opening?

14-40. Describe the procedure for grading aggregate and explain how cumulative percentages and fineness numbers are calculated.

14-41. A 500-g sample of fine aggregate is placed in a standard sieve and shaken. The following weights are recorded: sieve 4, 7.5; 8, 66; 16, 99.5; 30, 106.8; 50, 128.2; 100, 90.5; pan, 3.0. Calculate (a) the percentage of aggregate retained by each sieve, (b) the cumulative percentages retained, and (c) the fineness number.

14-42. A 500-g sample of fine aggregate is placed in standard sieve and shaken. The following weights are recorded: sieve 4, 9.2; 8, 60.0; 16, 100.5; 30, 108.8; 50, 126.2; 100, 89.4; pan, 5.9. Calculate (a) the percentage of aggregate retained by each sieve, (b) the cumulative percentage retained, and (c) the fineness number.

14-43. A 5000-g sample of coarse aggregate is placed in a standard sieve and shaken. The following weights are recorded: number 4, 15.5; $\frac{3}{8}$ in., 620.4; $\frac{3}{4}$ in., 922.6; 1$\frac{1}{2}$ in., 1050.0; 3 in., 1222.2; 6 in., 1084.5; pan, 96.6. Calculate (a) the percentage of aggregate retained by each sieve, (b) the cumulative percentage retained, and (c) the fineness number.

14-44. Repeat Problem 14-43 using the following data: number 4, 17.5; $\frac{3}{8}$ in., 610.4; $\frac{3}{4}$ in., 950.6; 1$\frac{1}{2}$ in., 1035.3; 3 in., 1155.4; 6 in., 1025.5; pan, 203.3.

14-45. Describe the slump test and how it is performed. Give several acceptable ranges of slump and state the uses of the concrete.

14-46. State two kinds of bulk specific gravity of an aggregate.

14-47. Describe the procedure used to determine (a) the bulk specific gravity of a coarse aggregate and (b) the percentage of absorption of moisture.

14-48. Describe the procedure used to determine (a) the bulk specific gravity of a fine aggregate and (b) the percentage of absorption of moisture of an oven-dried sample.

14-49. A coarse aggregate weighs 5000 g and has a submerged weight of 3130 g. The sample is surface dried, and when weighed again it is found to be 5062 g. (a) What is its specific gravity? (b) What is the percentage of absorption of the oven-dried weight?

14-50. A coarse aggregate weighs 5002 g and has a specific gravity of 2.7; it is immersed for 24 hr in water. When dried it weighs 5136 g. (a) What is its weight when suspended in water? (b) If the percent absorption is 1.5, what is the oven-dried weight of the moisture?

14-51. A fine aggregate of a 500-g saturated surface-dried material is placed into a 260-cm^3 flask. Assume that it takes 86 cm^3 of water to fill the flask. Calculate (a) the bulk specific gravity of the sample and (b) the percentage of moisture absorption if the oven-dried weight of the sample is 492 g.

14-52. A fine aggregate of a 500-g saturated surface-dried material is placed into a 292-cm^3 flask. The volume of the water required to fill the flask is 98 cm^3. Calculate (a) the bulk specific gravity of the sample and (b) the percentage moisture absorption if the oven-dried weight of the sample is 485 g.

14-53. List the various volumes that enter into the computations when one is determining a concrete mix.

14-54. Using Fig. 14-5, (a) calculate the water–cement ratio. (b) What is the ultimate compressive strength of a standard and high-strength cement that has the water–cement ratio? (c) Is the slump acceptable for this mixture? What is it? (d) Calculate the quantity of water, aggregate, and cement needed to produce 1 yd^3 of concrete for point *B*. (e) Assume that 1.5 per cent will need to be added for voids and lost water. What is the weight of the concrete? (f) How many bags of cement are needed? (g) How many gallons of water are needed?

14-55. Repeat Problem 14-54 for point *C*, Fig. 14-5.

14-56. A reinforced concrete beam 2 by 4 by 16 ft is to be poured. How much each of cement and fine and coarse aggregate are needed?

14-57. Assume a ratio of 1:3:5 of concrete for a large construction block of 4 by 6 by 8 ft. How much cement and fine and coarse aggregate are needed?

14-58. Explain the stacking of aggregate and the availability of air space. How is this air space utilized in concrete?

14-59. (a) What percentage of the available space is occupied by a mixture that has 10 per cent sand in the aggregate? (b) What percentage of the space is occupied by the aggregate? Refer to Fig. 14-7.

14-60. Given that an aggregate occupies 70 per cent of the available space, from Fig. 14-7 determine the *two* possible percentages of sand to gravel that could yield such an aggregate.

14-61. Repeat Problem 14-60 for a 75 per cent aggregate.

14-62. Assume a percentage of 12 per cent air space when the sand percentage is 35 per cent and that of the gravel is 65 per cent. The bulk density of gravel is taken to be 110 lb/ft³, and of sand, 105 lb/ft³. Calculate (a) the bulk volume of the gravel, (b) the bulk volume of the pores, (c) the bulk volume of the additional sand, (d) the total volume per 100 lb of mixture, and (e) the weight of the total mixture per cubic yard. The specific gravity of gravel is assumed to be 2.60, and of sand, 2.65.

14-63. Assume the same constants as those in Problem 14-62; repeat Problem 14-62 for a combination of 80 per cent sand and 20 per cent gravel.

14-64. Calculate the number of sacks of cement needed to pour 500 ft³ of concrete, assuming that $4\frac{1}{2}$ gal of water are needed for each bag of cement used. The bulk volume of gravel needed is 2.62 ft³, and of sand, 1.67 ft³. The bulk density of the gravel is taken to be 110 lb/ft³ and of sand to be 105 lb/ft³. The specific gravity of cement is taken to be 3.25.

14-65. What are some factors that determine the physical properties of concrete? List them and make a brief statement about each.

14-66. What effect does the water-to-cement ratio have on the strength of concrete?

14-67. Discuss the following physical properties of concrete: (a) compressive strength, (b) tensile strength, (c) flexure strength.

14-68. Why is normally reinforced concrete classed as a high-strength concrete? Explain.

14-69. Describe the process of making prestressed concrete. What is the purpose of prestressing the structure? Explain.

14-70. Explain the process for making poststressed concrete beams.

15 | Plastics

15-1 PHYSICAL PROPERTIES OF PLASTICS

Plastic materials are synthetic materials that, during some stage of their development, are deformable, and which may remain deformable or be subsequently hardened depending on their structure.

Plastics are classified into two general categories. *Thermoplastic* materials may be softened and upon cooling will harden again. *Thermosetting* materials will harden when heated, take a permanent set, and cannot be made plastic again. Examples of thermoplastic materials are polystyrene, polyethylene, nylon, Plexiglas, and Teflon. Examples of thermosetting materials are Bakelite, rubber, silicones, and epoxy.

In general, thermoplastic materials will not withstand high temperatures. Teflon is one exception. Thermosetting materials, in general, will not burn, but will char and disintegrate. Thermoplastic materials melt if heated to a high enough temperature. Thermosetting plastics are brittle, whereas thermoplastic materials exhibit varying degrees of ductility.

Since the melting or disintegration temperatures of plastics are much lower than steel, plastics creep at room temperature. This tendency to creep is present over the entire temperature range at which plastics are used, and therefore small changes in temperature may have marked effect on the physical properties of the material. The effect of temperature and strain rate on tensile strength must be carefully evaluated when plastics are to be used.

Plastics deform elastically, as do metals up to a given increase of length. At the yield point, there may be a drop in stress. In general, the softer the plastic, the lower the elastic limit and the greater the percentage of elongation. In some very soft plastic materials the yield point is absent. In general, the modulus of elasticity of plastics is lower than steel.

Changing the strain rate may change a ductile plastic into a brittle plastic. This causes a decrease in the elongation, which in turn causes an increase in tensile strength. A decrease in temperature has the effect of increasing the strain rate. Therefore, low temperatures make plastics brittle, even if they are worked slowly.

Impact tests are generally conducted on pieces that have been fabricated, rather than on test specimens. The reason for this is that fabrication itself may change the physical properties of the material. Izod tests and tensile impact tests are used to determine these properties. Plastics will also fail when loaded repeatedly below the yield strength of the material.

15-2 STRUCTURE OF PLASTICS

The *mer* is the basic unit of the chain that forms a *monomer* molecule. A *polymer* is a combination of many units that make up a large molecule. All three are shown in Fig. 15-1(a), (b), and (c), respectively. In Fig. 15-1 the tie bar that binds the hydrogen atom to the carbon atom represents two valence electrons. Thus in the mer in Fig. 15-1(a), each hydrogen atom is bound to the carbon atom by a pair of electrons. In Fig. 15-1(b), the carbon atoms are tied to each other by two pairs of electrons, indicated by the double tie bars. Multiple bonds generally signal that the monomer is unsaturated. That is, the molecule does not have the maximum possible number of hydrogen atoms tied to it. Under specific conditions of heat, pressure, and the appropriate catalyst, one bond from a double bond will transfer and form a bond with another monomer, which in turn forms a large molecule, or chain. The result is *polymerization*. Polymerization may occur naturally or artificially. The polymer shown in Fig. 15-1(c) is *polyethylene,* used in film and plastic squeeze bottles.

The principle is to generate polymers that have four bonds for each

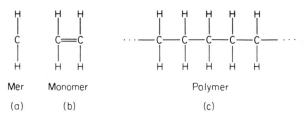

| Mer | Monomer | Polymer |
| (a) | (b) | (c) |

Figure 15-1

carbon atom, where each hydrogen atom has one bond to the carbon. Oxygen and sulfur share two bonds with other atoms, as shown in Fig. 15-2. Nitrogen shares three bonds. In addition, it is possible to replace one or more of the hydrogen atoms with chlorine, fluorine, benzene, etc.

It should be noted that in Fig. 15-2 the small molecules are monomers out of which are formed many of the plastics to be found in this chapter. Figure 15-2(a) is the *ethylene* configuration already mentioned. It is a colorless gas that is inflammable. It is used in the manufacturing of ethyl alcohol, styrene, and acrylic acid. In Fig. 15-2(b) one hydrogen atom has been replaced with an OH molecule to form the monomer *vinyl alcohol*. In Fig. 15-2(c) one hydrogen atom is replaced with a chlorine atom to form *vinyl chloride*. In Fig. 15-2(d) one hydrogen atom has been replaced with a benzene ring to form *styrene*. In Fig. 15-2(e) one hydrogen atom is replaced with an acetate configuration to form *vinyl acetate*. In Fig. 15-2(f) a hydrogen atom has been replaced with a CH_3 configuration to form *isopropylene*. In Fig. 15-2(g) two hydrogen atoms have been replaced by two chlorine atoms to form *vinylidene chloride*. Finally, in 15-2(h) all four

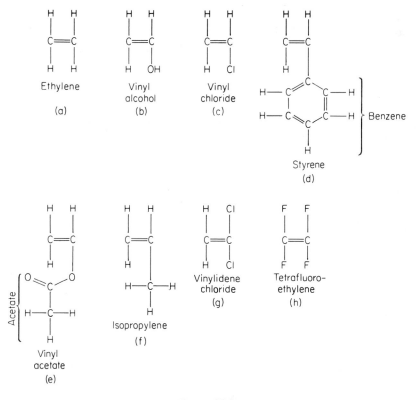

Figure 15-2

hydrogen atoms have been replaced with fluorine atoms to form *tetra-fluoroethylene.*

It is also possible to have different arrangements of the same atoms. Different structural arrangements of the atoms yield different properties. A different arrangement of the same compound is called an *isomer.*

15-3 SUMMATION OF MONOMERS

The next item to be studied deals with the summation of monomers into polymers made up of many mers. There are three basic methods by which polymerization takes place: *addition, copolymerization,* or *condensation.*

Addition polymerization takes place when several *similar* monomers join to form a chain. Thus in Fig. 15-1(b) the monomer is *ethylene,* which shows its mers joined by a double bond. Under proper conditions of heat and pressure, one bond is transferred from the monomer to join the ethylene monomers into a chain to form *polyethylene* [Fig 15-1(c)]. Mers that are joined by addition polymerization result in thermoplastic materials. The monomers vinyl chloride are linked by a transfer of one bond from the double chlorine bond. These double bonds are shown in Fig. 15-3(a). The polymerization takes place and produces the long chain of polyvinyl chloride shown in Fig. 15-3(b). The addition polymerization of styrene [Fig. 15-4(a)] is accomplished with a shift of one bond into the long chain polystyrene [Fig. 15-4(b)].

Copolymerization is the process that combines two or more *different* monomers by the process of addition polymerization. Many mono-

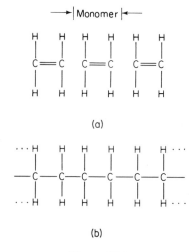

Figure 15-3

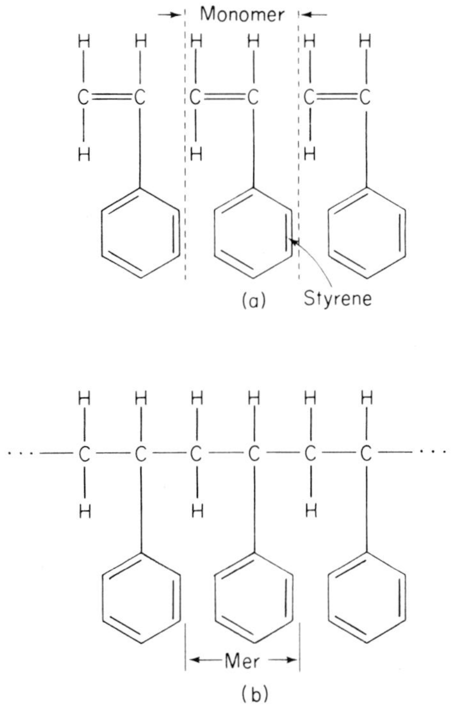

(a) Styrene

|←—Mer —→|
(b)

Figure 15-4

mers will not combine with themselves, but will combine with other monomers. These are called *copolymers*. In Fig. 15-2(e), acetate is linked to ethylene by replacing one of the hydrogens. This vinyl acetate monomer and the vinyl chloride monomer [Fig. 15-5(a)] transfer a bond to form a long-chain copolymer, as shown in Fig. 15-5(b). The properties of the end product of copolymerization are different than the properties of each of the constituent polymers.

Condensation polymerization takes place when long-chain molecules combine to form more complicated chains within their own compound structure, or when they combine with other compounds, and upon polymerization there is a residue. This residue is generally water. The materials that result from condensation polymerization may be either thermoplastic or thermosetting. See Fig. 15-6(a) (nylon) and Fig. 15-6(b) (Bakelite).

Long-chain structures are generally nonlinear and, if coiled as shown in Fig. 16-1(c), exhibit plasticity. When a force is applied to the material, the coils are straightened. The material is said to be plastic.

There are other materials in which this plasticity is greatly restricted after they are heated. Such is the case in plastics in which the structure

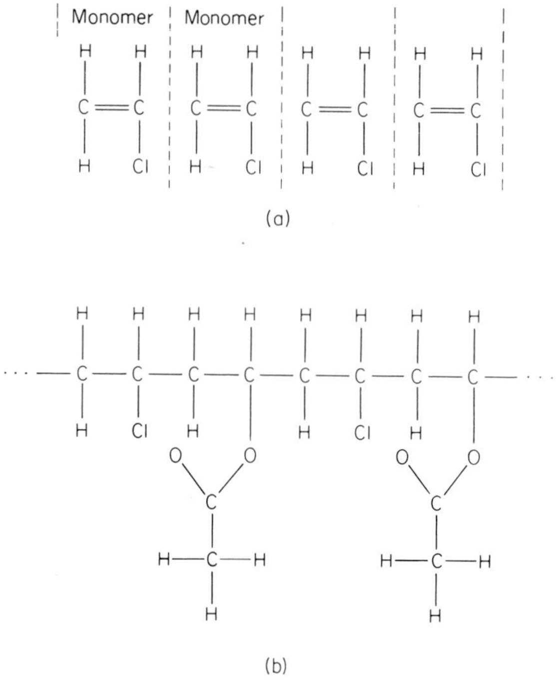

(a)

(b)

Figure 15-5

shows *cross-linking* of the molecules. Cross-linking is shown in Fig. 15-6(a) and (b). After a material is heated, the more cross-linkage exhibited by it, the greater is the loss of plasticity. It should be noted that plasticity and elasticity do not have the same meaning. Rubber loses its plasticity, but retains its elasticity when vulcanized. In general, cross-linking takes place *after* the material has been shaped, formed, and heated. If cross-linking occurs, the material is thermosetting. If it may be repeatedly heated and cooled without producing cross-links, it is thermoplastic.

Chain, or cross-link, bonding is called *primary* bonding. Other forces produce *secondary bonds*. The latter bonds result from the electrostatic forces within the atom due to the nonsymmetrical distribution of electrons around the nucleus. These bonds are weaker than the primary bonds. They may be broken when the material is heated and reform when the material cools. Cross-linked bonds, once broken, will not reform.

In other instances polymers may develop *branched structures*. Controlled branching, when carried out, affects the physical properties of materials by restricting the motion of layers of molecules by entangling these branches with the chains.

In yet other instances, even though plastics are basically noncrys-

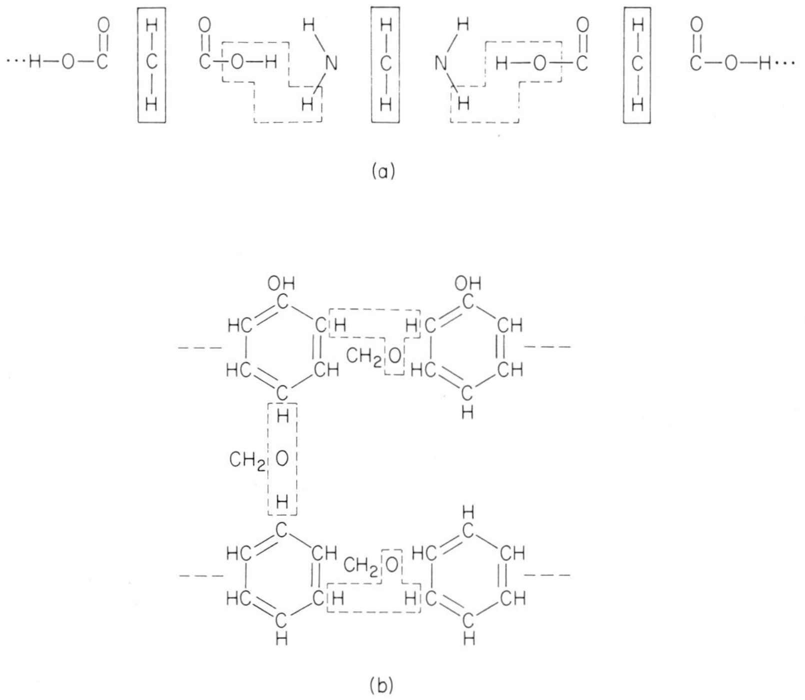

(a)

(b)

Figure 15-6

talline in structure, the molecules become aligned simply because of the way the chains intermingle. If they are aligned properly, the forces operate as though the structure were crystalline. Since they act like crystals, the physical properties that accrue to the material also parallel those of crystalline materials. Thus the temperature at which this plastic softens is increased. The softening temperature range is also narrowed. Also, change from a liquid to a solid state, or solid to a liquid state, is accompanied by significant dimensional changes. Increased crystallinity will also restrict deformation.

15-4 THERMOPLASTIC MATERIALS

The following materials are only part of a constantly expanding list of thermoplastic plastics. They are, as indicated, materials that may be softened and which will harden again upon cooling.

ABS, sp. gr. 1.04, is a copolymer that results from three monomers, acrylonitrile, butadiene, and styrene. They have tensile strength in the

5000-psi range. They exhibit good resistance to impact, even at lower than room temperatures. As the butadiene percentage is increased, the impact strength of ABS is increased, and its tensile strength is decreased. It also exhibits good toughness and resistance to thermal distortion. It has fair electrical insulating properties and good resistance to weather. Its main disadvantages are that it is inflammable, soluble in organic solvents, and affected by sunlight. ABS is used in automobile parts such as dashboards, and in liners for refrigerators.

Acetal resin, sp. gr. 1.40, has the chain formula shown in Fig. 15-7. It develops a long chain with a yield and tensile strength of about 9000 psi. It has high impact strength, excellent fatigue resistance, and dimensional stability. It has a low coefficient of resistance and good resistance to organic solvents. It burns easily, breaks down when exposed to radiation, and has poor resistance to weather. Acetal resin is used in carburetors, gears, and bearings. It is also used for pipes, joints, and fittings. Because of the finish that is achieved when these plastics are molded, the low cost of producing parts, and their high strength, they are used in many instances as replacements for metals.

Acrylic resins, sp. gr. 1.1, are colorless, highly transparent, and will not discolor or fade when exposed to light. They also exhibit excellent ultraviolet and white light transmission and optical homogeneity. They are also resistant to oxidizing agents and have good resistance to water. They have a tensile strength of about 8000 psi and a compressive and flexure strength of about 15,000 psi. Polymethyl methacrylate is shown in Fig. 15-8.

The acrylic plastics have low abrasion resistance and, unless spe-

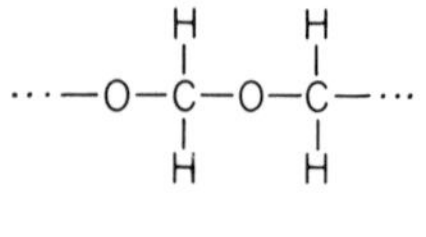

Figure 15-7

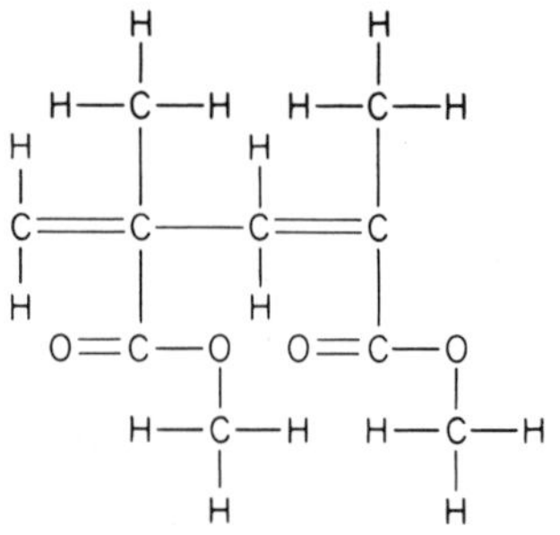

Figure 15-8

cially treated, they are subject to crazing when used as glass. They also soften at relatively low temperatures, approximately 200°F.

These materials are used in making Lucite and Plexiglas. They are used in laminated glass, transparent sheet, protective coatings, and molded products, such as costume jewlery, novelties, and frames for handbags. They are used in dental materials for dentures and for windows and shields in aircraft. Since they may be easily machined, they are manufactured in sheets, rods, and tubes. They may be injected or compression molded as a powder, cast, or molded into sheets.

Cellulose plastics, sp. gr. 1.1 to 1.3, are made up from long-chain molecules. The four common forms of the cellulose chains are ethyl cellulose, cellulose acetate, cellulose acetate butyrate, and cellulose nitrate. They are transparent and may be manufactured into very thin sheets and sold as cellophane. They may be used for packaging food. In another highly purified state, they are used for making high-grade writing paper. They are also used as a filler for plastics in such items as buttons, floor tile, and knobs and to thicken paints and cosmetics.

Ethyl cellulose is colorless, nonflammable, alkali-resistant, flexible, and stable to light. It is used for wrapping, molding plastics, and as a hardening agent in waxes. It is the plastic material that is used for dipping steel parts so that they will be protected from corrosion. It is easily stripped off when the parts are ready for use. Its strength is less than the cellulose acetates discussed below.

Methyl cellulose is a good emulsifier and as such is used in soap, waxes, glues, fats, floor wax, and shoe polish. It lowers the surface tension of the water used to produce strong emulsification.

Cellulose acetate, sp. gr. 1.3 to 1.4, has an amber color and is transparent. It is nonflammable and has good heat and light stability. It can be molded or made into delicate thin fibers for manufacture into rayon or other fabrics. As such, these fibers are flexible, strong, easily dyed, and mildew resistant. Its physical properties are as follows: tensile strength, 4000 to 8000 psi; compressive strength, 20,000 psi; a softening range of from 120 to 200°F. Its chain is shown in Fig. 15-9.

Cellulose acetate butyrate has the same characteristics as cellulose acetate, except that it has better stability to light and heat and better resistance to water absorption. It is used in coating materials, varnishes, and lacquers, electrical parts and tape, cable covering, and aircraft coating.

Cellulose propionate is used for the extrusion of high-impact parts.

Cellulose nitrate, sp. gr. 1.4, is used in the making of plastics, rayon, and lacquer. The low nitrates are very flammable and their fumes are poisonous, but they will not explode. This is the material from which celluloid is made. It has a tensile strength of about 7000 psi and a compressive strength of about 25,000 psi. It is a very tough material and may

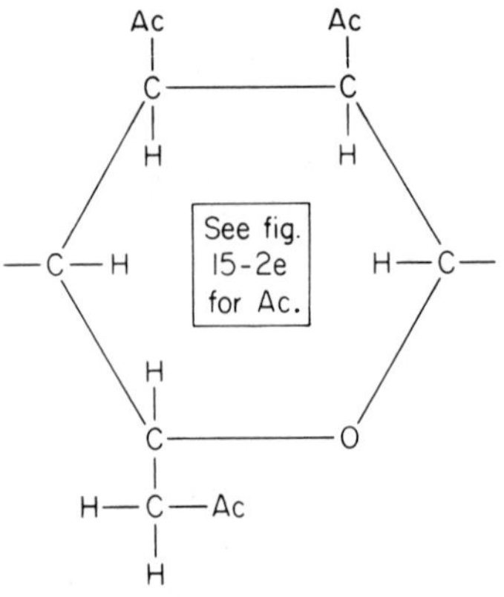

Figure 15-9

be easily dyed. They are marketed in sheet, tube, and rod forms. When they are used as lacquers, they are mixed with pigments and plasticizers to make them flexible and tough in thin film form. One cellulose nitrate polymer is shown in Fig. 15-10.

Fluoroplastics, sp. gr. 2.2, can be produced so that they will possess high heat resistance. They are resistant to almost all chemicals. Teflon is one such material. Its chain is shown in Fig. 15-11(a). The fluorine atom has replaced the hydrogen atom. Its dielectric strength is about 1500 V/mil. Its heat resistance is about 500°F. It melts at 594°F. It also possesses self-lubricating qualities and a very low coefficient of friction, which makes it useful as a bearing material. It may be used as electrical covering or tape because of its insulating qualities at high temperatures.

In another material of the fluorplastic series, one fluorine atoms out

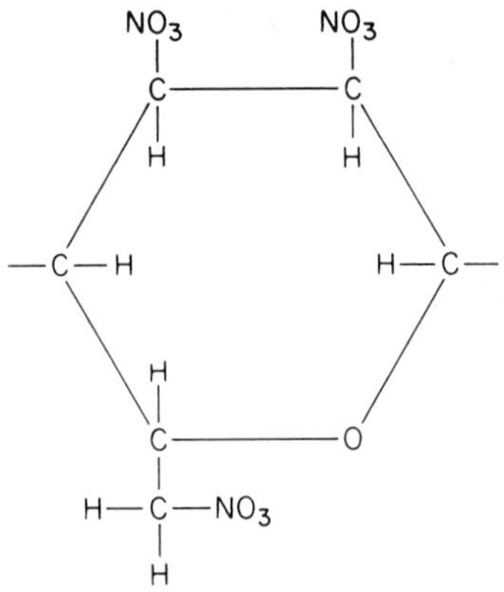

Figure 15-10

(a)   (b)

Figure 15-11

of every four is replaced by a chlorine atom [Fig. 15-11(b)]. This material may be used at temperatures as high as 300°F, and has a high dielectric strength. A somewhat similar material has resistance to heat up to 400°F and high compressive and tensile strength. It is used for gaskets, liners, diaphragms, and coatings. It may also be produced in a rubber form for acid- and heat-resistant hose, gaskets, protective coatings, and paints.

The method by which these materials must be processed is very expensive. Because of their high viscosity at elevated temperatures, they are powder compacted instead of being poured as a liquid.

Ionomers are plastics that develop bonding which is both ionic and covalent. They link in such a way that, even though they are stiff at room temperature, the bonds may be broken down thermally and the new linkages will cause the material to act as though it were a thermoplastic material. They have low densities, resist oil, and are flexible, tough, and transparent. They have very good low temperature properties, but should be used below 150°F. They are used for electrical insulation, tool handles, and housewares.

Polyamide plastic, sp. gr. 1.0 to 1.5, is a thermoplastic material formed by a long-chain condensation reaction. Shown in Fig. 15-12 is the chain for *nylon.* It is in long threadlike chains. It is tough, strong, and elastic. These threads are spun into cloth and are used in knitting. It is heat and wear resistant. It is lightweight, has a low coefficient of friction, and is chemical and abrasion resistant. Its tensile strength is affected by moisture content. The addition of 2 per cent moisture to nylon reduces its tensile strength from molded nylon by about one-half, from about 12,000 psi to 6000 psi for 6 per cent strain. The specific gravity of a nylon filament is 1.07 with a tensile strength of 53,000 psi. It softens at about 450°F. This material is used for brushes, tennis strings, fishing leaders, surgical sutures, and insulation for electrical wires. Nylon is also used for gears and bearings. It is sometimes impregnated with graphite to impart other qualities to the structure of the material.

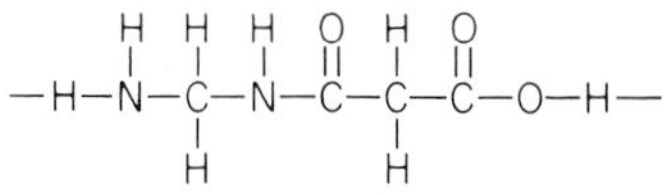

Figure 15-12

The *parylene polymers* are comparatively new plastics. They are produced as particles, as film, or in coating form. They have the characteristic of sticking to any material upon which they are sprayed. They polymerize immediately on contact with the surface being sprayed. They are currently used on fabric, paper, and metals that are subject to corrosion, and ceramics.

The *polyallomer and polybutylene* plastics are two crystalline materials that are thermoplastic and have high impact and fatigue strength. Polybutylene has very unusual elasticity properties. It can be stretched up to 100 per cent and recover completely when the forces are released. Both materials have excellent fatigue resistance and are free from cracks that develop from stress. They are used for gaskets, diaphragms, and washers.

The *polycarbonate,* sp. gr. 1.2, series of plastics has high strength and toughness in combination with ductility. It has excellent electrical insulation properties, heat resistance, and good dimensional stability. It is a transparent material. Its elongation under tensile stresses is better than 100 per cent. They are capable of supporting 2000 to 3000 lb/in.^{2} at temperatures as high as 300°F. They have excellent notched-bar impact strength. They also have high creep strength. The monomer is shown in Fig. 15-13. These materials are used in the manufacturing of hard hats, housings for hand power tools, unbreakable kitchen utensils, taillights for automobiles, nails and screens, instrument cases, and bearings.

Polyethylene, sp. gr. 0.95 (Fig. 15-14), is one of the softest and most flexible plastics. It has excellent resistance to water and moisture. It is a tough material, but has low tensile strength properties, about 2000 psi. It is chemical resistant and has good electrical insulation properties. The thermoplastic material has a melting point of about 240°F, whereas the thermosetting plastic material is stable at 350°F. It is, however, subject to becoming brittle if exposed to the ultraviolet rays of the sun. It will become thermosetting if exposed to radiation from high-energy electrons or ultraviolet light, or if cross-linked with other agents such as organic peroxide. Crystallization of polyethylene always makes it thermosetting. Amorphous polyethylene is thermoplastic. There are many varieties of polyethylene used for insulation, plastic bottles, pipes, containers, wrap-

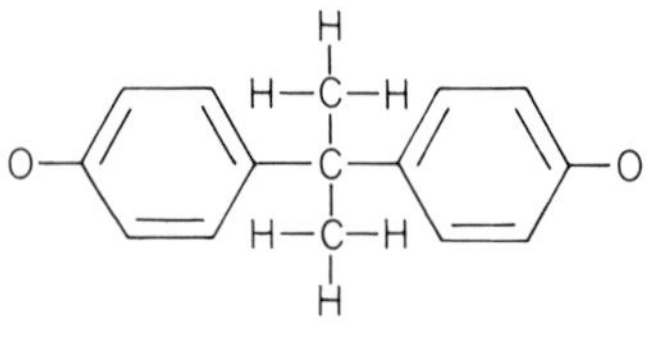

Figure 15-13

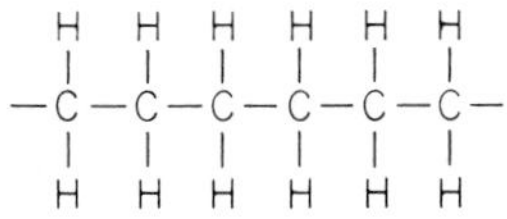

Figure 15-14

ping of pipe, as a foil for packages, coating for paper, and for injection molded toys.

Polypropylene, sp. gr. 0.9 (Fig. 15-15), has excellent fatigue strength and electrical insulating qualities. It has good resistance to distortion from heat and is chemically inert. Like polyethylene, it is flammable, subject to deterioration by the ultraviolet rays in sunlight, and brittle at low temperatures. Carbon black is used as a coating against such exposure. It is generally used with a filler material such as glass, asbestos, or wax. It is used in the automobile field, for radio and television cabinets, housewares, laboratory devices, and plastic piping.

Polystyrene, sp. gr. 1.05 (Fig. 15-16), is a plastic material that may be used as a simple polymer or as a styrene that has been modified with other monomers, rubber, or stabilizing agents to achieve special properties. It is made from ethylene.

It has a short time tensile strength of about 4000 to 8000 psi and a compressive strength of about 15,000 psi. It deflects at temperatures of about 150°F, will ignite at 900°F, and burns with a sooty black flame. It is tough at low temperatures, is dimensionally stable, has good water resistance, and will heat seal at temperatures from 60 to 80°F. It is resistant to alkali attack and to most acids.

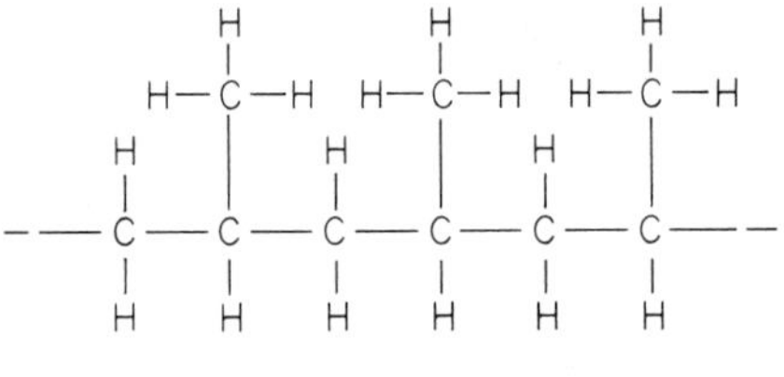

Figure 15-15

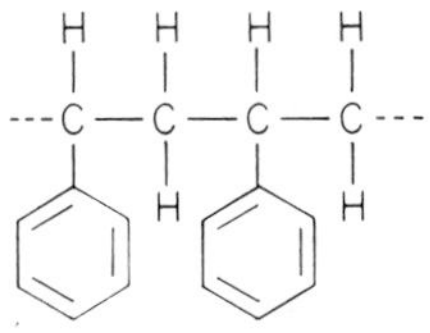

Figure 15-16

Like polypropylene and polyethylene, it is affected by sunlight and is subject to cracking unless treated. The polymers, such as polystyrene, polypropylene, and polyurethane, may be formed and used in packaging. Familiar trade names are Styrofoam and Ethafoam. This material is used in cushions, padding, containers, and cups. It may be manufactured as either a thermoplastic or a thermosetting material. It is used in refrigerator and appliance parts, toys, indoor lighting panels, radio cabinets and dials, in paints, cable wrapping, cosmetic containers, chemical bottles, and battery cases.

Vinyls, sp. g. 1.4 to 2.00, are widely used thermoplastic materials. In general they are flame, abrasion, chemical, and electrical resistant. They have good sound-absorbing properties. They are generally polymerized from ethylene derivatives.

Polyvinyl acetate boils at about 165°F. It is a clear liquid and clear plastic with a density of 1.19. It is soluble in benzene and alcohol, but is not affected by water or oils. This material is stable to ultraviolet light. It is a tough, hard material. It is used in paints and adhesives as a water-resistant nonstaining bonding material. It may be used to bond metallic and nonmetallic materials.

Polyvinyl chloride may be molded and extruded. It is used to coat wrapping paper and fabrics. The molded material has a specific gravity of 1.4 and a heat distortion temperature of about 165°F. In its unplasticized form it is used to make ducts, form heads, pipes, and garden hose. It is used to make an adhesive cement and in an extruded form to make webbing for seats. In another form it is used to make sheet wall covering and floor tile.

Polyvinyl chloride acetate is manufactured as a fine thread as thin as silk. It is elastic, strong, and waterproof. It is also acid and alkali resistant and may be woven into fine-mesh screening used for filtering liquid materials that have temperatures less than 300°F. See Fig. 15-5(b).

Polyvinylidene chloride is a clear colorless liquid made from petroleum and brine. It softens at about 280°F. In general, it is resistant to all chemicals and water and is used in tubing for handling these chemicals. It is sold under the trade name of Saran by the Dow Chemical Co. It is also used in gaskets for woven materials to be used in aircraft and automobile seats, fishing leaders, and tubing and couplings for refrigerators and air conditioners. It is used in film form as a packaging material. When woven into a fine mesh, it is used as noncorrosion screening.

Polyvinyl alcohol is a white odorless powder that is used in latex as a thickener, as coatings for paper and fabrics, and as adhesives. It is extruded into a tough acid-, alkali-, and heat-resistant material that is used as sheets or tubes. The material is not affected by oils or oily materials, and is resistant to penetration by gases. It is, however, affected by water and cannot be used in contact with aqueous solutions. It is also sensitive to heat.

Polyphenylene oxide and *polysulfone* are two additional thermoplastic materials. The major characteristic of polyphenylene oxide is its wide temperature range, −275°F to 375°F, through which its physical properties are useful. It is stable to acids, bases, and moisture. It also has good electrical properties. Polysulfone is also stable through a wide temperature range, −150°F to 300°F. Sulfur makes this plastic highly resistant to heat and oxidation. It is used in electrical tools and places where electrical stability is needed.

Reinforced plastics are made by adding fibrous materials to either thermoplastic or thermosetting resins. The reinforcing of plastics yields better properties such as tensile and impact strengths; improved resistance to moisture, sunlight, or ultraviolet rays; and improved resistance to corrosion. These materials, when added, combine with the monomers and therefore alter the physical properties of the plastics. Reinforcement can also take place by adding fillers, glass fibers, asbestos, cotton, nylon, paper, or metal.

Table 15-1 lists the property ranges for some of the thermoplastic plastic materials.

15-5 THERMOSETTING PLASTIC MATERIALS

Thermosetting materials result when cross-linking takes place between the molecules after plastics are heated. They become rigid and will not regain their ductility upon reheating. The following materials are considered thermosetting.

Allyd compounds, sp. gr. 1.2 to 2.5, form chains by condensation of water to produce an unsaturated polyester. When reacted with other monomers such as styrene, or one of the vinyls, cross-linking occurs. The resultant plastics are hard, thermosetting, and heat resistant. They have good dimensional stability and good electrical properties.

They are used in varnishes and caulking compounds. In one form, they are used as baked enamels for household equipment such as stoves and refrigerators and for automobile finishes. This material is resistant to oils, humidity, or light rays. In another form, they also find use in electrical parts.

Allyl plastics, sp. gr. 1.35 to 1.4, form another series of thermosetting materials produced from allyl alcohol or allyl chloride. The allyl ester is a clear liquid that will polymerize to form an allyl plastic. Castings are

Table 15-1   Physical Properties of Thermoplastic Plastics

	ABS	Acetal	Acrylic	Cellulose	Fluorocarb	Ionomers	Polyamides	Polycarbonates
Tensile St. x 10^3 psi	4 – 7	9 – 10	5 – 11	2 – 11	2 – 6	4 – 8	8 – 12	8 – 9
Compressive St. x 10^3 psi		14 – 18	4 – 18	2 – 35	2 – 7	0 – 20	7 – 13	12
Flexture x 10^3 psi	5 – 11	13 – 14	8 – 18	2 – 14	7 – 9	0 – 1	8 – NB	13
Elongation %		60 – 75	20 – 50	5 – 100	80 – 400	100 – 600	25 – 300	60 – 100
IZOD	2 – 10	1 – 3	1 – 4	0.5 – 12	2 – NB	6 – NB	1 – 3	15
Thermal Exp. x 10^{-5} in/in/°F	0.5	4.6	3 – 5	4 – 11	2 – 5	5 – 10	–	4
Thermal Cond. BTU/Hr/ft/°F	0.1 – 0.2	0.04 – 0.13	0.07 – 0.12	0.1 – 0.2	0.10 – 0.14	0.12	0.12 – 0.14	0.10
Burn rate in/min.	Es	1	1 – 1.7	SL	N	1	A	A
Sunlight	P	C	N	S	N	P	D	Eb
Acid (weak)	S	S	N	S	N	A	A	A
Alk. (weak)	N	N	N	S	N	N	N	S
Organic Sol.	SO	N	SO	SO	H	Ah	N	SO
Molding Char.	G	E	E	E	E	E – G	E	G
Machine Char.	G	E	G	E – G	E	Fr	E	Ex

Extracted from tables 14.2, 4, 5, 6, Materials Data Book, Earl E. Parker, McGraw Hill Publ. Co.
* These are ranges. For specific materials refer to a handbook.

Table 15-1 *(continued)*

Polyethylenes	Polypropylenes	Polystyrenes	Vinyls	Polyesters	Epoxies	Phenolics
1-5	4-10	2-12	0.5-12	20-60	14-60	5-60
3	6-14	2-16	1-14	15-50	30-70	20-50
1-7	6-15	4-20	4-18	10-90	20-100	10-90
15-900	3-700	1-140	2-400	-	-	-
1-NB	1-6	0.3-10	1-20	2-30	8-26	1-35
6-13	2-5	2-11	3-13	-	-	
0.2-0.3	0.07-0.14	0.05-0.2	0.08-0.17	-	-	-
1	SL	SL	A	SL	A	N
Eb	P	D	S	P	S	D
A	A	A	S	S	N	A
N	N	N	N	S	N	A
N	N	SO	SO	N	N	I
E-G	E	E-G	Fr-E	E	E	CA
E-Fr	E	Fr-G	P-E	G	Fr	F

KEY

A = attacked by oxid. acid
Ah = attacked by hydrocarbons
C = chalks
CA = cast
D = discolors
E = excellent
Eb = embrittles
Es = self-extinguishing
Fr = fair
G = good
H = swells in halogens
I = insoluble in most solvents
N = none
NB = no break
P = needs pigments for stability
Po = poor
S = slight
SL = slow
SO = soluble

clear and hard. They have compressive strengths of about 20,000 psi They have very good dielectric qualities, and dimensional and temperature stability.

This material is used for cooking utensils, and as optical glass for lenses, prisms, radar domes, and reflectors. It is also used for windows and clock dials. In one form it is clear, abrasion-resistant, and noncrazing. In the methallyl alcohol form, it is used in varnishes because of its chemical resistance and its ability to withstand temperatures in the 400°F range.

Amino plastics, sp. gr. 1.5 to 2.0, contain urea or melamine. The *urea monomer* is shown in Fig. 15-17(a), and the *melamine monomer* is shown in Fig. 15-17(b). When reacted with formaldehyde, the resulting monomer is polymerized by condensation. When heated it is further polymerized.

Urea formaldehyde plastics, sp. gr. 1.5, are produced by reacting urea, synthesized carbon dioxide and ammonia, with formaldehyde in the presence of a catalyst and mixing with a filler. The physical properties of this material are somewhat similar to the phenol formaldehyde plastics discussed later. They may be translucent, but will hold coloration well. They have excellent arc resistance. They are used in household fixtures, knobs, and handles. The formaldehyde molecule [Fig. 15-17(c)] and the urea molecule [Fig. 15-17(a)] polymerize as shown in Fig. 15-17(d).

Urea formaldehyde plastics are used as adhesives for bonding plywood to other types of board. They have tensile strengths of about 6000 psi, and compressive strengths of about 26,000 psi. When subjected to high temperatures for long periods of time, they lose their electrical properties and strengths. They will also fade.

Melamine formaldehyde plastics, sp. gr. 1.5, are complex organic materials that are usually mixed with fillers. They have better heat resistance and arc resistance, and are harder than the phenolic (discussed

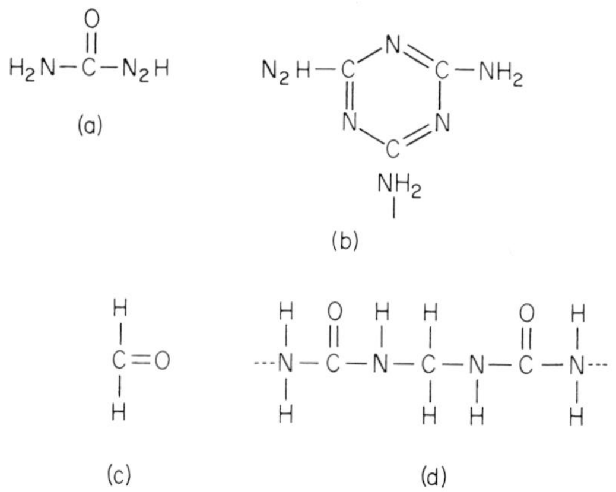

Figure 15-17

later) and urea plastics. They are water resistant. Because of this they are used as an adhesive bonding material for exterior plywood. They are used in electrical circuit breakers, ignition parts, and terminal blocks. Because they can be used in contact with strong detergents and hot water without being affected, they are used as dinnerware (Melmac). They resist acid and alkali attack and have good heat and color stability. They have a melting point of 670°F.

The amino plastics in general are hard, chemical and water resistant, rigid, and have excellent thermal stability.

Phenolic formaldehyde plastics, sp. gr. 1.3 to 2.0, are produced from the phenols and the aldehydes. Their most extensive use is in the manufacture of molded products.

The most common type is made from phenol and formaldehyde, generally in the presence of a catalyst. It is called Bakelite. Its basic structure is shown in Fig. 15-18(c). The formaldehyde molecule is shown in Fig. 15-18(a) and the phenol molecule in Fig. 15-18(b). Phenolic resins are ground to a fine powder, mixed with other materials as fillers to achieve the desired physical properties and coloration. When molded, it has a high finish. Cotton and pulverized cloth will make the end product shock resistant. For a material that will withstand heavy impact, shredded cord or canvas is used as a filler. Inorganic fillers such as mica, asbestos, and silica will produce a plastic which has good dimensional stability, a low coefficient of thermal expansion, heat resistance, and water absorption. These plastics have excellent electrical properties. Wood filler is used as a filler when a good general-purpose phenolic plastic is desired.

In general, molded phenolics have good resistance to mild solvents, but are severely damaged by strong acids and alkalies. They are brittle, transparent, and soluble in alcohol. They are insoluble in turpentine and water. They have tensile strengths of from 6000 to 8000 psi and high flexure strength of about 10,000 psi. Their notch impact strength is very low, ranging from 0.25 to 0.45 ft-lb/inch of notch. They also have the highest heat resistance of any of the phenolic plastics at about 300°F.

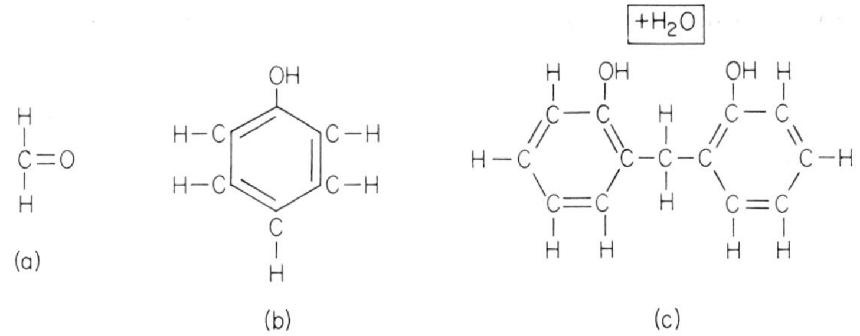

Figure 15-18

When wood fillers are used, their tensile strength may be as high as 12,000 psi and their compressive strength as high as 35,000 psi. Some of these plastics are laminated with linen, paper, or canvas. This produces a laminated Bakelite with a high compressive strength of about 40,000 psi and a tensile strength of about 12,000 psi. Examples of such Bakelite laminates are Micarta and Formica.

They are used as molded parts in radio cabinets, utensil handles, electrical insulators, clock cases, and tube sockets. The laminates are used for table and counter tops.

Epoxy plastics, sp. gr. 1.00 to 1.25, when combined with hardening agents, form cross-links. These plastics are very brittle, are chemical resistant, and have good electrical properties. They have a low absorption coefficient and very good dimensional stability. They cure at standard temperature and pressure. They have excellent bonding properties. However, their shock loading strength is poor. They are used as castings, encapsulating, or potting. When used as surface coatings for metals, they form an abrasion-proof, corrosion-resistant, hard surface. They are used as bonding agents in concrete, electrical insulation, as an aircraft adhesive, for tooling as foundry patterns, dies, jigs, and fixtures. They also find use in paints.

The epoxy resins are mostly thermosetting. They will react readily with curing agents to produce hardening. Many such agents are available. The alcohols, ammonia, water, and inorganic acids and bases are but a few.

The characteristic epoxy monomer and chain are shown in Fig. 15-19. Characteristic of these chains is the presence of oxygen atoms.

Furane plastics, sp. gr. approximately 1.00, are made from easily grown vegetable products such as cornstalks, peanut shells, straw, etc. They have very good resistance to acids and alkalies. They are excellent bonding agents for porous materials. They are used as bonding agents in grinding wheels, foundry sand, glued wood products, etc. In some forms, they are brushed on the inside walls of tanks as linings. This is especially true where high heat and chemical resistance is required. In the latter instance, they are self-curing.

Polybutadiene and *butadiene styrene plastics* will be discussed in

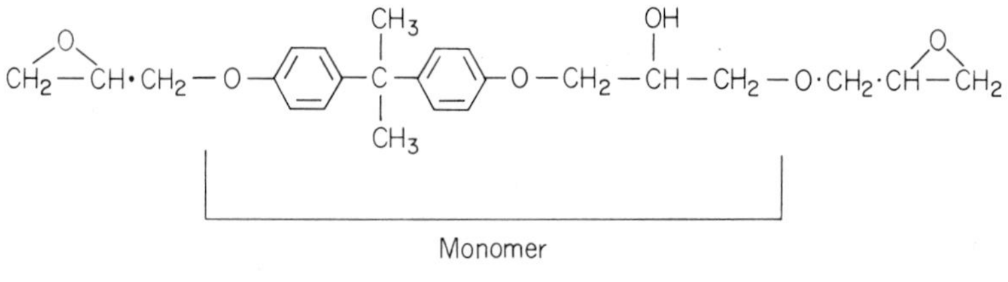

Monomer

Figure 15-19

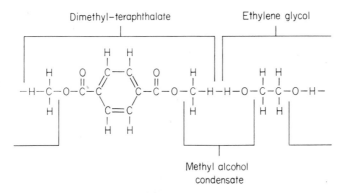

Figure 15-20

the section that deals with rubber. As thermosetting plastics they are unsaturated and cured by reaction with a catalyst. They are used for coating
cans, and for electrical- and moisture-resistant parts.

Polyester plastics, sp. gr. 1.00 to 1.5, are used widely under the trade
names of Dacron, Kodel, and Mylar. The chain for Dacron or Mylar is
shown in Fig. 15-20. This is a condensation polymerization where methyl
alcohol forms as a condensate instead of water. Monomers are sometimes
added to produce specific properties, such as color, flexibility, and strength.
They may also be added to reinforce the original plastic material. Fiber
glass is one such additive.

The plastic results from the reaction of an acid and an alcohol. The
thermosetting characteristics result from cross-linking. When polymerized, the resulting plastic is hard, dense, and has good toughness.

The reinforced polyesters are used in aircraft parts, nose cones, boat
hulls, skis, septic tanks, and helmets. The nonreinforced polyesters are
used in bowling balls, auto body patches, floors, table tops, and surface
coatings.

Polyamide plastics are high-strength thermosetting materials that
may be used at very low or very high temperatures without appreciable
loss of physical properties. They exhibit good electrical- and wear-resistant properties. They also have very low coefficients of friction. They are
used as adhesives and as binding material for laminates. They cannot be
molded and must be bought in the solid form.

Silicone plastics, sp. gr. 1.6 to 1.8, are long-chain polymers of silica
and oxygen with hydrocarbon radicals. Part of the long alternating silicon–
oxygen chain is shown in Fig. 15-21.

The polymerization of this organic plastic produces materials that
may be liquid, rubberlike, or very hard. They have high water resistance.
They are stable at temperatures with a range of about 500°F to as low as

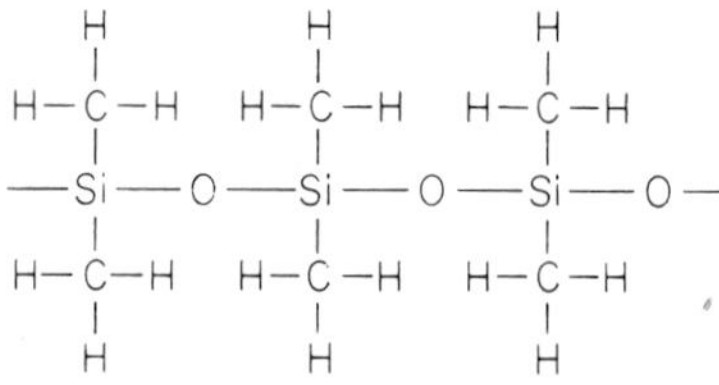

Figure 15-21

−70°F. Therefore, they may be used in high-temperature-resisting paints and varnishes. The liquid silicones when used with graphite are effective within the above temperature range. Since they are stable at their boiling temperature of about 750°F, and since they have a low vapor pressure, they may be used as hydraulic fluids. They are nonflammable and resistant to weather and to most chemicals.

They are used as textiles and in water-repellent paints. They are also used in terminal strips, insulating foam, and as laminates.

Polyurethane plastics, as indicated in Sec. 15-4, may be thermoplastic or thermosetting. These materials have a foamed structure. As such, they have very fine thermal and sound insulating properties. Flexible polyurethane foam is used in mattresses and as linings in clothing to be worn in severe winter weather.

They are used as foam slabs, adhesives, and as coatings. When used as coatings, they are tough, hard, and chemical resistant, and have been used in sulfuric acid surroundings. They are a good substitute for foam rubber. They are, however, harmed by high humidity, oxidize at low temperatures, and suffer ultraviolet degradation. They are discussed further in Sec. 16-4.

Table 15-2 lists the property ranges for some of the thermosetting plastic materials.

Problems

15-1. Define the term "plastic materials."

15-2. Define (a) thermoplastic and (b) thermosetting. (c) What is an elastomer?

15-3. Discuss some physical properties of thermoplastic and thermosetting materials.

15-4. Will plastic materials creep at room temperature? Explain.

15-5. What is the effect of temperature on the ductility of plastics?

15-6. Why are impact tests carried out with pieces that have been fabricated, rather than with test specimens?

15-7. Define and illustrate (a) the mer, (b) a monomer, and (c) a polymer.

15-8. What is the significance of (a) the single tie bar in Fig. 15-1(a), and (b) the double tie bar in Fig. 15-1(b)?

15-9. Describe the formation of a long chain by the process of shifting bonds during the process of polymerization.

15-10. Using Fig. 15-2, draw each of the following monomers: (a) styrene, (b) vinyl acetate, (c) isopropylene, (d) vinyl chloride, (e) vinyl alcohol, (f) tetrafluoroethylene, (g) vinylidene chloride, (h) ethylene.

15-11. Define and describe the term *isomer.*

15-12. (a) Describe polymerization by addition and give at least one illustration. (b) What is copolymerization?

15-13. Describe polymerization by condensation and give at least one illustrated example.

15-14. Describe the *coil* chain polymerization. What is the effect of such a chain on the elasticity of the material?

15-15. What effect does cross-linking have on the physical properties of plastics? Describe.

15-16. Relate cross-linking to thermoplasticity and thermoelasticity.

15-17. (a) How do primary bonds react when heated? (b) How do secondary bonds react when heated?

15-18. What is "branching"?

15-19. Discuss "crystallinity" as it relates to plastics.

15-20. Discuss the copolymer ABS.

15-21. Describe some properties of acetal resin.

15-22. Discuss the acrylic plastics and their uses.

15-23. List the cellulose plastics indicated in this chapter. Discuss at least four of them.

15-24. Which of the cellulose plastics are the following: (a) cellophane, (b) rayon, (c) celluloid, (d) Teflon, (e) nylon, (f) Saran.

15-25. Discuss the advantages gained when Teflon is used as an engineering material.

15-26. Discuss the fluoroplastic materials.

15-27. What is an ionomer? What is its special characteristic?

15-28. What are the special characteristics of nylon that make it a useful material? What is its polymer?

Table 15-2   Properties of Thermosetting Plastics

	Allyl	Epoxy	Furance	Melamine Formal	Phenol Formal	Phenolic
Tensile St. x 10^3 psi	5 – 10	5 – 30	3 – 4	5 – 13	3 – 18	3 – 9
Compressive St. x 10^3 psi	20 – 50	15 – 40	10 – 13	4 – 24	3 – 50	5 – 20
Flexture x 10^3 psi	8 – 20	8 – 60	6 – 9	10 – 23	6 – 60	5 – 18
Elongation %	–	1 – 70	–	0 – 6	0.5 – 2	1 – 2
Impact (IZOD)	0.3 – 15	0.2 – 25	–	0.2 – 6	0.2 – 8	0.2 – 5
Sp. Gr.	1.3 – 1.8	1 – 2	1 – 7	1.5 – 2	1.3 – 2	1.3 – 1.7
Thermal Cond. BTU/hr/ft/°F	0.12 – 0.6	0.1 – 0.7	–	0.17 – 0.27	0.1 – 0.5	0.1 – 0.2
Coef. Exp. x 10^{-5} in/in/°F	0.5 – 5	0.5 – 3.5	–	1 – 2.5	0.5 – 2	2 – 4
Optical Char.	TP – O	TP – O	O	TL – O	O – TL	O – TL
Burn Rate in/min.	Z – 0.3	SL – Z	SL	Ni	SL – Z	SL – Ni
Sunlight	S – N	S – N	–	S	D	D
Acid (weak)	N	N	S	N – S	S	S
Alk. (weak)	N – A	N	N	S – N	S	A – N
Organic Sol.	N	A	N	Ni – N	S	A – S
Molding Char.	E – Po	E	G	G – E	F – E	–
Machine Char.	G – E	Po – E	Fr	Fr – G	Po – G	Fr – E

Extracted from tables 14.7, 9, 10, 11, Materials Data Handbook,
    Earl and Parker – McGraw Hill Publishing Co.
* These are ranges. For specific materials refer to a handbook.

Table 15-2 *(continued)*

Polyester	Polyester Alkyd	Silicone	Urea–Formal	Urethane
1 – 13	5 – 50	3 – 35	5 – 13	1 – 10
3 – 6	15 – 50	10 – 18	25 – 45	20
8 – 23	5 – 80	8 – 30	10 – 13	12
40 – 300	0.5 – 2	100	0.5	300 – 1000
0.2 – 7	0.3 – 10	0.2 – 15	0.2 – 0.4	BENDS
1 – 1.5	1.3 – 2.3	1 – 2.8	1.5	1.2 – 2.5
0.1	0.2 – 0.6	0.09 – 0.28	0.17 – 0.23	0.12
3 – 5	1 – 1.5	0.5 – 16	1 – 2	6
TL	O – TL	TP – O	O	TL
Ni – 1.1	Ni	SL – Ni	Ni	SL
D	S – N	N	S	S
N	S – N	N	N	S
S	A – N	S – N	S	S
C	S – N	A	Ni	Ni
–	E	G – E	E	G
Fr – G	Po – G	Fr	F	F – E

KEY	
A	= attacks
C	= chlorine
D	= darkens
E	= excellent
Fr	= fair
F	= fast
G	= good
L	= low
N	= none
Ni	= nil
O	= opaque
Po	= poor
S	= slight
SL	= slow
TL	= translucent
TP	= transparent

15-29. What is the special characteristic possessed by parylene?

15-30. What are the outstanding physical properties of polyallomer and poly-butylene?

15-31. List the special characteristics of the polycarbonates.

15-32. Discuss the plastic polyethylene.

15-33. What are some of the special characteristics of the material polypropylene?

15-34. What is polystyrene? Draw its polymer. What are some of its uses? Discuss its physical properties.

15-35. List the plastics that may be foamed.

15-36. List the vinyl plastics discussed in this chapter. Give a characteristic of each.

15-37. What is Saran Wrap? What are some of its special characteristics? What are some of its uses?

15-38. What is the special use of the two plastics polyphenylene oxide and polysulfone?

15-39. (a) Why is it desirable to reinforce plastics? (b) List the materials generally used to reinforce plastics.

15-40. Compare the physical properties of the acrylic plastics with the vinyl plastics (see Table 15-1).

15-41. Repeat Problem 15-40 for cellulose and polyethylene.

15-42. Discuss the cross-link development of the alkyd compounds.

15-43. List the various uses of the alkyl plastic group.

15-44. Draw the monomers of urea and melamine. Draw the chain of both.

15-45. What is the effect of reacting formaldehyde with either urea or melamine?

15-46. Discuss the uses of the urea formaldehyde and the melamine formaldehyde plastics.

15-47. List some uses of the epoxy resins and give the reasons for such use.

15-48. Draw the monomer for epoxy resins.

15-49. What are some uses of the furane plastics?

15-50. Draw the mer, monomer, and polymer for Bakelite.

15-51. How are the phenolic formaldehyde plastics reinforced? Discuss them.

15-52. Discuss the properties of the phenolic formaldehyde plastics.

15-53. Discuss the lamination of Bakelite, the reinforcing materials used, and the effect upon properties. Give two examples of laminated Bakelite.

15-54. What are some of the uses of the reinforced polyester plastics? Of the nonreinforced polyester plastics? List three trade names for these plastics.

15-55. What are the special characteristics of the polyamide plastics?

15-56. What are the special characteristics of the silicon plastics? How are they used?

15-57. What is the special property of the polyurethane indicated in this chapter that makes it useful as an engineering material?

15-58. Compare the properties of the epoxy plastics with those of the polyesters (see Table 15-2).

15-59. Repeat Problem 15-58 for the urethane and phenolic plastics.

15-60. Repeat Problem 15-58 for the silicone and urethane plastics.

16 | Rubber and Organic Coatings

16-1 RUBBER: ITS SOURCE

Rubber is the gum resin which exudes from rubber trees and plants that grow in tropical climates in Asia, Ceylon, the Congo, Liberia, and Indonesia. Grafting is used to grow a tree that produces the best grade of rubber. A species that produces strong roots, but has low yield, is grafted to a high yield trunk. This in turn is grafted to a species which produces a foliage that has a high resistance to leaf disease.

Brazilian latex is dried over a fire into a dark solid mass, which is then shipped to processing plants. The latex is often shipped in its "milk-like" state. Castilla rubber is from Mexico and Panama. Euphorbia rubber comes from South Africa. Once processed they are alike. Mangabeira rubber is grown in the Amazon region. The latex is from the mangabeira tree. The grade is as good as the Brazilian Ceara rubber. A lower grade rubber, called Assam, comes from India and Malaya.

Crude rubber is sold as smoked sheet. In quantities of about 30 per cent, sulfur is used as a hardener. When lesser percentages of sulfur are used, the elasticity may be controlled accurately. If cured in vapor and neutralized with magnesium carbonate, it is suitable to be used as very thin sheet rubber.

Reclaimed rubber is the end product of ground old rubber that is treated with chemicals and heat. Foreign matter such as metal and other fibers are first removed. It generally is easier to process and has better resistance to aging than crude rubber.

Synthetic rubbers are materials with "rubber-like" qualities. They are highly resistant to oil, chemicals, heat, and aging. They are not as elastic as rubber and therefore are often mixed with rubber to produce a product with the characteristics of both. There are many types of synthetic rubbers: Buno rubber, a butadiene polymer; Methyl rubber, a dimethyl butadiene polymer; Neoprene, a chloroprene polymer; Koroseal, a vinyl chloride polymer; Vistanex, an isobutene polymer; and Thoikol, an organic polysulphide. In general, the monomer is manufactured first and then polymerized into rubber.

Buno rubber, a butadiene (76%) styrene (24%) combination was developed by the United States during the Second World War. The cold variety, which is processed at about 40°F, has good properties of elongation, resistance to cracking, abrasion, and heat buildup when in use. Its useful properties may be extended by processing the rubber in an oil or rosin emulsion. *Methyl rubber* is a Buno type rubber manufactured in Germany.

Neoprene is a rubberlike material that is stable in light and to aging. It has greater resistance to chemicals, oils, and heat than natural rubber. It has a specific gravity of 1.25 to 1.8 depending on the type and quantity of the fillers used. It is used in molded products, gaskets, belting, hose, and adhesives.

Koroseal is a synthetic rubber which has a higher strength and abrasion resistance than Buno rubber, but has a lower chemical resistance. It is used in cable sheathing, gaskets, and to impregnate fabrics.

Vistanex is a petroleum hydrocarbon which has elastic properties that are not quite as good as those of rubber. It is oil, chemical, heat, and wear resistant. When compounded with rubber it is used for hose, gaskets, insulation, and packing.

Thoikol is a rubber-like material which may be vulcanized and used for wire covering, coatings, or molded products.

It should be noted that the above discussion represents only a small sample of all the possible combinations of monomers which produce elastomers.

16-2 ELASTOMERS

Materials that recover after being deformed are called *elastomers.* These materials form chains that are deformed when a stress is applied. It is the configuration of the chain that deforms, rather than the material itself.

Figure 16-1(a) shows the mechanism of *bond lengthening* by which most materials stretch or elastically deform. Atomic forces hold the atoms in position. When forces are applied to both ends of the material, the bonds stretch. The particles may also elongate. When the forces are released, reaction forces pull the atoms back into position. Elongation is generally not over 1 to 2 per cent.

In other materials the bonds may be covalent and the binding forces may be operating at an angle when in an equilibrium state. In this case, when the forces are applied, the bonds are straightened. When the deforming force is released, the equilibrium bond angle is reinstated. The mechanism of *bond straightening* produces an elongation of about 2 to 5 per cent. This is shown in Fig. 16-1(b).

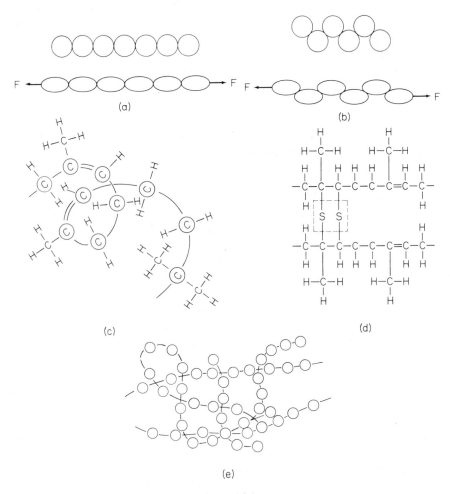

Figure 16-1

Rubber elongates up to 1000 per cent. Neither of the above mechanisms explains this very great difference in percentage of elongation. The schematic chain in Fig. 16-1(c) shows a polymer of rubber coiled into a helix because the unsaturated part of the chain is positioned on one side of the chain. This generates an unbalance of forces along the force line, and the chain coils until it finds an equilibrium force position. The equilibrium configuration is a helix. When *helical elongation* takes place, the material stretches like a spring. This type of mechanism would explain the high percentage of elongation in rubber and rubberlike materials.

Cross-linking of "weak" rubber with oxygen or sulfur prevents the polymers from sliding over each other to assume new positions that remain after the forces have been removed. This type of rubber will not elongate and take a permanent set. When cross-linked, the bonds give the polymers some stability — a link — so that its elasticity is increased considerably. The degree of "hardness" achieved by cross-linking can be varied from very hard brittle rubber to rubber that will stretch and recoup considerably. The *addition* of sulfur cross-linking, enclosed in a dashed rectangle in Fig. 16-1(d) is called *vulcanizing*.

Instead of cross-linking with oxygen and sulfur, as indicated above, a hydrogen atom may be removed at the end of a polymer and replaced by an oxygen or sulfur atom as a cross-link. This is a substitutional process. It needs energy input to break the hydrogen–carbon bond.

Branching is yet another method for entangling the chains in such a manner that the structure is capable of extensive movement without rupture or permanent set. See Fig. 16-1(e). External energy is required to cause branching in materials.

There are many types of synthetic rubbers manufactured. In general, natural rubber is stronger than synthetic rubber and does not generate as much internal heat as synthetic rubber. Wherever severe flexure is needed, natural rubber must be used.

Vulcanizing changes the properties of rubber drastically. Sulfur, as indicated, breeds cross-linking. The more cross-links, the harder the rubber. Accelerators and activators, such as fatty acids, increase the number of cross-links and speed up vulcanizing time.

Fillers, such as carbon black, clay, or zinc oxide, or the fibrous fillers, such as those made from metals, organic matter, and synthetic materials, change the physical properties of rubber by affecting its tensile strength, tear resistance, and wear and abrasion resistance.

Oxidation retarders are added to rubber to prevent aging and cracking. Other materials are added to rubber to preserve its plasticity during the molding or extruding process. Such plasticizers are oils, tars, and other petroleum products.

16-3 TYPES OF RUBBER

Figure 16-2 shows the various monomers used to make up the various synthetic rubber products.

Natural rubber is a resin from plants and trees that grow in tropical climates. Its monomer is shown in Fig. 16-2(a). The polymer of natural rubber is shown in Fig. 16-3(a). As indicated, it hardens with the addition of sulfur.

Natural rubber has outstanding flexure characteristics. Its specific gravity is about 0.9. The tensile strength of pure rubber is about 3000 psi. When carbon black is used to reinforce rubber, its tensile strength increases to 4500 psi. Its elongation is about 700 per cent for pure rubber and 600 per cent when reinforced with lamp black. Its efficiency of recovery is excellent, with a low flexure heat buildup. It has good tear resistance and excellent abrasion resistance. It has high resistance to permanent set.

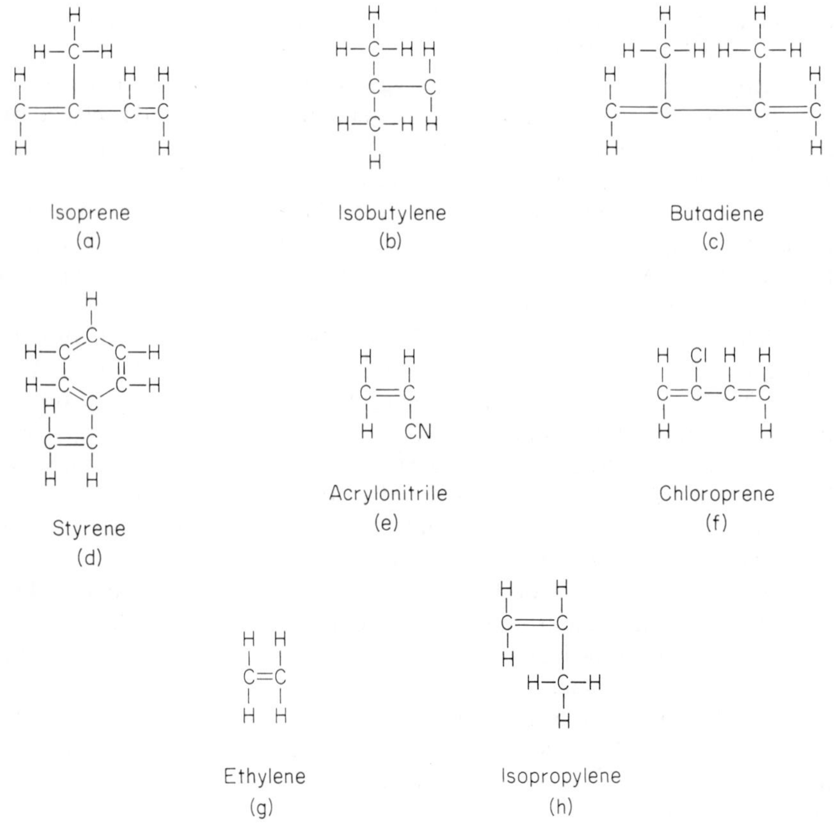

Isoprene
(a)

Isobutylene
(b)

Butadiene
(c)

Styrene
(d)

Acrylonitrile
(e)

Chloroprene
(f)

Ethylene
(g)

Isopropylene
(h)

Figure 16-2

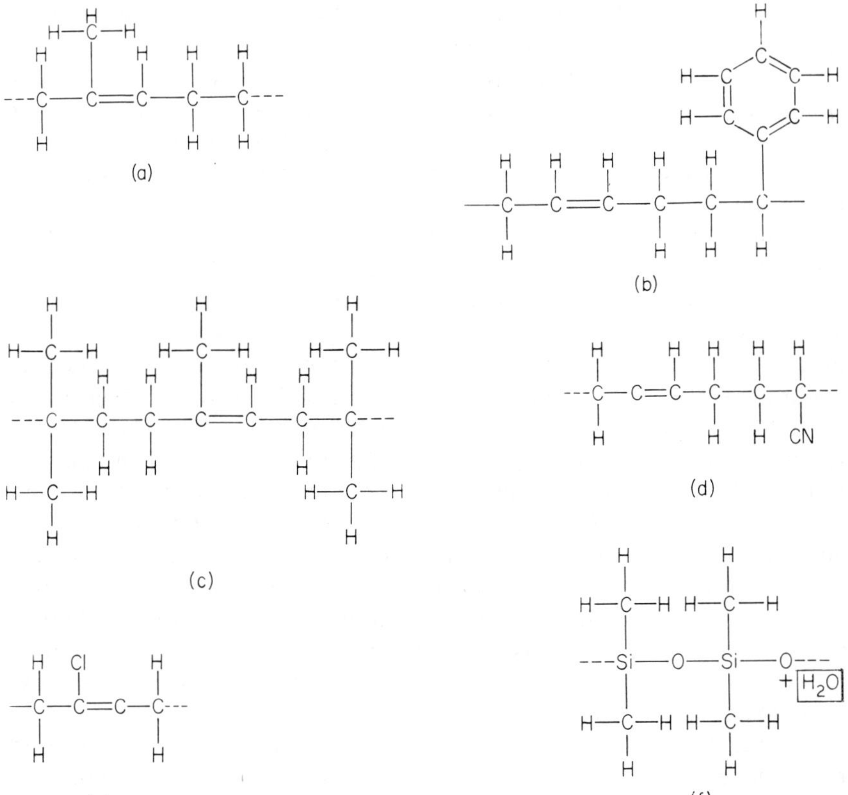

Figure 16-3

Its resistance to aging (sunlight, heat, oxidation, and shelf-aging) ranges from good to poor. It has poor resistance to gasoline products, and excellent resistance to attack by acids.

Reclaimed rubber is reprocessed rubber that is ground and chemically treated with all metallic and fibrous materials removed. It is then processed into sheets, slabs, or powder for reuse. Its physical properties are at least as good as fresh natural rubber.

Styrene–butadiene (SBR) represents the bulk of the synthetic rubbers used in the United States. The polymer generally used is about 78 per cent butadiene [Fig. 16-2(c)] to 22 per cent styrene [Fig. 16-2(d)]. The SBR polymer is shown in Fig. 16-3(b).

It may be polymerized cold at 40°F or hot at 125°F. The cold polymerized rubber has proved to be superior to natural rubber in automobile tires. The heat generated in heavy equipment tires has made natural rubber unsatisfactory for these applications. Polymerized rubber has high resistance to abrasion and weathering, and only fair resistance to hot tear. In its nonreinforced state it has a tensile strength of about 400 and a tension modulus of 200. In its reinforced state its tensile strength increases to as much as 3000 psi, its tension modulus to 2000, and its percentage of elongation becomes 500 per cent. This rubber is used in hoses, shoe soles, insulation, and flooring.

Butyl rubber is a copolymer of isobutylene [Fig. 16-2(b)] and butadiene [Fig 16-2(c)] or isoprene [Fig. 16-2(a)]. The polymer of butyl rubber is shown in Fig. 16-3(c). It has a structure that is capable of confining air because of its very low permeability to gases. It is therefore used in inner tubes, tubeless tires, and sealing compound. Its tensile strength is about the same as nonreinforced natural rubber. Its modulus of tension is about 300, and about three times higher when reinforced with lamp black. Its elongation when reinforced is about 300 per cent and when nonreinforced it is about 1000 per cent. It has good tear resistance and abrasion resistance. It has good-to-excellent resistance to aging. It has poor resistance to gasoline products, excellent acid resistance, excellent resistance to solvents in general, and fair resistance to hardening in cold temperatures.

Ethylene–propylene rubber, known as EPR, is polymerized from ethylene [Fig. 16-2(g)] and isopropylene [Fig. 16-2(h)]. These monomers produce a lightweight rubber that has good electrical properties and excellent resistance to aging and sunlight. Ratios of 35 to 65 per cent of either of the two monomers, with a catalyst to slow down the polymerization of ethylene, will produce a rubbery substance. This synthetic rubber has good resistance to abrasion and is used in tires.

Fluoro rubber is a product of butyric acid that has been converted to butyric alcohol and polymerized with acrylic acid to produce rubber. It may, under certain conditions, be polymerized with butadiene and be reinforced with carbon black. It may also be modified with bromine. In the acrylic polymer, it has high resistance to sunlight, oils, and solvents. With the bromine modifier, it has low gas permeability and high aging resistance. This rubber will not vulcanize but will withstand low- and high-temperature uses.

Nitrile rubber is a copolymer of acrylonitrile [Fig. 16-2(e)] and butadiene [Fig. 16-2(c)]. Its polymer is shown in Fig. 16-3(d). If the nitrile is increased, its oil-resisting properties are increased. Increasing the nitrile also decreases the flexure properties of the polymer. The nitrile groups may be reinforced with carbon black. It has excellent grease, solvent, and oil resistance. It is also a heat-sealing material. It has a specific gravity of about 1.00. In the pure state it has a tensile strength of about 600, and a tensile modulus of about 200. When carbon black is added, its tensile strength increases to 3500, its tensile modulus to 2500, and its elongation is about 600 per cent.

It is used extensively as an adhesive and to impregnate or as a coating for paper products. It is used as lining in hoses that are used to transport petroleum products, for lining in fuel tanks, for O-rings that must be resistant to these products, and for heels of shoes.

Polychloroprene rubber, also known as neoprene rubber, is a widely used rubber product. It is a polymer of chloroprene [Fig. 16-2(f)] and chlorobutadiene (not shown). Its polymer is shown in Fig. 16-3(e). It is resistant to petroleum products, sunlight, and abrasion. It is used in gaskets, gloves and clothing for protection, adhesives, hoses and hose linings, conveyor belts, and as an electrical insulation. It has excellent tensile properties even when not reinforced. The addition of carbon black will increase its tear and abrasion resistance. It has a specific gravity of approximately 1.25. It has a tension modulus of 400, which increases to 3000 when reinforced with carbon black. Its percentage of elongation is about 800 per cent. In one form it is resistant to crystallization at $-20°F$. In general, however, its major disadvantage is its high freezing temperature.

Polysulfide rubber, also known as thiokol or koroseal, is a condensation polymer. It has a low permeability to gases and excellent resistance to solvents. It is used as an adhesive for metals. It has a specific gravity of about 1.35. In its pure state it has a tensile strength of 300, a tension modulus of 450, and a percentage of elongation of 300. When reinforced with carbon black, it exhibits a tensile strength of 1500, a tension modulus of 1500, and a percentage of elongation of 500. Zinc sulfide, zinc oxide, or titanium dioxide are other reinforcing fillers generally used. It has poor-to-fair resistance to tear and abrasion. Its resistance to sunlight and oxidation is excellent. It has excellent resistance to petroleum products and hot water swelling, good resistance to acids, and poor resistance to heat from a flame or from cold temperatures. It is used as a caulking material, linings for fuel and acid tanks, gaskets, sealing materials, gasoline hoses, and cable coverings.

Polyurethane rubbers, as indicated in Sec. 15-3, may be classified as thermoplastic, thermosetting, or elastomers. As elastomers they may be used in hard rubber paints, as adhesives, or as tires. They are also used as heels on shoes, oil seals, diaphragms, gears, vibration mounts, gaskets, etc. The materials are mixed, then poured into a mold and cured. Their resistance to abrasion, tear and wear resistance, high percentages of elongation, and tensile strengths of up to 8000 psi make them suitable for the applications just mentioned. They are used for tires on slow-moving vehicles, such as lift trucks, because at high speeds they build up heat too rapidly at the contact point between the road and the tire. They are not suited for use as automobile tires. In spite of this, they have a low coefficient of friction, so that under special conditions they are used as bushings. They are also resistant to temperature changes from -100 to $+200°F$.

Silicone rubber results from the condensation polymerization of nitrile, phenyl, or fluorine groups. It is a long-chain dimethyl silicone, as shown in Fig. 16-3(f). It has oxygen alternating with silicon and two methyl groups attached to each of the silicon atoms. Carbon black is generally not used to reinforce the silicone chain, except when the polymer contains a vinyl monomer. Other fillers such as zinc oxide, iron oxide, or titania are used to reinforce the structure and thus change the physical properties of the material.

These rubbers are very stable, are unaffected by sunlight, have excellent electrical properties, are resistant to hot oils, and have the ability to remain flexible within the temperature range of -100 to $+500°F$. They have average tensile strengths at room temperature, with better tensile strengths as the temperature increases with a range of 300 to 600 psi. They are capable of 120 per cent elongation.

In general, to summarize, natural rubbers have good physical properties. They are not affected radically by the heat generated because of the deformation that takes place when a tire rolls on a pavement. Styrene–butadiene is a good general-purpose rubber, which, in most cases, can be substituted for natural rubber. Butyl rubber is solvent resistant and also has a low gas permeability, which makes it useful in items such as tire tubes. Ethylene–propylene has good resistance to abrasion, and is a good lightweight rubber. The fluororubbers have good high- and low-temperature resistance. The nitrile rubbers have excellent petroleum-product-resistant qualities. Polychloroprene has good electrical resistance and is resistant to petroleum products. The polysulfides are also solvent resistant and have good gas permeability. The polyurethanes are abrasion resistant, have good heat-resistant qualities, and have good wear resistance. The silicon rubbers have excellent low- and high-temperature flexural characteristics.

16-4   CLASSIFICATION OF RUBBER

Grades of rubber have been classified with ASE–ASTM numbers. Table 16-1 shows type R, which is a non-oil-resistant rubber, and type S, which is an oil-resistant synthetic rubber. These letters are used as prefix letters. The suffix letters are also shown in Table 16-1. They indicate various physical properties that may be expected from a rubber that has been assigned a particular number. The illustrated example that follows will show how to interpret these numbers.

Table 16-1   Classification of Rubber*

Prefix Letters

Type R (non-oil-resistant)
    RN = natural rubber or reclaimed
    RS = synthetic rubber
Type S (oil-resistant synthetic)
    SA = maximum oil resistant
    SB = good oil resistant
    SC = medium oil resistant

Suffix Letters

A = oven-aging requirement
B = low-compression set requirement
C = weather-resistant requirement
D = load-deflection requirement
E = oil-aging requirement
F = flexible at $-40°F$
FF = flexible at $-70°F$
G = maximum resistance to tear
H = maximum resistance to severe flexing
J = superior abrasion resistance
K = adhesion requirement
L = maximum resistance to moisture absorption
M = maximum inflammability resistance
N = maximum resistance to repeated impact
P = minimum staining effect of enamel and lacquer
Z = an additional laboratory test over and above those designated by
    the suffix letters listed above, life or performance requirements

Source: Kent, *Mechanical Engineering Handbook,* 12th ed., Table 3, pp. 5–6. New York: John Wiley & Sons.
*Classification of the compound is determined by the predominating basic elastomer therein.

Example 1

Given a number SB 425 ABEFL, interpret its meaning.

solution:

    SB = a good-oil resistant rubber
      4 = grade 40 durometer number
    25 = 2500-psi tensile strength
     A = oven-aging requirement
     B = low-compression set requirement
     E = oil-aging requirement
     F = flexible at $-40°F$
     L = maximum resistance to moisture absorption

16-5 ADHESIVES

The following are terms that will be used in this discussion of adhesives:

Adhesive materials are those used to join two surfaces. *Pressure-sensitive adhesives* are those that will cause adhesion simply by the application of pressure.

Cement is an adhesive material.

Adherend materials are the base materials that are to be joined.

Tack is that property of an adhesive which causes it to stick to the surface upon which it comes in contact and to form a bond with a second surface.

Set is the conversion of a tacky surface into a hardened state.

Cure is set that occurs as a result of polymerization or cross-linking.

Peel strength is the measured resistance of an adhesive to stripping.

Bond strength is the resistance to separation of joined surfaces.

Wettability refers to the spreading of the adhesive over the surface of the base material. This occurs when the forces between the adhesive molecules and the base metal molecules are greater than the forces between the adhesive molecules.

In general, two types of forces operate when an adhesive is used. There are forces operating between like molecules within the adhesive material and those operating between the adhesive material and the adherend material. If the forces within a material are greater than any of the other forces affecting the drop, the material will form a sphere, as shown in Fig. 16-4(a). It will generate a "force film" at the surface. We call this force film "surface tension." Such is the case when solder forms little spheres on the surface to be soldered.

If the forces between unlike molecules are greater than those between like molecules, the adhesive material will spread out over the surface of the adherend material. In the case of soldering, the solder coats the base material with a very fine film of solder. This is called "tinning," or "wetting." It is shown in Fig. 16-4(b).

Pressure-sensitive adhesives do not "wet" the surface to which they adhere. Pressure-sensitive tapes, such as Scotch tape, must have the op-

(a)   (b)

Figure 16-4

posite side of the tape from that to which the adhesive has been applied treated with a "release" coat. This is necessary so that the tape may be stripped from the roll without stripping the pressure-sensitive adhesive from it.

From the prior discussion, it should be evident that a good adhesive material must wet the surface of the base material with a very thin clean film. Ideally, the forces *within* the adhesive, *between* the adhesive and the adherend material, and *within* the adherend material should all be the same, so that one set of forces does not have an advantage over the others. Unfortunately, this is an ideal state.

In general, if the adhesive film is *very thin,* the weaker of the two sets of forces may be within the adherend material. In this case, if two pieces, which have been caused to adhere by an adhesive substance, are caused to be separated, the adherend material will tear away. If, however, the forces between the adherend material and the adhesive material are weaker than the forces within the adherend material, the adhesive bonds will break first. This weakness may result from dirt, corrosion, voids, or discontinuities. However, even if none of these imperfections exist, a thick layer of adhesive material for some instances means that the adhesive material will separate first.

There is a *best thickness* for adhesive materials. Films that are thicker or thinner than this best thickness yield weak bonds. Also, bond strength is proportioned to the *width* of the overlap, not the length of the overlap. Increasing the length of overlap decreases the bond strength of the joint.

16-6 TYPES OF ADHESIVES

Glues are animal gelatins mixed with water. Glues made from animal hides or bones are generally manufactured as flakes. Glues made from fish parts are generally liquid. *Mucilages* are made from vegetable glues.

The least expensive of all the adhesive materials are those which deteriorate when exposed to the atmosphere. Such materials are the *starch pastes.* These adhesives are usually mixtures of starch or dextrine with water and a strengthener such as glue, a resin, or gum. *Latex* pastes are used because they do not shrink as much as the starch-base glues and because the excess can be removed by rubbing. Many glues are made from tapioca flour. They are classified as *vegetable* glues and are used as an adhesive on such items as envelopes, labels, and stamps. Glues made from starches give good remoistening qualities. *Dextrine* has better remoistening qualities than starches, and when mixed with animal glue, various qualities may be achieved. Cassava roots are also ground into powders for use in vegetable glues for low-grade plywood. Corn and potato starches are also used as adhesives in low-grade plywood.

Tapes may require adhesives that need to be remoistened or those which do not need to be remoistened. The former type may be made from paper or cloth. They are coated with animal, starch, or dextrine glues. Those which need not be remoistened and rely on pressure are usually coated with combinations of rubber and resin.

Rubber cements may be made by using rubber that has not been cured and a chemical solvent. Synthetic self-curing rubber cements are also available. These latter rubber cements are cured by using heat and pressure or chemicals. The uncured rubber cements are waterproof and have good initial strength. Because they are not cured, they are subject to disintegration. The cured-type rubber cements are also waterproof, are stronger than the noncured cements, and last longer. Natural rubber has high bond and cohesive strength and good initial tackiness. It has good aging and pot life. Synthetic rubbers usually need tackifiers added.

Cellulose Adhesives.   The *pyroxylin adhesives* are made from nitro-cellulose (movie film) and a chemical solvent such as ether alcohol, called collodion. Usually plasticizers such as gums or resins are added. These cements are generally called the "household" cements. They will adhere to almost any surface, have very little initial tackiness, and are cured by evaporation of the solvent. They have poor resistance to heat and are flammable.

The *cellulose acetate* adhesives have good heat-aging properties. But their weather and adhesion properties are poorer than the nitrocellulose adhesives.

Cellulose acetate butyrate adhesive has about the same properties as the acetate adhesives, except that the butyrate adhesives have good resistance to moisture.

Other cellulose adhesives are available that are water soluble.

Plastic Cements.   *Plastic cements* are manufactured from the resins discussed in Chapter 15. These may be classified as thermosetting and thermoplastic cements. Thermosetting cements cannot be reheated after they have cured. Thermoplastic cements may be softened by reheating or with solvents.

Thermosetting Adhesives.   *Thermosetting adhesives* are made from epoxy, furane, melamine, phenolic, and urea resins.

Epoxy cements are used chiefly to achieve good bonding to glass, ceramics, steel, or wood.

Phenolic resins, when combined with formaldehyde and partially polymerized, are thermosetting adhesives. The bonding occurs when the adhesive is used with applied heat and pressure. When polymerization is completed, this adhesive produces excellent bonding. These plastic adhesives have recently been adopted for use in bonding abrasives to paper or cloth to make sandpaper and emery cloth. They are also used to bond metal to metal. These types of adhesives are furnished as thin films, in powder, and in liquid form. In the liquid form, they sometimes require catalysts. Curing occurs at room temperature. The resistance of the bond to severe moisture conditions is improved if the adhesives are cured at about 200°F. They are also known as resorcinol adhesives.

Urea formaldehyde adhesives, have a shorter curing period than the phenolics and require lower curing temperatures. They are not as strong as the phenolic adhesives. They are soluble in water, may be cured at 70 or 300°F, depending on the catalyst, or they may be fortified by the use of fillers. Curing time may range from a few minutes up to several hours, depending on the heat and pressure used. They have excellent durability under protected conditions. They disintegrate rapidly when exposed to severe moisture conditions or temperatures as low as 150°F. The addition of starch makes them waterproof.

Melamine formaldehyde adhesives are cured under pressure and heat only. They are expensive. In general, they have better durability than the ureas, but not as good as the phenolics. They are sometimes used in combination with urea resins to form a water-resisting glue for plywood.

Furan cements are usually made from furfural and polymerized with alcohol. In one form a silica powder filler is used. They are highly resistant to chemical attack and cure to a hard solid. In one form they are pressure-sensitive adhesives.

Polyurethane adhesives are used to bond wood, metal, or plastics. They are also used to increase the adhesion qualities of rubber.

Polyester adhesives are polymerized allyd resins and styrene. They vary from very hard setting adhesives to pliable gels. Their water and solvent resistance is good, but they are not heat resistant.

Silicon adhesives are used in conjunction with rubber to bond rubber itself, plastics, metal, or ceramics. They have a wide temperature use.

Thermoplastic Cements. *Thermoplastic adhesives* are made from cellulose, acrylics, and polyvinyls.

Polychloroprene (neoprene) adhesives are probably the most popular of the thermoplastic cements. They generally are liquid in form and are used with a filler as a strengthener, together with a curing agent. Sometimes an activator is added so that they can cure at room temperature. These ingredients may be mixed just prior to their use.

Polyvinyl alcohol adhesives are resealable. They may be used to bond leather, paper, and textiles. They are sensitive to water, although water-soluble grades are available.

Polyvinyl acetate adhesives are used as resealables in gummed tapes, shipping bags, tile cement, and cellophane. They may be used to bond porous materials or nonporous materials such as glass and metals. *Polyvinyl butyral* adhesives bond well with glass. Their chief use is in safety glass.

Acrylics, such as Lucite and Plexiglas, are used in solution form to bond these materials. Ethyl acrylic may be used in pressure-sensitive adhesives because it is flexible and always remains slightly tacky.

Rubber-based cements have a wide range of applications. Almost all rubbers are used in the manufacture of adhesives. In general, natural rubber has good tack and resiliency. The silicone adhesives will bond to Teflon and resist high-temperature applications. Polysulfide has excellent resistance to solvents. Acrylonitrile is used for metal bonding. Reclaimed rubber needs tackifiers. Polychloroprene has the best cohesive strength.

As a summary, a list of the types of adhesives is given in Table 16-2, and Table 16-3 is a list of plastics and their general uses as adhesives and sealants.

Table 16-2   Types of Adhesives

Chemically Reactive

Plural components	
Epoxy*	Polysulfide
Phenolic*	Polyurethane
Polyester	Silicone
Moisture cure	
Silicone	
Urethane	
Cyanoacrylate	
Heat activated	
Polybenzimidazole	Phenolic
Polymide	Polyvinyl acetals
Epoxy	Rubber
Nylon	Urethane (blocked)

Solvent or Water Based

Rubber: natural, reclaimed, synthetic	Phenolic
(butyl, S–B rubber, nitrile, neoprene)	Polyamide
Acrylic	Urethane
Cellulose esters	Vinyls (including polyvinyl acetate)

Hot Melt

Olefin copolymers, ethylene vinyl	Polyethylene (and polypropylene)
acetate, ethylene ethyl acrylate,	Polyvinyl acetate
ionomers, etc.	Polyvinyl ether
Phenoxy	Asphalt
Polyamide	Ceramics
Polyester	

Table 16-2   *(continued)*

Film

Epoxy (and epoxy alloys)	Polyamide
Nitrile alloys	Polyolefins
Olefin copolymers (ethylene acrylic acid, ionomers, etc.)	Polyvinyl butyral
	Polyvinyl chloride
Phenolic (and alloys)	

Pressure Sensitive

Rubber (all types)	Polyvinyl ether
Polyacrylate	Silicone

*Frequently used with elastomer such as nitrile rubber, neoprene, or vinyl. Combination is referred to as an "alloy."

16-7 ORGANIC FINISHES

The various types of organic finishes manufactured are paints, varnishes, lacquers, and enamels. Oils and synthetic resins are used as vehicles that harden by cross-linking. Therefore, the vehicle is generally thermosetting.

Paint is a mixture of pigment and a vehicle. *Varnish* contains no pigment. *Enamel* is a combination of paint and varnish. *Lacquer* is a quick-drying material in which the volatile materials evaporate quickly and leave a nonvolatile residue.

Coatings are used to protect surfaces from corrosion or damage. Protective coatings of polyvinyl or cellulose resins are vehicles that are deposited by hot dipping. When the part is ready for use, the coating is stripped off. The vehicle in this instance does not contain any pigment to be deposited.

Paint, however, is composed of a binder and pigment. The binder creates the hardened surface; the pigment provides the hiding power and color. The pigments used are white lead, red lead, titanium oxide, aluminum, lamp blacks, or Prussian blue. Since the vehicle produces the gloss, the ratio of pigment to vehicle controls the intensity of the gloss. The less vehicle, the more pigment, and the greater is the nonreflective power of the paint. Linseed oil and tung oil are used as vehicles; turpentine or other thinners as the solvent. Water paints contain a water emulsion as the vehicle.

The drying rate of paint is controlled by the type and amount of solvent used or by the rate of oxidation of the vehicle. Drying rate is proportional to evaporation time. Paint that is sprayed requires shorter evaporation time than paints that are brushed.

Water paints are those which use water as a thinner. They are generally odorless, quick drying, and easy to apply indoors. The binder is

Table 16-3 Plastics as Adhesives and Sealants

Type	Advantages	Limitations	Major usage
Acrylic Cellulosic Ethers Esters	Ultraviolet stability clear Good solubility	Heat resistance Cost, moisture sensitivity	Bonding of plastics, glass Bonding of leather, paper, wood, nonwoven fabrics
Cyanoacrylate	Fast setting, adhesion to many hard-to-bond plastics	High cost, restricted glue-line thickness and area	Specialty adhesive for "spot welding" surgical suturing
Epoxy	Versatility, high strength, solvent resistance	Peel strength, cost	Structural bonding (high peel alloys available), concrete repair construction industries
Olefin polymers (copolymers of ethylene with) a. Vinyl acetate b. Acrylates c. Acrylic acid or acid salt (ionomer)	Flexibility, ease of handling as films or hot melts, cost	Creep, heat resistance	Laminating, packaging, bookbinding
Phenolic	Low cost, heat resistance, weatherability	Brittleness	Plywood, abrasive wheels, structural bending
Polyamide	Flexibility, oil and water resistant	Cost	Can seam sealant, hot melt for shoes
Polyaromatic Polybenzimidazole Polyimide	Stability at 700°F with good low-temp. properties, "ladder polymer" structure	Cost, handling, high temps. required to form bonds	Aerospace applications, honeycomb sandwich assembly
Polyester	Flexibility, sharp melting point for hot melts	Cost, limited adhesion	Foils, plastics, shoe bonding
Polysulfide	Weatherability, wide temp. range	Strength	Elastomeric adhesives or sealant

Table 16-3 (continued)

Type	Advantages	Limitations	Major usage
Polyurethane	Cryogenic performance, versatility	Handling, heat resistance	Bonding of flexible to non-flexible substrates
Resorcinol	Room temp. setting	High cost	Marine plywood, tire cord adhesion
Rubber Natural Reclaimed Styrene–butadiene Neoprene Nitrile	Elastomeric, high "tack," weatherability (S–B), strength (neop.), oil resistance (nit.)	Low tensile and shear strength, poor solvent resistance	Pressure-sensitive tapes, household adhesive, laminating adhesives, footwear; nitrile forms alloys with thermosets for structural adhesives
Silicone	Performance at temperature extremes	Cost, strength	Sealants, pressure-sensitive tapes for aerospace electrical usage, coupling agents
Urea	Very low cost, room temp. curing	Moisture sensitivity	Plywood, furniture
Vinyl Polyvinyl chloride Polyvinyl acetate Polyvinyl acetals Polyvinyl alkyl ethers	Versatility, adhesion to glass, oil and grease resistance	Moisture sensitivity	Household glues (PVA), safety glass lamination (acetals), furniture, brake linings, footwear

Source: J. A. Clark, "Polymer Based Adhesives," *Technology Tutor*, Bek Publications (April 1971).

generally an oil and resin combination. In some instances the resin is rubber based. Because of the water, these paints are subject to freezing when exposed to cold temperatures and spoiling when exposed to excessive heat when stored.

Varnish is a mixture of a natural or a synthetic resin and a drying oil, such as fish, castor, soybean, cotton, or corn oil. Natural resins are gums or pitches such as hauri, batu, copal, manila, etc. Synthetic resins, such as the alkyd, epoxy, melamine, phenol, silicone, and urea, are all used in paints to achieve various physical property effects. *Alkyd* varnishes are hardened when heated. When used with an oil base, they produce a high gloss and resistance to chemicals, heat, and moisture, are flexible, and exhibit outstanding adhesion properties. *Phenol* varnishes have outstanding bonding properties when baked. They have good insulating properties and are resistant to salt water. Their resistance to chemicals in general is not as good as the alkyds.

Bituminous paints or varnishes have coal tar dissolved in mineral spirits and are used as a waterproofing paint. They have very low permeability and for that reason have high resistance to water, which, therefore, makes them good waterproofing paints. They are sensitive to sunlight and will shrivel and crack when exposed for even short periods of time.

Enamel results when pigments are added to varnish. Very smooth surfaces result after drying. If the enamels are of the synthetic resin variety, they harden by polymerization. They are largely replacing the oil-based enamels. They may be dried in air or by the application of heat.

The acrylic enamels have good color stability at high temperatures. The epoxy enamels are flexible, chemical resistant, and generally require baking. The phenolic enamels are chemical, water, and oil resistant. The coating is brittle. The alkyd enamels have good color and gloss retention. They are resistant to moisture and most chemicals. When polymerized with silicone, their heat resistance is increased. The straight silicone enamels have outstanding heat resistance. The urethane enamels are abrasion resisting, and the vinyl enamels are moisture resisting.

Lacquers form a film by evaporation of a vehicle. They are generally made from one of the cellulose resins dissolved in a solvent. Color pigments are added to achieve desired color effects. They may be modified with other resins. Acrylic resins make the lacquer water and chemical resistant. Vinyl resins make lacquers abrasion and oil resistant. Styrene–butadiene makes the lacquers water and chemical resistant and gives them very good adhesive qualities. Cellulose acetate makes them heat resistant.

Rubber-based coatings, such as the chlorinated rubber paints, are used as protection against attack from acids, alkalies, alcohol, most gasoline products, and salts. They will not resist attack by vegetable or animal greases or oils. Neoprene coatings are resistant to oils, alkalies, most acids, and salts. When the solvent evaporates, it leaves a coat that has the above characteristics. The sulfur–polyethylene resin rubber-based coatings are resistant to attack by the oxides and have good stability from temperatures of about 100 to +300°F.

Problems

16-1. Explain the purpose of grafting in the production of rubber.

16-2. What effect does sulfur have when added to rubber?

16-3. Describe the differences between crude, reclaimed, and synthetic rubbers.

16-4. Describe the process of elastic deformation by the mechanism of bond lengthening.

16-5. Describe the process of elastic deformation by the mechanism of bond straightening.

16-6. Explain deformation using the mechanism of helical elongation. Draw a diagram of this process.

16-7. What effect does cross-linking have on the elasticity of rubber? Use Fig. 16-1(d).

16-8. Describe substitutional cross linkage.

16-9. What is branching? How does it restrict permanent set?

16-10. What are the effects of the following on the physical properties of rubber: (a) flexure, (b) vulcanizing, (c) fillers, (d) oxidation, (e) plasticizing?

16-11. Draw the monomers for the following: (a) ethylene, (b) acrylonite, (c) isobutylene, (d) styrene, (e) isopropylene, (f) isoprene, (g) butadiene, (h) chloroprene.

16-12. Draw the polymer for natural rubber. Explain how the isoprene monomer is polymerized to form natural rubber.

16-13. Compare the physical properties of pure rubber and lampblack-reinforced rubber.

16-14. How is reclaimed rubber reprocessed for use?

16-15. Draw the monomers for styrene–butadiene rubber and show how they form the SBR chain.

16-16. Discuss the physical properties of nonreinforced and reinforced SBR rubber. What are some of its uses?

16-17. Draw the monomers for butyl rubber and show how they polymerize to form butyl rubber.

16-18. (a) What is the outstanding structural characteristic of SBR rubber? (b) To what uses can this rubber be put as a result of your answer in part (a)? (c) Discuss its physical properties in general.

16-19. (a) Draw the monomers of ethylene–propylene rubber. (b) Discuss its physical characteristics.

16-20. Discuss the physical properties of fluororubber.

16-21. (a) Draw the monomers of nitrile rubber. (b) Show and explain its polymer. (c) Discuss its physical properties and list some of its uses.

16-22. (a) Draw the polymer of polychloroprene rubber. (b) Discuss its physical properties. (c) What are some of its uses?

16-23. List some physical properties of the polyurethane rubbers. Discuss some uses of this rubber.

16-24. Draw the polymer for silicon rubber. Discuss its physical properties.

16-25. Given an ASTM number for rubber of S615FHJ, explain each digit and letter.

16-26. Repeat Problem 16-25 given the ASTM number SA518AGL.

16-27. Repeat Problem 16-25 for the number SC210CEN.

16-28. Define the following terms used for adhesives: (a) adhesive, (b) adherend, (c) tack, (d) cure, (e) set.

16-29. (a) Define *peel strength* and *bond strength*. (b) What is meant by the word "wettability"?

16-30. (a) Discuss the forces in operation when a "drop" forms on the surface of an adherend; (b) when the adhesive wets the surface.

16-31. Define and explain the term "pressure-sensitive" adhesive.

16-32. Describe the effect of a thin or thick adhesive layer upon the operating forces.

16-33. What is the effect of the width and length of overlap on the holding power of an adhesive?

16-34. (a) Discuss the three glues listed in this chapter. (b) Discuss their uses.

16-35. Discuss the properties of uncured and cured rubber cement.

16-36. Discuss the cellulose adhesives and compare their physical properties.

16-37. List the thermosetting and thermoplastic cements. What is the difference between the two types?

16-38. Discuss the epoxy resins as adhesives.

16-39. Discuss the phenolic resins as thermosetting adhesives.

16-40. Discuss the urea and the melamine formaldehyde adhesives.

16-41. List one characteristic for each of the following adhesives: (a) furfural, (b) polyurethane, (c) polyester, (d) silicon adhesives.

16-42. List the thermoplastic adhesives and give at least one characteristic use for each.

16-43. Table 16-1 categorizes the types of adhesives into five major divisions. List these divisions and explain the purpose for each.

16-44. If you are in need of an adhesive to be used with plywood, which two adhesives would you choose? Refer to Table 16-3.

16-45. Repeat Problem 16-44 if you are to bond glass.

16-46. Repeat Problem 16-44 if you are to repair furniture.

16-47. Which plastic adhesive is used in the manufacture of tires? (See Table 16-3.)

16-48. Discuss the role of the vehicle, pigment, and solvent in organic coating.

16-49. Describe water paints and their uses.

16-50. Define the term "varnish." List several and explain their uses.

16-51. What are the uses of the bituminous varnishes?

16-52. Define the term "enamel." List the outstanding feature of each enamel listed in this chapter.

16-53. Define the term "lacquer." What effect does the addition of resin have on the lacquers?

16-54. Discuss the rubber-based coating and their uses.

17 | Wood

17-1 WOOD STRUCTURE AND DEFINITIONS

The two broad categories into which wood may be classified are *softwood* and *hardwood*. These designations do not refer to mechanical hardness in the same sense as when the term is applied to steel. Softwoods come from trees that have needlelike leaves. Evergreens are such trees. Hardwoods come from trees that have broad leaves. Thus, according to these definitions, the wood from a softwood tree may actually have a greater mechanical hardness than the wood from a hardwood tree.

In most classes of trees, growth takes place by the formation of new fibers in the form of rings between the bark and the last existing ring. This type of growth adds rings year after year to the trunk and is called *exogenous* growth. When growth is such that the new fibers mesh with the old growth, it is called *endogenous* growth. Bamboo has such a growth.

Sapwood [Fig. 17-1(a)] is the outer portion of the tree, which contains the living cells and in which growth takes place. As the tree grows, the sapwood is converted to *heartwood* in the form of annular rings. As such, it ceases to contain living growth cells. Foreign substances are deposited in the heartwood, making it darker in color than the sapwood. It also becomes more decay resistant.

The grain direction and annular rings provide three axes of structural strength, as shown in Fig. 17-1(a). The radial axis is along the diameter of the trunk; the tangential axis is tangent to the annular rings; the longitudinal axis is parallel to the grain direction. All three structural axes are mutually perpendicular to each other. The strength of wood is very different along each axis.

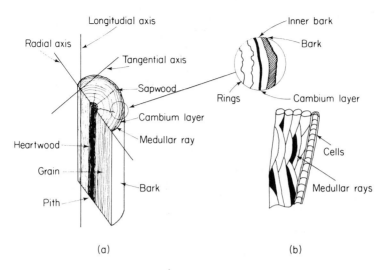

(a)   (b)

Figure 17-1

Growth takes place under the bark in the layer called the *cambium* layer. The cells in the cambium layer are long, *tubular cellulose fibers* held together by a natural resin called *lignin*. Radial cells, called *medullary rays,* are also found intertwined with these longitudinal cells in the cambium. They are shown in Fig. 17-1(b). The longitudinal cells run the length of the tree and carry food to the leaves. The medullary cells carry food from the bark to the inner portions of the cambium and also operate as storage repositories of the food.

Wood grain refers to the direction of the grain as it appears at one of the cut surfaces of lumber. A *straight-grained* board is shown in Fig. 17-2(a), where the grain runs parallel to the surface. Figure 17-2(b) shows a *cross-grained board.* In this case the grain runs at 90 degrees to the face of the board. Boards can also be cut from logs to produce *angle grains,* as shown in Fig. 17-2(c).

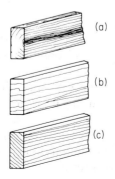

(a)

(b)

(c)

Figure 17-2

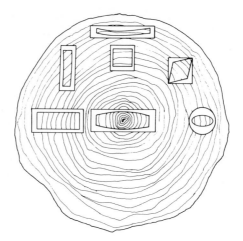

Figure 17-3

The degree to which wood warps is related to the direction of the grain with reference to the face of the board and to the method of drying. Figure 17-3 shows the cuts that can be made from a log and the resulting warpage that can be expected. Moisture content and its effect on wood will be discussed in Sec. 17-4.

The next two sections deal with the various softwood and hardwood trees. In general, the softwood trees are cedar, cypress, fir, hemlock, pine, redwood, and spruce trees. The hardwood trees are ash, basswood, beech, birch, cherry, chestnut, elm, gum, hickory, mahogany, maple, oak, poplar, walnut, and willow trees.

17-2   SOFTWOODS

Of the softwoods described in this section, pines are probably the most important for industrial uses.

Pines.   Pines are the most widely used softwoods. Western white, sugar, and lodgepole pine are found in the western and northwestern part of the United States. Western white pine is a uniform wood suitable for building construction, patternmaking, and boxes. Sugar pine is soft and easily worked. It is used in casting patterns, countertops, and blinds. Lodgepole grows thinly. Its trunk is not suitable for cutting into lumber. It is suitable for poles, ties, and construction.

Ponderosa pine grows in the western part of the United States. Since it is easily worked, it is used in general building construction, frames, paneling, and, after treatment, for railroad ties.

Southern yellow pine grows in the southeastern part of the United States. It is generally soft, but since it has good strength it is used in temporary construction and in poles, boxes, crates, and piling.

Firs. White and Douglas firs grows in the western part of the United States. White fir is generally soft, and is easily worked because of its uniform grain. It is a general-purpose wood used for general construction purposes. Douglas fir is strong and hard. It is used for building construction, plywood, furniture veneer, and boxes. Balsam fir is used for general building construction. It is also found in New England.

Spruce. Eastern, sitka, and Engelmann spruce trees are grown in the northeastern and the western parts of the United States. Eastern spruce possesses medium hardness and is easily worked. It is used in buildings, boxes, ladders, and general woodworking. Sitka spruce is strong and moderately shock resistant. It is used in general building construction, doors, siding, and in plane and boat construction. Engelmann spruce has a straight grain structure, is lightweight, and has moderate shrinkage. It is generally used in ties, studs, poles, mining timber, and flooring.

Hemlock. These trees are grown in the western and the eastern part of the United States. They are moderate in weight and hardness. They are used in building construction, doors, posts, piling, sashs, and wall and ceiling boards. They have a straight grain and small knots.

Cedar. Alaskan yellow, western red, and northern and southern white cedars are the main source of cedar in the United States. In general, cedars are light, soft, and are not very strong. They are generally moisture resistant, and have very good resistance to shrinkage even when environmental conditions change.

Alaskan yellow cedar grows in the western part of the United States and Canada. It is uniform in texture and has very good shrinkage qualities. It is used in furniture, cabinets, and internal finish. It has a bright yellow coloration. Port Oxford white cedar is also grown in the western part of the United States. It has a light yellowish-brown color with a uniform texture and a very pungent, spicy odor. It is used for the same purposes as the Alaskan yellow cedars, and for siding and shingles as well.

Western red cedar is grown in the northwestern part of the United States. It gets its name from the color of the heartwood. This wood will not withstand the forces required to drive piling. It is soft, has good shrinkage qualities, and is highly resistant to decay. It is used for siding, shingles, posts, and poles.

Northern white cedar grows along the Great Lakes and in most of the northeastern United States. This type of cedar has a strong odor. It is a very weak wood, is brittle and very soft. It has very good shrinkage qualities and excellent resistance to decay. It is used as shingles, posts, and poles.

The southern white cedar grows in the eastern coastal regions of the United States. It has essentially the same physical properties as northern cedar. This tree grows larger than the northern white cedar, and therefore can be cut into lumber and wooden objects. Both white cedars are used as linings in closets and trunks to repel insects.

Cypress.   This tree grows in the southeastern part of the United States. Coastal cypress has a dark reddish color. Inland cypress is a yellow-white. It has a very high resistance to decay. It is used in building construction for porches, sashes, siding, and paneling. It is also used to line tanks and in greenhouse construction.

Redwood.   This tree grows in the western United States. It grows very large and therefore is suitable for cutting into wide boards. These boards have uniform structural characteristics. It is a coarse-grain wood and has a dull red, almost mahogany, coloration. It has very high resistance to decay and is a durable wood, having low shrinkage. It is used for interior and exterior finishing in building construction, for structural timber, mill roofs, sashes, doors, siding, tanks, ties, posts, and shingles.

17-3   HARDWOODS

Oak.   In general, oak is divided into white and red oak. There are about 60 types of oak grown in the United States. Several of the oaks classified as white are white, post, swamp-white, and chestnut. Some of the oaks classified as red are red, southern red, swamp-red, pin, willow, water, and black oak. White oak is very hard and difficult to work because of its close-grained structure. An intermeshed growth in the pores of white oak makes it less permeable to moisture penetration. White oak has a somewhat better resistance to decay than red oak. Red oak is somewhat softer and has a coarser grain than white oak. This wood is used for flooring, furniture, trim, cross-ties, timber, and plywood.

Maple.   Sugar and black maples are classified as hard maples. Red, silver, and Oregon maples, even though they are classified as hardwood, are mechanically soft. Sugar maple is very hard, difficult to work, and very close grained. It is stiff and has good resistance to wear and shock loading. It has moderate shrinkage. Its grain is moderately straight, but is interrupted by knots and waves. It is found primarily in the northeastern United States in the Appalachian region. It has a reddish-brown coloration. Black maple has properties very similar to those of sugar maple. These maples are used for flooring, agricultural tools and implements, handles, and furniture.

Red, silver, and Oregon maples are not as strong, stiff, or as hard as sugar maple. Red maple is heavier and stronger than silver maple. These are used for wood trimming and interior trim.

Ash.   The most commonly used ash are the white, green, and black. White and green ash grow principally in the eastern parts of the United States. White ash is very hard, stiff, and strong. It also possesses high resistance to shock loading. It is used for handles, ladders, farm implements, baseball bats, and furniture. Black ash is softer, is easier to work, and is lighter in weight than white ash. It is used in furniture, veneer, and containers.

Basswood.   This tree is grown in the eastern United States around the Great Lakes. Its color is an off-white. It has a straight grain. It is a lightweight wood, easy to work, and not very strong. It is used in the manufacture of woodenware, crates, baskets, and furniture.

Beechwood.   This tree is grown in the eastern United States. It has a white-to-tan coloration. It is a heavy wood.

Hickory.   The true hickories are the shagbark, bigleaf shagbark, mockernut, and pignut hickories. They are found in the eastern United States. The sapwood is generally white, and the heartwood is red. It is hard, heavy, strong, and shock resistant. Its chief fault is that it shrinks considerably when dried. It is used in automobiles for spokes and rims, for tool handles for hatchets, axes, picks, and sledges, and for golf-club shafts, ladders, and gym equipment.

The *pecan hickory* woods include nutmeg, water hickory, and pecan. They are found in the eastern part of the United States. It is a white wood with a red heartwood. Otherwise, it is used for the same purposes as the true hickories. Its properties do not rank as high as the true hickories.

Mahogany.   This tree grows in the West Indies and in other countries with tropical climates. The heartwood is reddish-brown. It is strong, durable, and has an open structure. It shrinks equally in all directions and therefore makes an excellent pattern material. It is used for cabinets, furniture, veneers, and interior finish.

Poplar.   This is one of the most widely used trees for lumber. It grows in the eastern part of the United States. It is soft, has a uniform texture, and is easily nailed, glued, or painted. The sapwood is white, is very strong and hard, and will polish easily. It is used in floors, kitchenwear, railroad ties, handles, and furniture.

Birch.   This wood is straight grained, is very strong, has good shock resistance, and is heavy. Yellow birch is the most prevalent of the birch woods. Sweet birch has a brown tinge with a mixture of red. Paper, sweet, and yellow birch grow in the Great Lakes region of the northern United States. It is used for interior finishing, veneers, musical instruments, furniture, plywood, and woodenware.

Cherry.   Cherrywood, called black cherry, grows in most of the territory east of the Mississippi River and Texas. It is strong and moderately hard, has low shrinkage and good shock resistance, and is close grained. It has reddish-brown coloration. It is used for paneling, furniture, and interior trim.

Elm.   The three types of elm grown in the United States are slippery, rock, and American. American and slippery grow east of the Rocky mountains. Rock elm is grown in the north central states. American and slippery elm are classed as mechanically soft. They have good strength and bending qualities. The rock elm is heavy and hard, and has better physical properties than do the soft elms. It has a coloration similar to walnut. It is used for heavy construction, tool handles, boxes, and crates.

Gum.   Sweet gum grows in the swamplands of the southern United States. Its heartwood color is reddish-brown. The sapwood is white. It is durable, strong, and has uniform texture. It will warp badly unless properly dried. It is used for furniture, lumber, veneer, boxes, and crates.

Chestnut.   This wood is grown in the southern Appalachian Mountains. The sapwood is a light brown; the heartwood is a dark tan. It is straight grained and highly resistant to decay. It is a moderately light wood. It is used in cabinets, furniture, caskets, boxes, crates, and the core stock in panels.

Walnut.   This tree is found in most of the states east of the Mississippi and Texas. It is grown in abundance in the upper Mississippi states. The heartwood is brown, and the sapwood is white. Black walnut is stiff, hard, and strong. It glues, nails, and works easily. It is used in plywood and panels. It is also used for gunstocks, cabinet works, interior finishes, and floors.

White walnut, or butternut, does not have the properties that black walnut has. It does have a beautiful grain structure. It is therefore used in paneling and for decorative purposes.

Willow. Black willow has high strength and excellent resistance to shock loading. It is used for lumber, veneer, furniture, and in flooring.

Other popular woods, generally imported, are teak, lawan, ebony, and crabwood.

17-4 MOISTURE AND ITS EFFECT ON WOOD

Moisture is found inside the fibers and outside the cell fibers as free moisture. The *free* moisture evaporates first, and the wood is then said to be at the *fiber saturation* state. In this condition, the moisture content of the wood should be about 25 per cent of the kiln-dried weight. Green wood moisture contents may range up to 200 per cent.

Moisture content is determined by cutting test pieces from the width of a board, about 1 in. long, and well enough away from the end of the board to avoid the possibility of end drying. The sample is weighed and recorded as the *original weight*. It is then placed in a furnace at 212°F and allowed to "cook" until all the moisture has evaporated. The extent of evaporation is ascertained by repeated weighing. A constant weight indicates that all the moisture has been driven off. The test samples are removed and weighed, and the results are recorded as the *dry weight*. The moisture content is calculated by using the equation

$$m = \frac{W_m - W_d}{W_d} \times 100 \qquad \begin{aligned} m &= \% \text{ moisture} \\ W_m &= \text{weight before drying} \\ W_d &= \text{weight after drying} \end{aligned}$$

Example 1

Determine the moisture content of a wood sample that weighs 220 g before drying and 180 g after drying.

solution:

$$m = \frac{220 - 180}{180} \times 100 \qquad \begin{aligned} W_m &= 220 \text{ g} \\ W_d &= 180 \text{ g} \end{aligned}$$

$$= 22.2\% \text{ moisture}$$

The amount of moisture in green wood varies, depending upon too many factors to list here. The amount of moisture in *air-dried* wood depends upon the type of wood, the protection it has while drying, the size of the wood, the moisture in the atmosphere, and the length of the drying time. As indicated, kiln-dried wood is considered completely dry when the moisture content does not exceed 25 per cent. Air-dried wood generally has a moisture content of 15 per cent. Woods used outdoors should have a moisture content not greater than 10 per cent; for indoor use the moisture content should not be greater than 5 per cent.

Wood is said to be *hygroscopic*. That is, it takes on moisture when exposed to it, and it gives up moisture when it is exposed to dry conditions. When atmospheric conditions are stable, wood will reach a moisture level that is in equilibrium with these conditions.

The quantity of moisture taken on or given off is related to whether or not the wood is sapwood or heartwood, and to the type of grain, the density, the size, and the species. End grains will absorb and give up moisture more readily than the side grains. The rate at which moisture is taken on or given up can be controlled, but not eliminated, by painting, varnishing, creosoting, soaking in linseed oil or wax, or by kiln drying.

17-5 DEFECTS IN WOOD

When wood absorbs moisture, it swells. When it gives off moisture, it contracts. These are the mechanisms that cause volume changes and therefore warpage in wood. The volume change is greatest at right angles to the axis of the tree, or crosswise to it. Cross-grained lumber will shrink more than straight-grained lumber of the same species.

Radial shrinkage in wood is less than tangential shrinkage. This was shown in Fig. 17-3. The medullary rays, which run radially, restrict the shrinkage in that direction. The wood is free to shrink tangentially, since there is nothing to restrict it in this direction. Thus, from Fig. 17-4(a), quarter-sawed lumber will shrink less in width than thickness. Quarter-sawed lumber will twist, warp, or check as readily as plain-sawed lumber as shown in Fig. 17-4 (b).

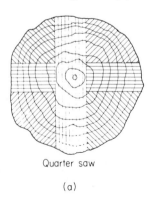

Quarter saw

(a)

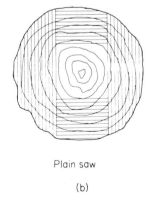

Plain saw

(b)

Figure 17-4

However, given the same conditions, plain-sawed lumber is easier and cheaper to cut and dries more readily than quarter-sawed lumber.

The difference in the amount of radial shrinkage as contrasted with tangential shrinkage is what causes wood to distort and separate. This is called *checking*. This checking defect takes place along the medullary rays, as shown in Fig. 17-5(a). Therefore, checking generally appears along the edges of quarter-sawed lumber and along the faces of plain-sawed lumber.

Shake is a defect that takes place between the annular rings. It is shown in Fig. 17-5(a).

Figure 17-5(b) and (c) shows lumber that has been subjected to stresses which cause it to "cup." If drying takes place more rapidly at the outer surfaces of a board, then, internally, it will shrink, but the external fibers will set in a stretched condition. This happens because the internal fibers, which are still "wet," restrain total external shrinkage. If sawed in this condition, the boards will cup so that the concave sides are toward the back side of the tree [Fig. 17-5(b)]. If drying continues before sawing, the inside will dry normally. As a result, the outside is still in an expanded condition, which restricts the shrinkage on the inside of the log. If the log is sawed in this condition, the cupping will be toward the center of the trunk [Fig. 17-5(c)].

Knots result from the limbs or branches that have been trimmed off the trunk. These may extend deep into the trunk, because the layers, or rings, of new wood grown each year bury them deeper into the ever-growing trunk. The appearance of the knot depends upon its position at the time the board was cut. Knots are defects in wood. The grain is forced to go around them, and therefore stress lines and cracks are created. The interrupted straight-line direction of the fibers has a greater effect upon the tensile properties of the board than upon the compressive properties. Thus, beams should be oriented with their knots near the top rather than near the bottom of the beam.

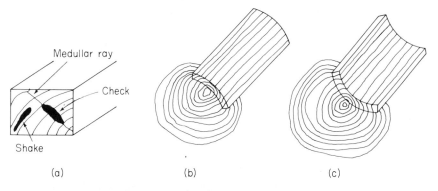

Figure 17-5

Rot is a decaying process, caused by bacteria, that causes the wood to separate. *Pitch pockets* are defects caused by an abnormal accumulation of resin.

17-6   GRADING AND SIZING OF LUMBER

Softwoods are generally graded as dense select structural, select structural, dense construction, construction, and standard. Grades are assigned that are based on certain well-defined grading techniques. The grader examines the wood for such items as splits, and checks bending stress, grain, slope of grain, torn grain, knots, etc. The size, location, and severity of the items as listed above are carefully observed and recorded, and a grade is assigned.

Softwoods are sometimes graded as timber, shop, and yard lumber. Timber is used when strength requirements are important, such as in beams. The cross section is generally over 2 in. thick. Shop lumber is used in the manufacture of wood models, patterns for molds, etc. Finishing grades are used as finishing lumber in trim, doors, etc. Low letters and numbers designate the better grades of lumber with fewer knots.

Hardwoods are graded as firsts, seconds, select, number 1 common, number 2 common, sound wormy, number 3 common. Again, certain well-defined criteria have been established, such as the size of the boards and the length and width of the imperfections. Warp, cup, and knots allowed are spelled out.

Lumber is sized according to the board foot. One board foot is defined as a board 1 in. thick by 12 in. wide by 1 ft long. In Fig. 17-6 the thickness T in inches times the width W in feet times the length L in feet equals the board feet B. Thus

$$B = \frac{NTWL}{12}$$

T = thickness, in.
W = width, ft
L = length, ft
B = bd-ft
N = number of boards

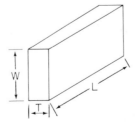

Figure 17-6

Example 2

Calculate the number of board feet in a piece of lumber $\frac{3}{4}$ in. thick by 8 in. wide by 16 ft long.

solution:

The board feet are

$$B = \frac{NTWL}{12} = 1 \times \tfrac{3}{4} \text{ in.} \times \tfrac{8}{12} \text{ ft} \times 16 \text{ ft}$$

$$= 8 \text{ bd-ft}$$

$T = \frac{3}{4}$ in.
$W = 8$ in.
$L = 16$ ft
$N = 1$

Example 3

A pile of 4 in. by 8 in. by 20 ft lumber contains 175 boards. Calculate the board feet in the pile.

solution:

The board feet in the pile are

$$B = \frac{NTWL}{12}$$
$$= 175 \times 4 \times \tfrac{8}{12} \times 20$$
$$= 9333 \text{ bd-ft}$$

$T = 4$ in.
$W = 8$ in.
$L = 20$ ft
$N = 175$

Example 4

Calculate the number of pieces when an order reads 1500 board feet of 2 by 10 in. by 20 ft long.

solution:

From the equation

$$B = \frac{NTWL}{12}$$

$T = 2$ in.
$W = 10$ in.
$L = 20$ ft
$B = 1500$ bd-ft

solve for N.

$$N = \frac{12B}{TWL} = \frac{12 \times 1500}{2 \times 10 \times 20}$$
$$= 45 \text{ boards}$$

17-7 PROPERTIES OF WOOD

As indicated, wood is composed of cellulose, lignin, and smaller quantities of inorganic materials. The latter comprise about 6 per cent of the total, combined with about 22 per cent lignin and 72 per cent cellulose. Chemically, about half of wood is carbon, about 40 per cent is oxygen, and the remainder is hydrogen, nitrogen, and ash.

Even though the chemical properties of wood do not change much, the physical properties vary from species to species. These properties may vary within a species, or be different in different directions within one piece of wood. As already indicated, knots, shakes, checks, and other imperfections will change the properties of wood.

Cell structure also has a marked effect on the physical properties of wood. The specific gravity of the materials that make up wood is about 1.54. This should make the density of wood more than 50 per cent greater than water. It is evident that wood is lighter than water. This is due to the presence of a large number of air spaces. The thickness of the fiber cell walls and the size of the cells have an influence on the amount of air and the specific weight of the various species.

The strength of a particular wood species is related to the fiber in the wood. The strength will vary directly with the specific gravity, because the specific gravity is an indication of the amount of fiber present in the wood. The strength also varies inversely with the number of pores present in the wood.

Beams may be subjected to flexural loading. Since this type of loading subjects the outer fibers to tension and compression, such beams fail in these fibers. They may fail by shear at the center fibers, or at the support points at the end, or across the fibers at the point of loading.

Moisture will *decrease* the mechanical properties of wood by a factor as high as 50 to 1 up to the point of saturation. Below the saturation point, an increase in moisture decreases the strength, but *may* increase the amount of deformation that a piece of wood can take before rupturing. If the saturation percentage is taken to be 30 per cent, Table 17-1 shows the percentage of increase in physical properties for each percentage of *decrease* in moisture content.

Table 17-2 lists the physical properties of lumber.

17-8 PRESERVATION OF WOOD

Wood, because it is an organic material, is subject to attack by bacteria and insects. It is also subject to destruction by burning.

Table 17-1   Changes in Properties of Wood

Properties	Percentage of increase in property due to 1% decrease in moisture below fiber saturation point (30%)
Static bending proportional limit	5
Static bending fracture stress	4
Modulus of elasticity	2
Impact bending, height of drop causing failure	0.5
Compression proportional limit parallel to grain	5
Compression crushing strength parallel to grain	6
Compression proportional limit perpendicular to grain	5
Shearing strength parallel to grain	3
Tension fracture stress perpendicular to grain	1.5
Hardness	3

Source: Parker, *Materials Data Handbook,* Table 11.2, p. 234. New York: McGraw-Hill Book Company.

Fungus.   Fungus attack takes on various forms. This type of destruction may cause discoloration without damaging the structural properties of the wood. But when fungi cause decay or rot, the timber is destroyed structurally. Certain fungi attack the wood between the fibers and consume the nourishment before it gets to its destination. This type of fungi growth is generally found near the surface of the wood.

Other fungi live by attacking the fibers or lignin. Under proper temperature conditions, they thrive on wood fiber, moisture, and air. Seasoned wood, under dry conditions, is the best protection against fungi growth. When possible, the wood should be treated with chemicals, creosote, or zinc chloride. Wood submerged in water will not be affected by fungi, because the air that they require to sustain life is absent.

Insects.   Three types of insects may do serious damage to wood. They are *subterranean* and *drywood* termites and the *marine borer.*

The *subterranean termite* bores tunnels in the ground to the wood, which is their food. Like fungi, they need moisture, oxygen, and warm temperatures to survive. They then bore into the wood and eat the fibers and lignen internally, leaving a hollow shell. The damage is not visible from the outside and therefore is not detectable until after the timber has failed. This type of termite has been known to build tubes up the side of stone and block foundation to get to the wood. They have been known to migrate from the soil inside hollow tile to the wood. In some cases they

Table 17-2 Physical Properties of Dry* Lumber

Species	Class	Sp. wt. (12%) Green lb/ft³	Dry lb/ft³	sp. gr. Green lb/ft³	Dry lb/ft³	Tension Parallel to grain lb/in.²	Shear Horizontal to grain lb/in.²	Compression Parallel to grain lb/in.²	Compression Perpend. to grain lb/in.²	Mod. of elast. $E \times 10^6$
Ash										
White	H	48	41	0.54	0.58	2050	105	365	1450	1.5
Green	H	46	38	0.50	0.55	1450	130	220	850	1.1
Black	H	52	34	0.45	0.49					
Bass	H									
Beech	H	55	45	0.56	0.64	2200	185	365	1600	1.6
Birch										
Yellow	H									
Sweet	H	55	44	0.57	0.63	2200	185	365	1600	1.6
Cedar										
Yellow	S	45		0.35		1600	130	185	1200	1.2
Red	S	37	37	0.44	0.47	1600	130	185	1200	1.5
White	S	24	23	0.31	0.32	1300	120	145	950	1.0
Cherry	H									
Chestnut	H									
Cypress	S	50	32	0.57	0.63	1900	150	220	1450	1.2
Elm										
Slippery	H	54	34	0.57		1600	150	185	1050	1.2
Rock	H					2200	185	365	1600	1.3
American	H					1600	150	185	1050	1.2
Fir										
White	S	38	35	0.41	0.44	1100	100	130	750	0.8
Douglas	S					2200	130	235	1450	1.6

Table 17-2 (continued)

Species	Class	Sp. wt. (12%) sp. gr.				Tension	Shear	Compression		Mod. of elast. $E \times 10^6$
		Green lb/ft³	Dry lb/ft³	Green	Dry	Parallel to grain lb/in.²	Horizontal to grain lb/in.²	Parallel to grain lb/in.²	Perpend. to grain lb/in.²	
Gum	H					1600	150	2200	1050	1.2
Hemlock	S	50	28	0.38	0.40					
East						1600	100	220	950	1.1
West						1900	110	220	1200	1.5
Hickory										
True	H	63	51	0.65	0.73	2800	205	440	2000	1.8
Pecan	H	55	48							
Mahogany	H	54	40	0.44	0.48	2200	185	365	1600	1.6
Maple										
Sugar	H	44								
Black	H	48								
Red	H	41								
Silver	H									
Oregon	H									
Oak		64		0.57	0.63					
White	H					2050	185	365	1350	1.5
Red										
Pine										
White	S	35	27	0.36	0.38	1300	120	185	1000	1.0
Sugar	S	52	25	0.35	0.36	1600	120	160	1050	1.1
Lodgepole	S	39	29	0.38	0.41	1300	90	160	950	1.0
Ponderosa	S	45	28	0.38	0.40	1600	120	160	1050	1.2
Yellow	S	55	41	0.47	0.51	2200	160	235	1450	1.6

Table 17-2 (continued)

| Species | Class | Sp. wt. (12%) sp. gr. | | | | Tension | Shear | Compression | | Mod. of elast. |
		Green lb/ft³	Dry lb/ft³	Green lb/ft³	Dry lb/ft³	Parallel to grain lb/in.²	Horizontal to grain lb/in.²	Parallel to grain lb/in.²	Perpend. to grain lb/in.²	$E \times 10^6$
Poplar	H	28				1450	130	160	1050	1.2
Redwood	S	50	28	0.38	0.40	1750	100	185	1350	1.2
Spruce										
Eastern	S	34	28	0.38	0.40	1600	120	185	1050	1.2
Sitka	S	33	28	0.37	0.40	1600	120	185	105	1.2
Engleman	S	39	23	0.31	0.33	1100	100	130	800	0.8
Walnut										
True	H	58	38	0.51	0.55					
White	H									
Willow	H									

Source: Parker, *Materials Data Handbook.* New York: McGraw-Hill Book Company.
*Shrinkage from green to oven dried, except where otherwise noted.

will follow cracks in foundations, using these cracks as passageways to the wood.

The control of subterranean termites may be accomplished in either of two ways:

1. By properly designing the structures to prevent growth of termite colonies.

2. By a planned program of extermination once they are discovered.

Foundations should be properly designed so that they will not readily crack. Copper flashing should rim the foundation. All wood should be kept well away from the ground. All wood scrap should be disposed of so that it does not become a breeding ground for termites.

If a building has already been constructed and termite infestation is suspected, regular inspection of the wood and foundation should be made. This can be done by "sounding" the wood with a hammer or piercing the wood with a sharp instrument. Piles of discarded termite wings call for inspection of and treatment of the soil with poison, such as chlordane.

Drywood termites are found in the southern part of the United States. These termites do not require moisture to survive. Wood preservatives and poison powders may be used to control them.

The *marine borer,* in the larva stage, bores into the wood at the moisture line, turns, and bores a series of tunnels parallel to the direction of the grain. Using the wood as food, it destroys the trunk internally. Many types of sheathing have been used to prevent the growth of the larvae.

There are other types of insects, such as ants, that bore into the wood and use it for nests rather than food.

17-9 PLYWOOD AND LAMINATES

The term *plywood* generally refers to the end product of laminated thin sheets of wood placed so that the grain of each sheet runs perpendicular to the adjacent sheets. All sheets are glued so that they form a laminated structure. Waterproof glues and plastic adhesives may be used to bind them. This structure has better shear strength than a single sheet of wood of the same thickness and made from the same type of wood. The flexure strength is also better than regular wood. Its resistance to splitting, warping, or cracking is very high when compared to nonlaminated wood.

Plywood may be a combination of many types of wood, or it may be made by laminating layers of one type of wood. These layers are made by either cutting the logs into thin sheets or by rotating the logs and peeling off thin layers of wood. These thin sheets are called *veneers.* Three processes for making veneers are shown in Fig. 17-7(a), (b), and (c).

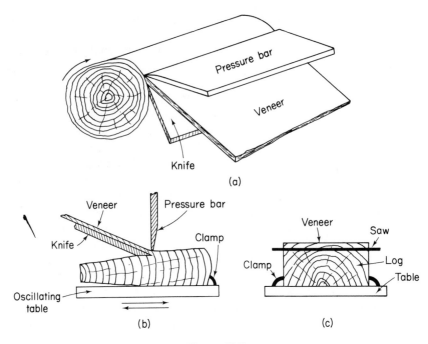

Figure 17-7

Before the veneer is sliced from the logs, the logs are cut into desired lengths and the bark is stripped off; then the veneers are peeled. Large imperfections are cut out of the sheets. The sheets are then sorted and dried in ovens. Small imperfections are cut from the veneers. Submarine-shaped patches are fitted and glued into position. The sheets are coated with an adhesive and laminated so that the grain direction is at an angle of 90 degrees to its neighbor. Either hot or cold pressure pads are applied to the laminates either to bond the glue (water resistant) or to set the thermo-setting resins (waterproof). The edges are trimmed, and one surface or both surfaces are sanded as desired.

Softwood plywood is graded by each company that manufactures it. Thus plywood may have various *odd* numbers of layers. They are manu-factured generally in 4- by 8-ft sheets and have thicknesses of $\frac{1}{4}, \frac{1}{2}, \frac{5}{8}$, and $\frac{3}{4}$ in. Other thicknesses and lengths may be obtained by special order. Sheathing grades are usually $\frac{5}{16}$ in. thick. They are graded from the best grade, "both-sides-good," to a grade referred to as "marine."

Hardwood plywood has three grades. The various types of hardwood veneers are graded, and the plywood is then made. The plywoods are graded into three classes: *red,* which is the best, *blue,* which is intermedi-ate, and *black,* which is the poorest grade. Various combinations of these three colors are painted on the plywood. The combinations indicate not only various grades of the veneers but also various types of adhesives used in making the plywood. Grades generally reflect the resistance to moisture.

Because of the orientation of the successive layers, plywood will not split from nailing and has high resistance to impact forces. In both instances the forces are distributed throughout the successive layers. Splintering will occur primarily from change in moisture conditions and the effects of this moisture on the different layers.

Other laminates are made by laminating lumber into beams, girders, arches, and columns. These glued lengths of wood are much stronger than similar sections of solid lumber. The lengths of wood are glued together with the grains running parallel to each other. Figure 17-8 shows one such beam.

17-10 HARDBOARD

Insulating fiberboard is made from shredded wood, asbestos, or sugarcane. The fibrous material is shredded either with a grindstone or from chips. In the latter case, the chips are soaked before they are shredded. The shredded materials are mixed with water, sized, and deposited on a drum to a uniform thickness. The water is removed from the fibers by a vacuum of about 15 psi. This "board" is pressed to a solid uniform sheet, cut, and put into drying ovens.

Sheathing fiberboard has its sides and edges impregnated with asphalt. It is manufactured in 4-ft widths, and 8- or 9-ft lengths. The thickness ranges from $\frac{7}{16}$ to $\frac{5}{8}$ in. It is used for moisture control as an exterior building material. It also has good sound and insulation qualities.

Insulating fiberboard is used for roofing insulation, ceiling tile, insulating paneling, or decorative paneling. It is manufactured in thicknesses of from $\frac{1}{2}$ to 1 in., in lengths from 4 to 16 ft, and in 4-ft widths.

Chipboard is made from wood wafers that are 0.010 to 0.050 in. thick by $1\frac{1}{2}$ in. square. The wafers are sorted according to thickness, dried, and coated with a polymer adhesive. They are transported to a forming

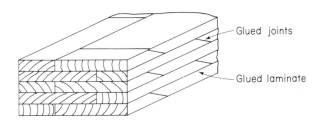

— Glued joints

— Glued laminate

Figure 17-8

table, where the thin wafers are distributed first, followed by two layers of thick wafers, then followed by another layer of thin wafers. This laminated board passes between two hot pressure rollers, which polymerize the adhesive and form a solid board. The finished boards are 4 ft wide and from 8 to 16 ft long. They are manufactured from $\frac{1}{4}$ to $\frac{3}{4}$ in. thick. They may be used as exterior sheathing or for interior construction.

Hardboard is boards made from wood chips that are processed by subjecting them to heat and pressure. This separates the wood fibers and the lignin from the wood. The fibers and lignin are mixed, hot pressed, and formed into hardboard.

Standard-grade hardboard is smooth on one or both sides. It has a high density, 60 lb/ft³, and a light brown coloration. It is available in 4-ft widths and 4-, 6-, 8-, 10-, 12-, and 16-ft lengths. It is manufactured in thicknesses of $\frac{1}{8}$ through $\frac{5}{16}$ in.

Tempered-grade hardboard is dark brown, has a density of 70 lb/ft³, and is harder than standard grades of hardboard. It is made from standard hardboard that is impregnated with a polymer. It is then backed to set the polymer. It is waterproof and may be used for exterior construction. It is available in sizes comparable to standard board.

Low-density hardboard does not have the qualities that standard and tempered hardboards possess. It has a density of 55 lb/ft³. It is processed by tempering and pressing patterns into the surfaces so that it imitates leather or the grain of wood. It is manufactured in widths of 4 ft, lengths of 4 to 10 ft, and thicknesses of $\frac{3}{16}$ and $\frac{1}{4}$ in.

Gypsum board is made from a core of gypsum sandwiched between two layers of heavy paper. It is fireproof and is used for interior walls and ceilings. This core is manufactured from a mixture of fibers and gypsum. It is manufactured in thicknesses of $\frac{3}{8}$, $\frac{1}{2}$, $\frac{5}{8}$, and 1 in., lengths of 4 to 12 ft, and 4-ft widths.

Asbestos cement board is asbestos fiber reinforced with portland cement that is treated with steam. This produces a very hard, fire-resistant board. This type of board is used on walls and roofs. These boards are sometimes laminated with a core of fiberglass or polystyrene. The surface may also be treated with a baked polymer paint. Special fastening procedures and devices must be resorted to when these boards are used in construction. They are manufactured in a standard 4-ft width, from 6- to 12-ft lengths, and thicknesses of $\frac{1}{8}$, $\frac{1}{4}$, $\frac{3}{8}$, and $\frac{1}{2}$ in.

Strawboard is a laminate of a core of processed straw sandwiched between two layers of tough paper. Insulating strawboard is manufactured 2 in. thick, 4 ft wide, and 5 ft long. It weighs slightly less than structural board. *Structural* strawboard is also made 2 in. thick and 4 ft wide, but can be purchased in 6- to 10-ft lengths.

Plastic foam board is made from polystyrene or polyurethane polymers, foamed, pressed into logs, and cut into boards. It has high compressive strength and very high insulation qualities. Adhesives or cements are used to bond the boards into position. They may be used as cores for lamination with steel, wood, or plastic sheets. They are manufactured in 4- to 12-ft lengths, 1- to 2-ft widths, and $\frac{1}{2}$- to 3-in. thicknesses.

Corkboard is made from a mixture of ground cork and resin. The mixture is baked and pressed into boards. It may have widths of from 1 to 3 ft, a length of 3 ft, and thicknesses of from 1 to 6 in. It is used to insulate walls and ceilings against cold, heat, and sound. Its insulating qualities are about one-third better than wood, and it has about 10 times better sound absorption qualities than brick. It is also made into cork tile, which is simply smaller squares of corkboard.

Paperboard is made from ground paper that is pressed into board $\frac{3}{16}$ to $\frac{1}{4}$ in. thick, or from layers of hard paper, corrugated, and sandwiched between two thick paper backings. These boards generally are 4 ft wide by 6 to 8 ft long.

Mineral fiberboard is made from a core of fiberglass or rock wool backed by a thick paper on either or both sides of the board. These boards are used for insulation. They measure 2 by 4 ft, and are $\frac{1}{2}$ to 2 in. thick.

17-11 PAPER AS AN ENGINEERING MATERIAL

Paper is a cellulose material made by removing the lignin and rolling the cellulose into thin sheets. Its monomer is shown in Fig. 17-9. The first papers appear to have been made for writing purposes in ancient Egypt and China. Later, paper was made in Damascus for use in the Near East and Europe.

There are many types of paper in use today. The type of paper produced depends upon the source of the cellulose and the way in which it is processed. If softwood is used, it is pulverized and cooked with chemicals, such as calcium sulfite, caustic soda, sodium sulfate, or magnesium bisul-

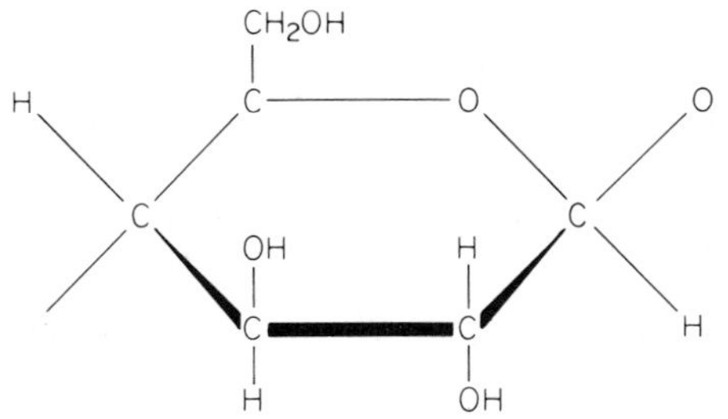

Figure 17-9

fite. If hardwood is the source of the cellulose, the chemicals used are soda ash and sulfite solution. The lignin is dissolved and removed, leaving the cellulose.

Wrapping paper is made from a mixture of pulps. Kraft paper is a heavy paper used for wrapping or in the building trades. Manila paper is made from hemp. Paper used in books is made from soda and sulfate pulp. Blotting and filter paper is made from bulky wood fibers. Filter paper, used to filter chemical gases, is made from thin fibers of glass pressed into thin sheets. It is also used as fiberglass paper for electrical insulation.

Writing paper is made from cotton products such as cotton or lignin rags. Typewriter paper is made from 80 per cent white rags. Fine ledger paper is 100 per cent white rags.

Oatmeal paper is used mainly for wallpaper. Cartridge wax paper is used as a waterproofing material, waxed on one side.

Transparent papers are very thin and are made by continuous rolling. These are the papers used in envelopes. They are glassine, onionskin, tissue, and tracing papers. Glassine paper is made from sulfite pulp. The better grade is called glassoid paper. Onionskin papers are made transpaper by hydration. Tissue characteristics depend upon the processes used in its manufacture. If loosely made, it has fine absorbent qualities. If tightly made, it may be used as a wrapping paper. Tracing paper is used in drawing and designing rooms. It is a thin rag paper that has been treated with a polymer. It takes ink or pencil, and erases easily.

Papers may be strengthened for wet strength, waterproofed, or treated for antistick qualities. Other types of papers are used for electrical insulation and in capacitors, as laminating paper in laminated plastics, as flameproof paper, and as metallized paper.

Building-trade papers are made from heavy kraft papers that may or may not have a rosin coating. The following types of paper are important in the engineering field:

Insulating paper made from asbestos fiber is used for insulating high-temperature vessels or pipes. In heavier weights it is also used as a building material. Wood fiber insulating paper is made from ground wood, and is used for insulating floors, roofs, and ceilings.

Roofing paper is made from wood chips that have been subjected to treatment with steam and shredded. It may be mixed with rag pulp for strength. It is used as roofing felt or as a heavy roofing covering.

Sheathing paper is manufactured in two forms. As plain sheathing paper, it may consist of a chemical pulp mixed with wastepaper, or it may be made from tough kraft paper. The latter may be either asphalt-coated or impregnated.

Fireproofing paper is made in two forms. It may be made from matted asbestos fibers similar to insulating paper. It may also be woven into threads, which are then rolled into paper. Since asbestos is fireproof, the paper is fireproof.

Vapor-barrier paper may be made from a thin core of asphalt sandwiched between two sheets of kraft paper, or wood pulp, treated and rolled into paper. Another vapor-barrier paper is waxed kraft paper. A third vapor-barrier paper is made from a thin copper foil, a layer of asphalt, and a sheet of kraft paper. All three types of paper have one purpose, which is to prevent the passage of moisture.

Concrete-form paper is made of strong, thick kraft paper rolled into spiral tubes. These tubes are used as forms for pouring concrete into columns or cores. Others are used as forms for pouring ribbed concrete slabs, which are made in the form of boxes, and which have been treated so that they will be abrasion resistant.

Wallpaper is made from bleached soda pulp and ground wood. Fillers, sizing, and coating are used to give this enough strength to keep from tearing when water paste is applied.

Cushion paper is similar to the wood-fiber insulating paper. It is used under slate roofs, carpets, and linoleum.

Laminating paper is used in the manufacture of laminated plastics. It is a high-strength kraft paper that has been treated with a polymer. Several sheets are then pressed together into one single sheet.

Envelope paper is used as a covering for gypsum board. Some of the insulating materials are inclosed in envelopes made from kraft treated or untreated paper.

Problems

17-1. Define softwood and hardwood. Give at least three examples of each.

17-2. What is exogenous growth? What is endogenous growth? Give an example of each.

17-3. What is sapwood? Heartwood? Explain fully.

17-4. Draw the three axes of structural strength of a tree trunk. Why should the strength of the trunk be different along each of these axes?

17-5. The long tubular cellulose fibers and lignin are very important components of wood. Discuss each fully.

17-6. What is the purpose of the medullary rays?

17-7. (a) What is the cambium layer? (b) Lignin? (c) Tubular cellulose fibers? Describe the function of each.

17-8. What is the difference between a straight-grained and a cross-grained board? Is wood ever cut with an angle grain?

17-9.   What is the effect of the grain direction on the degree to which a board will warp?

17-10.   Draw the cross section of a tree trunk and show how it can be cut to produce straight, cross, and angle grains in lumber.

17-11.   Analyze the seven cuts made in Fig. 17-3 and explain the forces that will cause the lumber to warp as shown.

17-12.   According to our definitions of softwood and hardwood, see if the leaf structure of the trees listed at the end of Sec. 17-1 fits the definitions.

17-13.   List the pine trees discussed in this chapter and at least two uses for each.

17-14.   List the fir trees in this chapter and their uses.

17-15.   List the spruce trees and their uses.

17-16.   What are two advantages of hemlock as a lumber?

17-17.   Discuss the types, physical properties, and uses of cedar.

17-18.   What outstanding advantage does cypress have over most other trees?

17-19.   Discuss redwood as a building material.

17-20.   List the types of white oak and red oak mentioned in this chapter. Discuss the characteristics of each.

17-21.   List the types and discuss the physical properties of maple as a building material. What are some of the uses of maple?

17-22.   Discuss the properties of the three types of ash found in the United States.

17-23.   Basswood, beechwood, hickory, willow, and pecan hickory are five woods listed as hardwoods. Discuss their physical properties and uses.

17-24.   List the properties and uses of mahogany, poplar, cherry, gum, and chestnut trees.

17-25.   List the various birch trees grown in the United States. Name one characteristic for each.

17-26.   What are the three types of elm listed in this chapter? List their physical properties.

17-27.   Discuss the uses of walnut as a building material.

17-28.   What is fiber-saturated wood?

17-29.   How is the moisture content of wood determined?

17-30.   Determine the moisture content of a wood sample that weighs 260 g before drying and 20 g after drying.

17-31.   Given a moisture content of 120 per cent and a dry weight of 148 g, calculate the green weight of the wood before drying.

17-32. Given a moisture content of 35 per cent and a green weight of 245 g, calculate the dry weight of the wood.

17-33. How many factors can you list that are responsible for the moisture content in green wood? List them.

17-34. What are the factors that control the air drying of wood?

17-35. How does the moisture content of air-dried wood differ from that of kiln-dried wood?

17-36. What is meant by the word *hygroscopic* as applied to wood?

17-37. What factors determine the amount of moisture taken on or driven off by wood? How is it controlled?

17-38. Which lumber will shrink most, cross- or straight-grained? Why?

17-39. List the characteristics of plain- and quarter-sawed lumber. Why do they exist?

17-40. Discuss checking and shake in lumber.

17-41. What causes cupping in lumber? Explain.

17-42. What are knots in lumber? How should a beam be oriented when it has knots? Why?

17-43. What causes rot? Pitch pockets?

17-44. How is softwood graded? List the criteria.

17-45. Describe the meanings of timber, shop, and yard lumbers when these terms are used to grade lumber.

17-46. How is hardwood graded? List the criteria.

17-47. Define the board foot. List the units for each dimension.

17-48. How many board feet are in a board 1 ft thick by 18 in. wide by 20 ft long?

17-49. A pile of lumber contains 425 boards. Their dimensions are 2 by 4 by 14 ft. How many board feet are in the pile?

17-50. An order reads 8000 board feet of 4- by 8- by 16-ft beams. How many beams should be delivered?

17-51. An order calls for 6750 board feet. If there are 280 boards that measure 2 by 6, how long are the boards? Assume that they are all the same length.

17-52. A carpenter is to use vertical planking $\frac{1}{4}$ by 4 in. in a room that is 20 ft long by 16 ft wide and has a ceiling 8 ft high. How many board feet does he need to enclose the entire room? Assume two windows, 2 ft wide by 48 in. long, and one door, 3 ft wide by 7 ft high.

17-53. (a) What are the percentages of cellulose, lignin, and inorganic materials in wood? (b) What percentage of wood is oxygen? (c) Carbon?

17-54.   How do you account for the fact that, even though the chemical composition of wood does not change much from species to species, the physical properties change radically?

17-55.   Explain the fact that the specific gravity of wood fibers is greater than water and yet wood will float in water.

17-56.   (a) What is the relationship between the strength and the specific gravity of wood? (b) Between the strength and the number of pores present?

17-57.   What is the effect of flexural loading on beams?

17-58.   What effect does moisture have on the mechanical properties of wood?

17-59.   Discuss the changes in properties of wood when related to the moisture content as stated in Table 17-1.

17-60.   From Table 17-2, compare the physical properties of white ash, red cedar, rock elm, Douglas fir, and maple.

17-61.   How do the physical properties of oak compare with redwood? (Refer to Table 17-2.)

17-62.   If you needed an 8- by 8-in. horizontal beam to support a floor would you use an oak or a spruce beam? Explain your answer. (Refer to Table 17-2.)

17-63.   If you needed four long timbers to support a water tower, would you use oak, yellow pine, or redwood? Explain the advantages and disadvantages of each. (Refer to Table 17-2.)

17-64.   Describe the two types of fungi that will attack wood. Describe how they operate to damage wood.

17-65.   Describe the feeding and migration habits of the subterranean termite.

17-66.   How may subterranean termites be controlled?

17-67.   How do the habits of the drywood termite differ from those of the subterranean termite?

17-68.   How does the marine borer destroy trees?

17-69.   Discuss the structure and physical properties of plywood.

17-70.   Discuss the three methods described in this chapter that are used to make veneer sheets for plywood.

17-71.   Discuss the making of plywood from the log stage to the finished product.

17-72.   How is softwood plywood graded? Hardwood plywood? Explain.

17-73.   Why is plywood a desirable construction material?

17-74.   Describe the reason that plywood resists splitting.

17-75. In your own words, explain why laminated wood (not plywood) is more desirable for structural use than solid wood. Check your library for other types and uses of laminated wood and list them.

17-76. How is insulating fiberboard made? List the sizes and uses of sheathing and insulating fiberboard.

17-77. How is chipboard made? List the standard sizes of chipboard.

17-78. How is hardboard made? What are the sizes available?

17-79. How does tempered hardboard differ from regular hardboard?

17-80. What are some uses to which low-density hardboard may be put?

17-81. How is gypsum board made? What are the standard sizes?

17-82. Discuss the manufacture, use, and standard sizes of asbestos cement board.

17-83. List the chief use of strawboard, paperboard, and mineral board.

17-84. Discuss the making of plastic foam board.

17-85. How is corkboard made? What are its chief qualities?

17-86. What is the basic structure of paper made from wood?

17-87. List 10 types of paper and their uses.

17-88. List and discuss the transparent papers to which this chapter refers.

17-89. Discuss the 10 building-trade papers listed in this chapter.

Index